THE EMERGENCE OF LIFE

Addressing the emergence of life from a systems biology perspective, this new edition has undergone extensive revision, reflecting changes in scientific understanding and evolution of thought on the question "what is life?" With an emphasis on the philosophical aspects of science, including the epistemic features of modern synthetic biology, as well as providing an updated view of the autopoiesis/cognition theory, the book gives an exhaustive treatment of the biophysical properties of vesicles, seen as the beginning of the "road map" to the minimal cell – a road map, which will develop into the question of whether and to what extent synthetic biology will be capable of making minimal life in the laboratory. Fully illustrated and accessibly written, *The Emergence of Life* challenges the reader directly with provocative questions, while also offering suggestions for research proposals taken directly from the author's bench experience. Dialogues with contemporary authors including Humberto Maturana, Albert Eschenmoser, and Harold Morowitz make this an ideal resource for researchers and students across fields including bioengineering, evolutionary biology, molecular biology, chemistry, and chemical engineering.

PIER LUIGI LUISI is Professor Emeritus at the Swiss Federal Institute of Technology in Zurich, Switzerland (ETHZ), where he developed his professional career, notably initiating Cortona Week in 1985. He has also held the position of Professor in Biochemistry at the University of Rome 3. He has authored more than 500 peer-reviewed papers as well as a number of books, recently including *The Systems View of Life* with Fritjof Capra (2014).

THE EMERGENCE OF LIFE

From Chemical Origins to Synthetic Biology

PIER LUIGI LUISI

University of Rome 3

CAMBRIDGE
UNIVERSITY PRESS

CAMBRIDGE
UNIVERSITY PRESS

University Printing House, Cambridge CB2 8BS, United Kingdom

One Liberty Plaza, 20th Floor, New York, NY 10006, USA

477 Williamstown Road, Port Melbourne, VIC 3207, Australia

314-321, 3rd Floor, Plot 3, Splendor Forum, Jasola District Centre, New Delhi - 110025, India

79 Anson Road, #06-04/06, Singapore 079906

Cambridge University Press is part of the University of Cambridge.

It furthers the University's mission by disseminating knowledge in the pursuit of education, learning and research at the highest international levels of excellence.

www.cambridge.org
Information on this title: www.cambridge.org/9781108735506

First published 2016
Second edition 2016
First paperback edition 2018

A catalogue record for this publication is available from the British Library

Library of Congress Cataloging in Publication data
Luisi, P. L., author.
The emergence of life : from chemical origins to synthetic biology / Pier Luigi Luisi,
University of Rome 3.
Second edition. | Cambridge : Cambridge University Press, 2016.
LCCN 2015049383 | ISBN 9781107092396
LCSH: Life – Origin.
LCC QH325 .L85 2016 | DDC 576.8/3–dc23
LC record available at http://lccn.loc.gov/2015049383

ISBN 978-1-107-09239-6 Hardback
ISBN 978-1-108-73550-6 Paperback

Contents

Acknowledgments

This book has been written immediately after the book with Fritjof Capra, *The Systems View of Life: A Unifying Vision* (2014), and the many fruitful discussions I had with Fritjof have helped to shape and enrich the new edition of the present book. I would also like to express my gratitude to Michel Bitbol, Rossella Mascolo, Michele Lucantoni, Angela Spaltro, and Pasquale Stano for their comments and continuous encouragement. The help of Pasquale Stano has been very important for the discussions in Part IV of this book. I would like to emphasize the collaboration with Angelo Merante, who was with me also in the first edition: he has been not only responsible for the many illustrations of the book, but also for the general organization of the various chapters, and for important critical comments and enrichments to the text.

Introduction

The first edition of this book, published in 2006, was written about 10 years ago; a partial update was done in the Japanese edition, in the 2010 Spanish edition, and in the 2013 Portuguese edition. A more complete English update was deemed necessary. Not that something dramatic has taken place since the first edition: the origin of life remains an unsolved question. At the last 2004 ISSOL meeting in Nara, Japan, a discussion took place on the question "what is life?," and this was conducted in similar terms and emphasis as the conversation 20 or 30 years back. Of course, in all these years, although not solving the core question, some particularly important research papers have appeared, and the corresponding update will be presented, to the best of my knowledge, in this second edition. However, this is not the main reason for this second edition.

Life on Earth is based on ordered sequences of proteins and nucleic acids, and on their mutual ordered interactions. And the solution to the quest for the origin of life is the answer to the question, of how this order came about. There are approaches to the origin of life that ignore this simple consideration and start from already pre-constituted ordered systems. This is the case of the (original) RNA-world, starting from a self-replicating RNA. This is a highly ordered, functional macromolecule, and to explain the origin of life from this ordered state would be akin to constructing a house roof first. The same can be said for those researchers who advocate viroids, or pristine forms of ribosomes, as the starting base to explain the origin of life. First, you should explain how this highly ordered state came about – and if you do so, perhaps, yes, you then have paved the way to explain the origin of life. To be clear on that: the research on the three areas mentioned above is often of the highest quality, and corresponds to the best pages of modern science. But in my view it will help very little to explain the origin of life on Earth.

The other approach to the origin of life starts from the opposite direction, namely from the disorder of monomers or low molecular weight compounds. The keywords here are hydrothermal sources, marine or volcanic smokes, endogenous prebiotic molecules, or molecules coming from space. Again, beautiful research is done on these areas, but you can have all the low molecular weight compounds of this world, in any quantity you wish, and you will never be able to make life. You cannot expect that a cathedral arises by simply having all the possible bricks. With 20 different bricks, you can construct a 100-bricks-long wall in an almost infinite number of ways; the question is to discover the principle that permits the construction of only a few ordered sequences.

This book starts from the consideration that we are still a long way from having a solution to this pristine problem of the origin of ordered structures (not that this is the only one in the origin of life). It attempts to analyze why this is so, indicating that one main reason may lie in the present-day reductionist (nucleic acid-centered) philosophical thinking in the field; and it then tries to propose some way to eliminate this shortcoming, emphasizing a systemic view of life and a corresponding systems view approach to its origin.

Part I of the book examines the various aspects of the bottom-up approach to the origin of life, as generally presented in the literature. This part also contains stringent criticism of the prebiotic RNA-world as origin of life, with an invitation to look at the origin of life from "ground zero." The discussion is enriched by a series of conversations with distinguished authors in the field, such as David Deamer, Albert Eschenmoser, Gerald Joyce, Doron Lancet, Harold J. Morowitz, Eörs Szathmáry, Sandra Pizzarello, and Nobelist Ada Yonath.

Part II is a detailed account of the theory of autopoiesis, as due to Maturana and Varela, who picture the cell as an open molecular system capable of self-maintenance, due to a regeneration of the components from within. This systems view is in sharp contrast with the reductionist, DNA- or RNA-centered visions of life according to which life is the result of the behavior of a single molecular species. Here, it is instead the system's organization of the internal web interactions that may cause cellular self-reproduction (eventually leading to Darwinian evolution) – and Darwinian evolution, being the result of that organization, cannot be seen obviously as the *prima causa* of life.

Also considered is the interaction of the living with the environment, which leads to the important notion of cognition with its epistemic and ecological aspects. Particularly in this part of the book, but then also as a general background framework, philosophical and biological aspects are strictly interwoven with each other, with the intention of showing that philosophy and biology should not be seen as two distinct disciplines, but as an integrated unity of the systems view of life: a message that should be given to all our students. Part II also includes the conversation with Humberto Maturana and – as a tribute to Francisco Varela's thought – the conversations with philosophically minded authors such as Amy Cohen Varela and Evan Thompson. Other enrichments of discussion on these arguments are a conversation with Denis Noble and a Side

Box concerning future developments of research on autopoiesis by Luisa Damiano.

With Part III, the book moves towards biological complexity, and here the two complementary notions of self-organization and emergence – from simpler molecular systems, as micelles and vesicles, to more complex structures (organized protein systems, ribosomes, viruses, and so on) – are presented in two distinct chapters. The following chapter concerns the most salient emergent property, namely self-replication/reproduction. Two side boxes, the first one concerning self-organization mechanisms in Hydra and the other focusing on the sciences of complexity – written by distinguished authors Giorgio Venturini and Stuart Kauffman – offer interesting points of view about the richness of the arguments handled in this section of the book.

The first chapter of Part IV concerns the world of surfactants and lipids, in particular the procedures for solute incorporation and questions about overall and local concentration. The other two chapters are focused on the world of vesicles, in particular their physical and chemical properties, with emphasis on their capability of entrapping biopolymers – whereby they are seen as the best models for the shell of biological membranes.

The information regarding the self-organization and self-reproduction of vesicles will be the basis for the last section of this book, Part V, which is devoted to synthetic biology and the attempts to construct the minimal living cell in the lab. First, general questions about synthetic biology – as today's most celebrated and ambitious laboratory approach to make new or alternative forms of life – are handled. The opening discussion about the epistemology of synthetic biology is enriched by two interesting conversations with relevant authors Paul Freemont and Sarah Lau. Then, a review of the experimental work carried out in my laboratory and others towards the synthetic biology of the minimal cells, which includes the new and unexpected finding of the spontaneous solute overcrowding in vesicles.

The general idea, the red thread that pervades this book, is to provide a unitarian view of life and its origin that departs from a reductionist, nucleic-acid-centered view, to favor instead a systems approach, in which the cellular organization, and its cognitive interaction with the environment, gives the basis for an understanding and, possibly, the reconstruction of life.

Part I

Approaches to the origin of life

Chapter 1

Setting the stage

Introduction

The origin of life on Earth was never a widely discussed issue until the middle of the nineteenth century. In those times, people were pious and good-natured and believed firmly that God has created the world and all living forms, once and for all. It would have been a blasphemy to doubt such a simple truth. And the blasphemy arrived in the form of a book written in 1859 by a British scientist named Charles Darwin, who refused the idea that living forms were fixed, saying instead that they were changing with time, and that they were evolving from a common ancestor. The blasphemy was very convincing and spread rapidly in the scientific community, also contaminating fields beyond biology. Darwin – adding one blasphemy to the other – arrived to postulate, in a letter written to Joseph Dalton Hooker on February 1st of 1871, that life might have originated from natural causes:

... It is often said that all the conditions for the first production of a living organism are now present, which could ever have been present. But if (and oh what a big if) we could conceive in some warm little pond with all sorts of ammonia and phosphoric salts, light, heat, electricity &c. present, that a protein compound was chemically formed, ready to undergo still more complex changes, at the present day such matter would be instantly devoured, or absorbed, which would not have been the case before living creatures were formed.

Thus the idea of the warm little pond, later called prebiotic soup or similar names, was born. And generally, in that time philosophers and biologists were increasingly accepting the idea of an origin of life based on natural laws.

At the time of Darwin, in parallel to the belief in the divine creation, there was also the idea of the spontaneous generation (abiogenesis) of simple life forms. The rationale was that God, in his grandeur, could not have had the time and will to think about the creation of such primitive forms of life. Thus, ants, flies, beetles, mosquitoes, and even rats would originate spontaneously from decaying

3

meat or dirty laundry in the wet cellar.[1] The work of Redi (already in 1668), and by Spallanzani one century later, aimed at providing evidence against abiogenesis, was not convincing to the community of the time. But the work by Louis Pasteur was: people after him accepted that life does not arise by spontaneous generation; life arises only from life itself. This statement, by a trait of irony, seemed to give more strength to the belief that life can only come from God.

1.1 The secular view on the origin of life

However, around the middle of the nineteenth century, the time was ripe for the idea of the secular origin of life. Thus, while Darwin's theories were spreading, the German biologist and philosopher Friedrich Rolle was writing in 1863:

The hypothesis of an original emergence of life from inanimate matter [...] can at least offer the advantage of explaining natural things by natural pathways, thus avoiding the invocation of miracles, which are actually in contradiction with the foundations of science.

Darwin himself was not directly interested in the origin of life, but some of his contemporary scientists popularized his views of a natural origin of life, most notably Ernst Häckel. He stressed that there is no difference in quality between the inanimate and the animate world (*Anorgane und Organismen*) and that, therefore, there is a natural and continuous flux from the one to the other (Häckel, 1866). This "continuity principle" had been advocated also by the already cited Rolle (1863) and by William Thierry Pryer (1880) and is part of the modern view of life.

Proceeding with the historical discourse, let us consider a surprising definition given by Friedrich Engels (yes, the same Engels of Karl Marx's memory), written in *Dialectics of Nature* (1883):

Life is the mode of existence of protein bodies, the essential element of which consists in *continual metabolic interchange with the natural environment outside them*, and which ceases with the cessation of this metabolism, bringing about the decomposition of the protein.[2]

This is indeed surprising, given the early date and the fact that Friedrich Engels certainly was not a biologist, and that at this time nobody had a clear notion of what "protein bodies" really meant (although Darwin himself had

[1] Famous is the experiment introduced by Jean-Baptiste van Helmont (1577–1644). He suggested that mice are spontaneously generated from wheat. Van Helmont believed it was human sweat which provided the generating principle of life and hence his experiments needed dirty shirts as well as wheat germ and 21 days of fermentation after which the vapors from the shirt with the vapors from the seeds would generate live mice. Van Helmont was surprised to find that such mice were exact replicas of natural mice originating from mouse parents.

[2] A few years before, Engels gave an almost similar definition: "*Life is a mode of existence of protoplasm* and consists essentially in the constant renewal of the chemical constituents of this substance. Protoplasm is here understood in the modern chemical sense and comprises under this name all substances analogous to the white of an egg, otherwise called protein substances" (Engels, 1877). Later, he updated again his definition: "Life is the existence form of proteic structures, and this existence form consists essentially *in the constant self-renewal of the chemical components of these structures*" (Engels, 1894).

spoken of a "protein compound"). Moreover, he pointed out the constant self-renewal of the chemical components of these proteic structures.

We had to wait over 50 years to have a more scientific rendering of Engels' concept. Let's consider a definition written by Perret in the early 1950s, and reiterated by John Desmond Bernal in 1965:

Life is a potentially self-perpetuating system of linked organic reactions, catalyzed stepwise and almost isothermally by complex and specific organic catalysts which are themselves produced by the system.

Bernal discusses this concept in more detail in his other books (Bernal, 1951; 1967; 1971).

In jumping from Engels to Bernal, we should not forget a big name in between them, Alexander I. Oparin. The question of the origin of life became a scientific issue only with the publication of Oparin's books. However, the first one, *Proiskhozhdenie Zhizni* (Origin of Life), published in Russian in 1924, was largely unnoticed until it was translated after 1950. His second book on this subject, *Vozniknovenie zhizni na zemle* (The origin of life on Earth), published in Russian in 1936, expanded and modified his earlier views in some important ways[3]. It was published in English in 1938 as *The Origin of Life*. The second edition of this book (1941) was published in English in 1953.[4] The reworked and much enlarged third Russian edition (1957) was translated in English in the same year as *The Origin of Life on the Earth*.

He subsequently wrote *Life: Its Nature, Origin, and Development* (1961). The Russian edition was published in 1960, in which he gave a description of life

---

[3] In his book of 1936, Oparin expressed his views in the form of a dialectical materialist analysis, explicitly citing Friedrich Engels. He described life as a naturally emergent stage in the evolution of matter, one in which physicochemical laws had been supplemented by the "purely biological" laws of natural selection and metabolism. Oparin drew more heavily on the current international literature in astronomy, geochemistry, organic chemistry, plant enzymology, and about the chemical evolution of the biosphere. Stanley L. Miller and H. James Cleaves wrote: "Careful reading of Oparin's 1924 pamphlet shows that, in contrast to common belief, at first he did not assume an anoxic primitive atmosphere. In his original scenario he argued that while some carbides, that is, carbon-metal compounds, extruded from the young Earth's interior would react with water vapor leading to hydrocarbons, others would be oxidized to form aldehydes, alcohols, and ketones [...] Oparin's ideas were further elaborated in a more extensive book published with the same title in Russian in 1936. In this new book his original proposal was revised, leading to the assumption of a highly reducing milieu in which iron carbides of geological origin would react with steam to form hydrocarbons. Their oxidation would yield alcohols, ketones, aldehydes, and so on, which would then react with ammonia to form amines, amides, and ammonium salts. The resulting protein-like compounds and other molecules would form a hot, dilute soup, which would aggregate to form colloidal systems, that is, coacervates, from which the first heterotrophic microbes evolved" (Miller and Cleaves, 2007).

Moreover, Oparin drew upon studies on colloidal coacervation, arguing that the formation of coacervate droplets by the electrostatic attraction of organic soils in the early seas provided a key requirement for the emergence of life: chemical pools separated by a membrane from the surrounding medium. Such droplets could selectively assimilate materials, and collect and accumulate catalysts and promoters that would accelerate chemical reactions. Although most of these coacervates were short-lived, Oparin believed that those with the fastest rates of reaction, the most stable internal configurations, and the ability to grow and divide most rapidly, would begin to undergo natural selection, leading to more organized forms and eventually to primitive living systems.

[4] I highly recommend reading the *Introduction* to the 1938 English edition of Oparin's book, written by Sergius Morgulis, editor and translator, as it is still one of the best analyses on the naturalistic essence of life and the progress from non-life to life.

based on six properties: (1) capability of the exchange of materials with the surrounding medium; (2) capability of growth; (3) capability of population growth (multiplication); (4) capability of self-reproduction; (5) capability of movement; (6) capability of being excited. He also added some additional properties, such as the existence of a membrane (a cardinal principle for him); and the interdependency with the milieu (Oparin, 1961). An enumeration of properties, as Oparin does, appears to be the preferred way of getting around the problem of giving a definition in a nutshell, and modern examples of this are given by Koshland (2002) and by Oró (in Schopf, 2002).

However, the list of properties of life may be extremely long and subjective, and the real point is to find the unique feature of life that gives rise to the list of properties (this is what we will do in the next part of this book).

The definition of life is a rather thorny question. About that, let me add that the term *definition* – with its strong ontological flavor – is perhaps too ambitious: the term "operational description" probably catches better the epistemic and pragmatic aspects of the question. As Primas says in a different context (1998):

... by contrast an operational description refers to empirical observations obtained by some pattern recognition methods which concentrate on those aspects we consider as relevant.

Actually, most of the "definitions" of life given in the literature comply with the above operational description.

There are plenty of them; for example, one may refer to those listed in the monographs by Folsome (1979), Chyba and McDonald (1995), or in a book edited by the late Martino Rizzotti (Rizzotti, 1996). See also Popa (2004). In addition, a few dozen definitions of life are given in over 40 pages by a corresponding number of authors in the book edited by Palyi *et al.* (2002). Out of this vast repertoire, I would like to mention some of the least traditional; for example, Alec Schaerer (2002) approaches the conceptual conditions for conceiving and describing life, including the aspects of language, cognition, and consciousness. Or the paper by Kunio Kawamura (2002), who approaches the origin of life from the angle of "subjectivity," referring to the philosophical work by Imanishi (for me, there are strong ties here with the view of autopoiesis, which we will explore in depth later in this book). This author provides a view of life from the classic Japanese philosophical view, with the notion of *shutaisei* (subjectivity). And, still in the same book, you will find the Vedanta view of life (Apte, 2002) as well as that of the Russian Orthodox tradition (Arinin, 2002). There are questions about life raised by other authors and researchers: "Is life reducible to complexity?" (Abel, 2002); "When did life became cyclic?" (Boiteau *et al.*, 2002); "Does biotic life exist?" (Valenzuela, 2002).

I mention these less traditional points of view not because we learn much about the origin of life from them, but rather to emphasize that, by asking the question, we are necessarily dealing with a broad spectrum of perspectives and diverse human cultures.

One cannot end this part on life definitions without mentioning the so-called "NASA definition of life." Originally, this was simply an operational perspective used by the Exobiology Program within the National Aeronautics and Space Administration – a general working definition. However, people working on the origin of life often use this definition –actually proposed earlier by Horowitz and Miller (1962) – which is as follows (Joyce, 1994):

Life is a self-sustained chemical system capable of undergoing Darwinian evolution.

This operative definition is one of the most popular, and probably it is so because it is based on a pragmatic operation. If NASA astronauts are going to find in some distant planet a colony of bacteria that behave just like a terrestrial colony of bacteria, they are going to communicate back that yes, they have found life. However, it is at this point that one would ask them, "Yes, but what is life?"

In fact, the above definition in my opinion is not very useful, nor correct from an epistemic point of view. It applies to populations (Darwinian), and is completely silent if you consider a single living organism at a time – a flying bird, a swimming fish, an oak tree, or a product of synthetic biology. Particularly in the last case, but also for an object found in a distant planet, the genetic background may be unknown or technically impossible to establish. We need a local, "here and now" life criterion to discriminate between the living and the non-living without waiting for evolution or reproduction.

The multiplicity of views presented in literature is thus impressive, and also impressive is the number of books devoted to the origin of life, which I listed to the best of my knowledge (see Side Box 1.1).

The popularity of the NASA definition among the scientists studying the origin of life reflects the obvious prejudice that the molecular mechanism of nucleic acids must be the main basis for defining life. Accordingly, this would bring life and evolution to equivalence.

Considering the overlap between evolution and life, one may recall the distinction made by Szathmáry (2002) between the units of life and the units of evolution. The author emphasizes that the two domains (life and evolution) may partly overlap, but that they should be considered as two distinct realms. Other authors emphasize the same concepts (Lewontin, 1970; Maynard-Smith and Szathmáry, 1995; Okasha, 2006).

Before looking deeper at Oparin's ideas, however, we need to present the scenario of the origin of life in terms of basic data.

Side Box 1.1

Books on the origin of life

Bastian, H. C. (1872). *The Beginnings of Life*. London.

Pryer, W. (1880). *Die Hypothesen über den Ursprung des Lebens*. Berlin.

Leduc, S. (1907). *Les Bases Physiques de la Vie*. Paris.

Osborn, H. (1918). *The Origin and Evolution of Life*. London.

Oparin, A. I. (1924). *Proiskhozhdenie Zhizni*. Moscow. (In Russian). Translated into English as: The Origin of Life, in *The Origin of Life*, J. D. Bernal (ed.), London, 1967, and Cleveland, 1967.

Haldane, J. B. S. (1929). The origin of life. *Rationalist Annual*, 148: 3–10.

Haldane, J. B. S. (1929). The Origin of Life. In *The Origin of Life*, J. D. Bernal (ed.), London, 1967, and Cleveland, 1967.

Oparin, A. I. (1936). *Vozniknovenie Zhizni na Zemle*. Moscow and Leningrad. Translated into English as: Oparin, A., *The Origin of Life*, New York: MacMillan, 1938.

Vozniknovenie Zhizni na Zemle. (1941). 2nd Edition (In Russian). Translated into English as: Oparin, A., *The Origin of Life*, New York: Dover Publ., 1953.

Vozniknovenie Zhizni na Zemle. (1957). 3rd Edition (In Russian). Translated into English as: Oparin, A. I., *The Origin of Life on the Earth*, Edinburgh: Oliver and Boyd, 1957, and London – New York: Academic Press, 1957.

Bernal, J. D. (1951). *The Physical Basis of Life*. London: Routledge and Paul.

Haldane, J. B. S. (1954). The Origin of Life. *New Biology*, 16: 12.

Schrödinger, E. (1956). *What is Life? And other Scientific Essays*. Garden City, NJ.

Brachet, J. (1957). Origin of life on the Earth, in *Rep. Intern. Symp.*, 361, New York: Pergamon, 1959.

Oparin, A. I. (1960). *Zhizn', Ee Priroda, Proiskhozhdenie i Razvitie*. Moscow. 2nd Edition, 1968. English translation as: *Life; Its Nature, Origin, and Development*. Edinburgh and London: Oliver and Boyd, 1961; New York: Academic Press, 1961.

Oparin, A. I. (1964). The chemical origin of life, in *American Lectures in Living Chemistry*, S. Kugelmass (ed.), Charles C Thomas, Springfield, IL.

Crick, F. (1966). *Of Molecules and Men*. University of Washington Press.

Oparin, A. I. (1966). *Vozniknovenie i Nachal'noe Razvitie Zhizni* (Moscow), English translation as: *Genesis and Evolutionary Development of Life*. New York, 1968.

Bernal, J. D. (1967). *The Origin of Life*. Cleveland, OH: World Publishing Co.

Margulis, L. (ed.) (1970). *Origins of Life*. I. New York: Gordon & Breach.

Bernal, J. D. (1971). *Ursprung des Lebens*. Lausanne: Editions Rencontre.

Margulis, L. (ed.) (1971). *Origins of Life*. II. New York: Gordon & Breach.

Rutten, M. G. (1971). *The Origin of Life by Natural Causes*. New York: Elsevier.

Fox, S. W. and Dose, K. (1972). *Molecular Evolution and the Origin of Life*. San Francisco: Freeman.

Orgel, L. E. (1973). *The Origins of Life*. New York: Wiley.

Miller, S. L. and Orgel, L. E. (1974). *The Origin of Life on Earth*. Englewood Cliffs, NJ: Prentice Hall.

Wickramasinghe, N. C., with Fred Hoyle. (1978). *Lifecloud: The Origin of Life in the Universe*. London: J.M. Dent.

Ponnamperuma, C. (1981). *Comets and the Origin of Life*. Dordrecht, The Netherlands.

Cairns-Smith, A. G. (1982). *Genetic Takeover and the Mineral Origin of Life*. Cambridge University Press.

Day, W. (1984). *Genesis on Planet Earth: the Search for Life's Beginnings*. New Haven, CT: Yale University Press.

Cairns-Smith, A. G. (1985). *Seven Clues to the Origin of Life*. Cambridge University Press.

Dyson, F. (1985). *Origins of Life*. Cambridge University Press.

Shapiro, R. (1986). *Origins: a Skeptic's Guide to the Creation of Life on Earth*. New York: Summit.

Fox, S. W. (1988). *The Emergence of Life*. New York: Basic Books.

de Duve, C. (1991). *Blueprint for a Cell: The Nature and the Origin of Life*. Burlington, NC: Carolina Biological Supply Co.

Eigen, M. and Winkler-Oswatitisch, R. (1992). *Steps Towards Life*, Oxford University Press.

Morowitz, H. J. (1992). *Beginning of Cellular Life*. Yale University Press.

Margulis, L. and Sagan, D. (1995). *What is Life?* London: Weidenfeld and Nicholson.

Rizzotti, M. (ed.) (1996). *Defining Life*, Publ. University of Padua.

Thomas, P. J., Chyba, C. F., and McKay, C P. (eds.) (1997). *Comets and the Origins and Evolution of Life*. New York: Springer-Verlag.

Brack, A. (ed.) (1998). *The Molecular Origin of Life*. Cambridge University Press.

Davies, P. (1999). *The Fifth Miracle: The Search for the Origin and Meaning of Life*. New York: Simon and Schuster Paperback.

Dyson, F. (1999). *Origins of Life*, 2nd edition. Cambridge University Press.

Fry, I. (1999). *The Emergence of Life on Earth: A Historical and Scientific Overview*. London: Free Association Books.

Maynard Smith, J. and Szathmáry, E. (1999). *The Origins of Life*. Oxford University Press.

Fry, I. (2000). *Emergence of Life on Earth: A Historical and Scientific Overview*. New Brunswick, NJ: Rutgers University Press.

Varela, F. J. (2000). *El Fenómeno de la Vida*. Dolmen Ensayo.

Wickramasinghe, N.C., with Fred Hoyle. (2000). *Astronomical Origins of Life: Steps towards Panspermia*. Dordrecht, NL: Kluwer.

Willis, C. and Bada, J. (2000). *The Spark of Life*. Cambridge, MS: Perseus Books Group.

Zubay, G. (2000). *Origins of Life on the Earth and in the Cosmos*. Cal. Academic.

Blum, P. (ed.) (2001). *Archaea: Ancient Microbes, Extreme Environments, and the Origin of Life*. New York: Academic Press.

Schwabe, C. (2001). *The Genomic Potential Hypothesis, a Chemist's View of the Origins, Evolution and Unfolding of Life*. Georgetown: Landes Bioscience.

Day, W. (2002). *How Life Began: the Genesis of Life on Earth*. Cambridge, MA: Foundation for New Directions.

de Duve, C. (2002). *Life Evolving, Molecules, Mind and Meaning*. Oxford University Press.

Schopf, J. W. (ed.) (2002). *Life's Origin, The Beginning of Biological Evolution*. California University Press.

Davies, P. C. W. (2003). *The Origin of Life*. Penguin Science.

Ganti, T. (2003). *The Principles of Life*. Oxford University Press.

Popa, R. (2004). *Between Necessity and Probability: Searching for the Definition and Origin of Life*. Berlin: Springer-Verlag.

Ribas de Pouplan L. (ed.) (2004). *The Genetic Code and the Origin of Life*. Kluwer Acad.-Plenum Publ.

Hazen, R. M. (2005). *Genesis: The Scientific Quest for Life's Origins*. Joseph Henry Press.

Shiller, B. M. (2005). *Origin of Life: The 5th OPTION*. Trafford.

Yockey, H. P. (2005). *Information Theory, Evolution, and The Origin of Life*, Cambridge University Press.

Luisi, P. L. (2006). *The Emergence of Life: From Chemical Origins to Synthetic Biology*. Cambridge University Press.

Teerikorpi, P., Valtonen, M., Lehto, K., Lehto, H., Byrd, G., and Chernin, A. (2009). *The Evolving Universe and the Origin of Life. The Search for Our Cosmic Roots*. New York: Springer Verlag.

Wickramasinghe, J., Wickramasinghe, C., and Napier, W. (2009). *Comets and the Origin of Life*. Hackensack, NJ: World Scientific.

Lane, N. (2010). *Life Ascending: The Ten Great Inventions of Evolution*. New York: W. W. Norton & Company.

Deamer, D. (2011). *First Life*. University of California Press.

Egel, R., Lankenau, D.-H., and Mulkidjanian, A. Y. (2011). *Origins of Life: The Primal Self-Organization*. Berlin: Springer-Verlag.

Rutherford, A. (2013). *Creation: The Origin of Life – The Future of Life*. London: Viking/Penguin.

Rutherford, A. (2014). *Creation: How Science Is Reinventing Life Itself*. New York: Current.

Mazur, S. (2015). *The Origin of Life Circus*. New York: McNally Jackson Books.

Gerald, M. C. with Gerald, G. C. (2015). *The Biology Book: From the Origin of Life to Epigenetics – 250 Milestones in the History of Biology*. New York: Sterling Publishing.

1.2 A few accepted facts

Let us begin by saying that, according to the big-bang theory, the age of the universe is estimated to be 13.8 Gya (billion years ago), while the origin of the solar system, and our Earth, dates to 4.5–4.6 Gya. They say that at this time our planet was more or less a fireball, which reached a certain geological and thermal stability about 4

Gya, at the time of a faint young sun. This is also the age of the first rocks on our planet, as well the age of the oceans. Evidence of oxygen production is at 2.7 Gya, but bacterial photosynthesis was operative long before the atmosphere originated – that is, until 0.6 billion years passed and much more oxygen was being produced, a development for which we do not know the reason.

The most ancient microfossils have been described by J. W. Schopf from the Apex Chert in Marble Bar, in western Australia (Schopf, 1993, 1998; Schopf and Klein, 1992). They are dated 3.465 Gya. The microfossils of Swaziland (Sud Africa) are more or less of the same period. The so-called Gunflin Chert of North America (1.9–2.3 Gya) and the microfossils of the Belcher Group in Canada were the first to be ascribed to the Precambrian period. Readers can find more information about microfossils at http://www.uni-muenster.de/GeoPalaeontologie/Palaeo/Palbot_2011/forschung.html.

Although Schopf's interpretations have been criticized (Brasier *et al.*, 2002), the general consensus today is still that unicellular organisms existed on our Earth at 3.4 to 3.5 Gya.

If these unicellular organisms already had a full-fledged genome, one is tempted to suggest that life cannot have started right away with such a complexity, and that therefore, the "origin of life" must be older than 3.5 Gya. In this regard, consider a quite recent paper by Bell et al. (2015). These authors have analyzed inclusion of graphite in zirconium crystals, and state that the carbon isotope ratio is "consistent with a biogenic origin and may be evidence that a terrestrial biosphere had emerged by 4.1 Gya, or ~300 My earlier than has been previously proposed."

The general view can be schematized as in Figure 1.1, which indicates main events that took place in our past on a time scale. Among these, the position of the plausible transition to life can be located between 3.5 and 4 Gya.

Where should we set the scenario for the origin of life? As we will argue later, the aqueous environment of the ocean is the most accepted scenario, often in the form of tides, which may increase the solute local concentration when withdrawing, with a possible localized scenario such as hydrothermal pools or small lagoons, or even coastal lakes. If life started only in one place and only once, it must have been a place from where it could then rapidly expand and "infect" large parts of the Earth.

1.3 Oparin's view, and its implications

As already mentioned, the view of the origin of life by a natural process, in conformity with the natural laws, was given by Alexander Oparin (1924; 1953; 1957), the brilliant Russian chemist who was influenced both by Darwinian

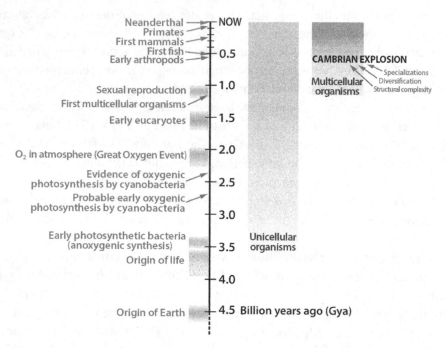

Figure 1.1 A simplified time scale of main events in the origin of life. These "steps" were among Gould's arguments in favor of contingency, as discussed later on in this chapter (Section 1.4). The plausible transition to life can be located between 3.5 and 4 billion years ago. The early eukaryotic organisms appeared around 1.6 Gya and the first simple multicellular organisms circa 1.2 Gya. Note also the Cambrian explosion around 540 million years ago. Prior to the Cambrian event, living organisms were simple, small, and, mostly, unicellular. Maybe, some complex, multicellular organisms gradually became more common in the years preceding the Cambrian, but it was not until this period that mineralized – that is, readily fossilized – organisms became common. Anyhow, the – more or less rapid – diversification of a wide variety of complex multicellular life forms in the first Cambrian produced the first representatives of modern phyla.

theories and by dialectical materialism. J. B. Haldane (1929; 1954) put forward a similar view on the origin of life, coming from a quite different context. Accordingly, there is a natural and spontaneous increase of molecular complexity, governed by the natural laws, up to the point in which spherical compartments – the first cells or protocells – were formed, which could make copies of themselves.

Oparin's view, which modern biology generally takes for granted, appears in most college textbooks, specialized literature, and mass media. The background of the left panel of Figure 1.2 is the already mentioned "continuity principle" (Oparin, 1924; Orgel, 1973 and 1994; de Duve, 1991; Eigen and Winkler-Oswatitisch, 1992; Morowitz, 1992; Crick, 1996). This sets a gradual continuity from inorganic matter to organic molecules and from these to molecular

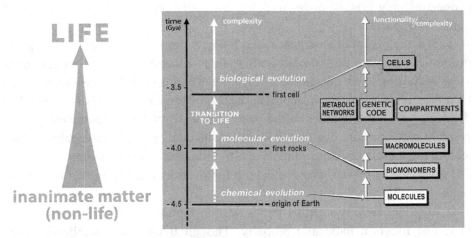

Figure 1.2 *Left panel*: according to Oparin and most of his followers, life on Earth originated from the inanimate matter via a continuous increase of molecular complexity and functionality. *Right panel*: the ideal representation, on a simplified and arbitrary time scale, of the continuous increase of complexity and functionality in the natural processes leading to the origin of life.

complexes, up to the onset of cellular life, with no qualitative gap between each stage. In this sense, then, the view expressed in the left panel of Figure 1.2 is the modern version of a kind of spontaneous generation, although on a sluggish time scale.

The left panel of Figure 1.2 appears as a simple-minded cartoon, but its acceptance has strong implications. The first one is that life is no longer a divine creation, but is instead due to natural laws. This brings immediately a conflict with the traditional religious view, particularly in our Christian world. This is an important issue, to which we will return in the next section.

There are other implications from Figure 1.2. Consider, for example, that the idea of a spontaneous increase of order and complexity – in a prebiotic world still deprived of the action of enzymes and nucleic acids – may appear at first sight as something against the thermodynamic common sense, which instead would suggest that natural processes bring about an increase in disorder and entropy. And, in fact, Oparin's proposition in this sense is a very bold one. We will learn, however, in the course of this book, that there are indeed natural processes that spontaneously produce order, and are even attended by an increase of entropy. This is an aspect of the concept of self-organization, and we will devote one chapter to that in Part III of this book.

However, the increase of molecular complexity alone would not be sufficient; we need also an increase of functionality. In this sense, the representation of Figure 1.2 has another important hidden message: that when a higher hierarchic structure is formed from constituent smaller parts, novel properties may arise – novel in the sense that they are not present in the constituent parts.

This is the notion of emergence, and the new properties are called emergent properties, indeed a central point in the whole apparatus: the anti-reductionist view, according to which – to use a well-known expression – "the whole is more than the sum of its parts." In addition, these new properties generally cannot be deduced with the properties of the parts. The construction of complexity is a creative phenomenon, one that will permit the quality of our molecular architecture to increase up to the ultimate quality: life itself. Given the importance of this point, we will dwell specifically on it for an entire chapter, in Part III of this book.

In addition, there is a further implication in Figure 1.2. The picture says, in fact, that life is a molecular property. Though it may be complicated and embellished by the notion of emergence, finally, according to Oparin, what we need to make life is "only" molecules and their interactions. It is at this point that the chemist can say: "Well, if it is so, let me just learn the basic reactions, and then I can make life in the laboratory." This is the idea that fuels prebiotic chemistry and the entire field devoted to the reconstruction of the origin of life. We will devote a substantial portion of Part I to this idea, starting from the famous Stanley Miller experiments in the next chapter.

Thus, we see that the simple-minded scheme traced in the left panel of Figure 1.2 contains *in nuce* the entire book, or almost all of it.

However, before turning toward chemistry, let us consider the philosophical implication of Figure 1.2. One is within the science camp and concerns the dichotomy between determinism and contingency; the other is between science and religion.

1.4 Determinism and contingency in the origin of life

The first of these two controversies can be expressed by the question of whether the origin of life is causally determined by the laws of physics and chemistry, or whether instead it is due to a unique, bizarre event resulting from contingent parameters – something that in the old nomenclature would be called *chance*.

The dichotomy between determinism and contingency is a classic theme in the philosophy of science (see, for example, Atmanspacher and Bishop, 2002) and in this chapter it will be considered only in the restricted framework of the origin of life (see also Luisi, 2003a).

Contingency, in this particular context, can be defined as the simultaneous interplay of several independent and concomitant factors to shape an event in a given space/time situation. For example, a tile falling on your head from a roof as you walk down the street below it can be seen as a chance event. But, in fact, it is due to the concomitance of many independent factors, such as the speed at which you walked, the sidewalk that you decided to use, the state of the roof, the presence of wind, the density of the tile's material, and so on.

In most of the epistemological literature, the word "contingency" has aptly replaced the terms "chance" or "random event" and, in fact, it has a different meaning, as I will demonstrate more directly later in this part. Note that each independent factor has a deterministic nature: the poor condition of the roof predictably determines some tiles sliding off and falling down, and the law of gravity does the rest. In this respect, the image that comes to mind is the one used by Maturana and Varela (1998) when discussing the subject of biological evolution. Consistently with Kimura's views on evolution, they use the metaphor of water falling from the top of the hill: the flow of water is due to gravity. However, the actual path is determined by the accidents on the ground – the trees, grooves, and the rocks encountered along the way, so that the actual downhill flow of water is a balance between the forces of determinism and contingency.

Particularly looking at evolution, one tends to consider a compromise between these two main factors: the compelling constraints of natural laws, and the vagaries of the environmental pressure. The invention of the wings, or of the flagellum, may be due to contingency, but then the wings and the flagellum have to respect the thermodynamics of the animal body, the laws of mechanics, the law of gravity, and so on.

The main point about contingency is that it is determined by many factors, each with an unknown statistical weight, and therefore, the event as a whole is unpredictable. If the contingent conditions are changed – perhaps only one of them – the result will be quite different. For instance, in the example given above, had you walked faster, the impact with the tile would not have happened.

Relative to the origin of life, the implications of this are profound. Through analyzing the main events in the evolution of life (see Figure 1.1), Stephen Jay Gould (1989) wrote his famous sentence, which has become an icon in the evolutionary biology:

... run the tape again, and the first step from prokaryotic to eukaryotic cell may take twelve billion years instead of two...

We can consider other well-known examples of contingency in the biological evolution. First of all, go back to *c.* 3 Gya, when only unicellular, bacteria-like organisms characterized life on our planet. At a certain point, one mutation in a particular colony of bacteria started the production of oxygen. It was not an obligatory event; this mutation might have happened millions of years later, or not happened at all. Then, most things that have occurred throughout the course of biological evolution might have been dramatically different.

Let us consider the "Cambrian explosion," when, there was a "relatively fast" passage from simple (still, mostly unicellular) organisms to a wide variety of complex and differentiated organisms. Again, it was not an obligatory event. It might have happened millions of years later, or not happened at all. And, as in the previous example, most developments in biological evolution might have been considerably different.

Another example is the asteroid, which 65 million years ago struck our planet, producing the long winter that eliminated the dinosaurs – animals that had dominated the planet for over 100 million years and might have lived happily for another 100 million years, or perhaps even longer.

The extreme consequence of this contingency view is that the onset of multicellular organisms, including humankind, may have not yet arisen or may never arise. This is contingency in its clearest form, which was also shaped by Jacques Monod's belief (Monod, 1971). Monod reiterated that the human species, being a product of contingency, might just as well not have come into existence, as a corollary to the famous notion of "being alone in the universe."

Compromises and balance between contingency and determinism may become very difficult in the question of the origin of life. The deterministic view assumes that the laws of physics and chemistry have determined the obligatory series of events leading from inanimate matter to life – that each step is causally linked to the previous one and to the next one by the laws of nature. As in all strictly deterministic situations, the state of a system at any point in time determines the future behavior of the system – with no random influences. In contrast, in the non-deterministic scenario governed by contingency, it is not possible to predict future behavior, and instead of a linear causal line of development, we may have a zigzagging of the pathway, whereby each step is determined anew by the set of parameters operating in that space in that very moment, and determined also by the specific structure of the evolving unit (structural determinism, as we will discuss later in the book).

These two views seem to be two extreme points of view, and the compromise on the origin of life appears difficult or impossible. Life is either obligatory, or due to contingency.

Considering first the deterministic point of view, we can refer to H. J. Morowitz, who in his well-known book on the emergence of cellular life (1992, p. 12), states:

We have no reason to believe that biogenesis was not a series of chemical events subject to all of the laws governing atoms and their interactions.

On this, we may all agree. Then he also adds something that is more critical to accept (p. 3):

Only if we assume that life began by deterministic processes on the planet are we fully able to pursue the understanding of life's origins within the constraints of normative science.

And he concludes with a plea against contingency (p. 13):

We also reject the suggestions of Monod that the origin requires a series of highly improbable events...

However, according to the late Christian de Duve, the extreme pioneer of the "absolute determinism" of the origin of life (2002, p. 55):

... I favor the view that life was bound to arise under the physical-chemical conditions that surrounded its birth...

Then he adds something much stronger (p. 298):

... It is self-evident that the universe was pregnant with life and the biosphere with man. Otherwise, we would not be here. Or else, our presence can be explained only by a miracle...

Interestingly, de Duve, a few pages earlier, writing about the evolution of life, asserts (p. 289):

[Evolution's ...] main mechanism [is] by natural selection acting on accidental genetic modifications devoid of intentionality. The finding of molecular biology can leave no doubt in this respect.

This complex and apparent set of contradictions testifies to the inherent difficulties of modern scientists in having a clear-cut view of the situation.

In this respect, the following statement by de Duve (1991) is also very interesting, and in keeping with the above quotation by Morowitz:

The science of the origin of life has to adopt the deterministic, continuity view – otherwise it would not be possible to adopt a scientific method of inquiry...

As I mentioned, the idea that life on Earth is the result of a deterministic pathway of highly probable and perhaps inevitable events, is frequent in the literature. In this regard, I would like to make a general point.

One cannot invoke natural laws, and then oblige them to follow an absolute determinism, namely an obligatory, very long and clearly complex pattern. Secondly, an obligatory pathway to life based on natural laws (thermodynamics and the like) would be tenable only if the pathway to life would correspond to a path of a minimum of energy – the most probable pathway – which is certainly not so in the case of the origin of life, or at least, whether it is so remains unknown.

I exchanged with de Duve, with due respect, a few letters on these arguments, an exchange that did not bring us any closer to clarity. He mentioned that few other important authors share his view – which is certainly true. In fact, for me this represents a problem, in the sense that I cannot conceive how a man of science can share the citation given above on the universe being pregnant with humankind. I can only explain it by assuming that these people hold in their heart – not necessarily at a conscious level – that life is a gift given by some transcendent divinity. I know that my colleagues who are for de Duve's view would not accept this badge of "crypto-creationists" – and I believe in their good faith. However, the unconscious is something one is not conscious of – particularly in a society dominated for so many centuries, and still today, by Christian creationistic religion.

Some authors criticize the claim of the inevitability of life on Earth; for example, Szathmáry calls it the "gospel of the inevitability" (2002), and Lazcano (2003) has similar words. This "inevitability" view has its counterpart in the notion that contingency is the basic creative force for shaping the molecular and evolutionistic constructs on Earth (which de Duve, 2002, dubbed "the gospel of contingency"). It should be said that de Duve accepts contingency, but in a context other than the origin of life (de Duve, 2002).

The contingency view on the origin of life and biological evolution is not new; actually, it is an old icon in the history of science. We have mentioned already Jacques Monod with *Chance and Necessity* (1971), his colleague François Jacob with *The Possible and the Actual* (1982), and the books by Stephen Jay Gould, who is perhaps the most cited author on contingency in biological evolution (see for example, Gould, 1989).

Actually, de Duve's statement concerns not simply life at the level of prokaryotes, but rather reaches up to the level of humankind. To me, to reach this level, we should consider that the origin of life is not one single event, but instead several different ones in a discontinuity scale (refer again to Figure 1.1). The first event was from organic matter to the first prokaryotes, certainly the most crucial of all. However, almost equally unpredictable was the jump from prokaryotes, probably from Archaea (Williams *et al.*, 2013; Spang *et al.*, 2015) to early eukaryotes, with the insertion of mitochondria (Lane, 2009) and chloroplasts, of algae or plant ancestors (Archibald, 2009, and references therein). There are also further critical steps, such as the jump from unicellular to multicellular organisms, and the jump from asexual to sexual reproduction, which were not a sequitur. And so on.

Contingency assumes that a cocktail of occasional and unpredictable environmental factors shaped each of these steps. There is a great distance from this concept to de Duve's "gospel of inevitability."

The same position of the inevitability of the origin of life on Earth is taken by the novel form of creationism, called intelligent design (ID). This movement does not negate the importance of the natural laws, but rather assumes that they have been guided by an Intelligent Designer. To be fair, one should add that Christian de Duve, in his latest book, *Singularities* (2005) takes a stand against intelligent design.

Before proceeding further, I would like to share the thoughts of Albert Eschenmoser, in what will be the first "conversation" hosted in this book, one whose message should accompany us for the rest of these pages.

Conversation with Albert Eschenmoser

 **Albert Eschenmoser** was born in Erstfeld (Uri), Switzerland, on August 5, 1925, received his Ph.D. at the Swiss Federal Institute of Technology (ETHZ) with Leopold Ruzicka in 1951 and served as professor of organic chemistry at the ETHZ (1960–1992). In the early 1960s, Eschenmoser began work on what was the most complex natural product to be synthesized at the time – vitamin B_{12}. In a remarkable collaboration with his colleague Robert Burns Woodward at Harvard, the two teams – engaging about 100 coworkers in the course of 12 years – accomplished two syntheses of the molecule in 1972. The work is considered a landmark in the history of organic chemistry. The Eschenmoser fragmentation, the Eschenmoser sulfide contraction, and Eschenmoser's salt are named after him. After the B_{12}-project, Eschenmoser embarked on an extended investigation of the etiology of the nucleic acid structure and demonstrated that not only RNA and DNA, but also oligomer systems such as threose nucleic acids and pyranosyl-RNA, constitute informational oligomers also belonging to his production.

Among his many honors (the list has more than 30 entries) are the Davy Medal (Royal Society, London, 1978), the Tetrahedron Prize for Creativity in Organic Chemistry (1981), the ACS Cope Award (1984), and the Wolf Prize in Chemistry (1986). Eschenmoser is a member of the German Academy Leopoldina (Halle), Academia Europaea (UK), and the Pontifical Academy (Vatican), as well as a foreign member of the American Academy of Arts and Sciences (Boston), the National Academy of Sciences (Washington), the Royal Society (UK), the Academy of Science in Göttingen, and the Croatian Academy of Sciences and Arts (Zagreb). His work, and his personality, were instrumental to graduate more than 140 doctoral students, and an equal or greater number of young post-doctoral coworkers from Switzerland and abroad were his students: the best numerical indication of a fruitful action within science at large and within the field of organic chemistry in particular.

PLL: You have been following the field of the origin of life for many years. Do you feel that we are closer now to the understanding of the chemical process leading from non-life to life? Or do you still see big conceptual and/or experimental problems?

ALBERT ESCHENMOSER: We must differentiate between *understanding* and *knowing*. In essence, we think to *understand* the thermodynamic and kinetic circumstances that were the drivers of the chemical process that led from non-life to life, but – of course – we neither *know* the chemistry, nor the location of the process. Understanding the thermodynamics is far from *knowing* how to synthesize a complex chemical structure, not to speak of a process such as biogenesis. I nevertheless believe that we are continuously "getting closer" by discovering chemical facts that are bound to be relevant to that search for knowledge: It is potentially relevant to know, that the basic building blocks of life are elementary chemical molecules, elementary in the sense that they are formed whenever excessive energy hammers on chemical matter. It is potentially relevant to know that under a variety of conditions that can lead to the formation of the four pentose diastereoisomers, it is *ribose* that is formed fastest. It is potentially relevant to know that informational base pairing is by no means a property exclusive to the structure type of the natural nucleic acids, but that rather a variety of other backbone structures can support that property which is so fundamental to life. It is potentially relevant to know that in vitro-evolution of RNA can lead to self-replicating RNA sequences and that informational oligomers containing non-natural backbones are evolvable toward specific properties. It is this kind of gradual accumulation of potentially relevant facts that has to continue. To think of, and to explore experimentally the chemistry of a potentially vast number of chemical scenarios that *could* have given rise to life's origin is the goal we can and have to pursue. The vastness of that number is and will remain to be one of the major problems.

What a paradox: billions of dollars are spent for operations that aim at finding out whether there is water on Mars. Because if there is water, then there might be, or might have been, life on the planet. Is it not amazing, considering the degree of our ignorance of how life arose, how undiscerning and loose such a conclusion sounds as co-justification for spending billions? Would it in that very context not be logical to invest also billions for boosting experimental research on the chemistry of life's origin? Such that the gap that yawns between the detection of water and the conclusion that there might have been life on Mars would be filled by as much knowledge as is knowable on the chemistry of life's origin and on prerequisites of biogenesis that go beyond the requirement of water? Whether there exists life elsewhere in the universe and whether life exists as a consequence of intrinsic necessity or of contingency are questions so fundamental that origin of life research in laboratories of chemistry and biology would justify a CERN-type initiative.

Would the money that would come from such an initiative help? Needless to emphasize that funding worldwide mostly flows to research that promises to be useful and not to work on problems of "natural philosophy". Given the degree of freedom and security that money provides, more chemists would join the field, fascinated by the depth of the challenge, top scientists among them attracted by the top problem of chemistry as a science in the 21th century: the creation of artificial chemical life in the laboratory. Creating artificial chemical life and searching for the origin of the life we know are two sides of the same coin. I may be forgiven for self-quoting what I had written in a recent review:

"We'll never be able to know" is a truism that leads to resignation with respect to any experimental effort to search for the chemistry of life's origin. But such resignation runs radically counter to the challenge imposed upon chemistry as a natural science. Notwithstanding the prognosis according to which the shortest path to understanding the metamorphosis of the chemical into the biological is by way of experimental modeling of "artificial chemical life," the scientific search for the route nature adopted in creating the life we know will arguably never truly end. It is, after all, part of the search for our own origin.

PLL: What do you suggest to younger people engaging in the field of the origin of life?

ALBERT ESCHENMOSER: What I would propose to young chemists entering the field is: Focus in your research on problems that promise to have dual relevance, namely, work on questions that can lead – to be sure – to results relevant to the origin of life problem, but not only: these results should at the same time constitute valid scientific contributions to contemporary chemistry. Relevance to the origin of life problem is *potential* relevance, and it may turn out to be short-lived; a result that is acknowledged as contribution to contemporary chemistry will be understood and appreciated by many, it will consolidate your career.

PLL: Are you a believer? And does this create some problem with the secular view, that life on Earth comes from inanimate matter?

ALBERT ESCHENMOSER: First, I think you should refrain from asking in a scientific context whether a scientist is a "believer." It sounds like you would "look down" on people whose world view is based on religion. This, we should not, and I think I know you well enough to know that you do not. The answer to the second part of your question of whether I have "some problem with the secular view that life on Earth comes from inanimate matter" is an unambiguous and unconditional: no, I do not. To me, the view of Science on our universe and its potentials is immensely beautiful and satisfying – irrespective of the uncounted intra- and extrapolations it implies and in which we have to "believe." Your question reminds me of my revered teacher Leopold Ruzicka (1887–1976), born and grown up as a Catholic in Croatia, who wanted as a boy to become a priest, yet as a scientist became an outspoken agnostic, a member of the papal academy as well. As a retiree after his career as natural products chemist he developed an interest in the origin of life problem and began in this context studying the new literature on biochemistry with the enthusiasm and intensity so characteristic of him. . . He became overwhelmed by looking as an organic chemist at all those chemical "miracles" in the world of biochemical reactions, and he made no secret out of it. Once in a private discussion he said to me: "Die Biochemie, die wäre eines lieben Gottes würdig!" At that time still too young to be the agnostic I am today, I replied: "Ja, Herr Professor, aber der Konjunktiv in Ihrer Feststellung, der ist das Problem."[5]

1.5 The question of creationism and intelligent design (ID)

We have mentioned that Oparin's view of the prebiotic evolution to life was destined to find opposition from the Christian believers. The same happened almost one century earlier with the theory of Darwinism. It is not always a frontal opposition: most of the theologians and church doctors are, in fact, ready to accept the view of evolution, but add hastily that the conditions to generate life from the inanimate matter, or to go from one species to another, have been given by God. This kind of statement *per sé* would not be a problem. The real problem for science, at large, lies in the fact that something has been created with the name of Creation Science, or Scientific Creationism. This is a branch of creationism that attempts to provide *scientific* support to the Genesis creation narrative. Thus, the world is being created in seven days *ex nihilo* within the last 10,000 years), challenging the geological and astrophysical evidence for the age of Earth and of the universe, challenging the relevance of fossils, and of course all forms of Darwinian evolution. It is also apparent that the anti-Darwinian movement comes not so much from research theology, but rather from sidekick zealots (see, to this respect, the short editorial by Holden, 2005). As for myself, I would be more sympathetic towards the creationists' camp if experimental evidence were to be provided. It is not difficult to conceive what this should be: simply find *equally old* fossils of horses, dinosaurs, hominids, trilobites, snails, mammoths, turtles, cyanobacteria, and swordfish – as implied in the notion of creation in a single act.

As long as this simple evidence is not forthcoming, it is probably safe to be scientifically very skeptical about the creationistic view (in this sense, it is almost funny that the creationists lament some gaps in the theory of evolution).

There is nothing new in the arguments of ID, particularly in their main point, that things of life are too complex for having arisen spontaneously, and therefore, they must have been designed in a predetermined way by a transcendent mind.

Actually the notion of ID has a classic, famous old reference. This is in the writing of William Paley, the Anglican priest who became well known for

---

[5] The translation from the German goes: "Biochemistry would be worthy of a Dear God!" The answer: "Yes, professor, but the problem is your putting it in conditional form."

having introduced one of the most famous metaphors in the philosophy of science, the image of the watchmaker (Paley, 1802):

... When we come to inspect the watch, we perceive [...] that its several parts are framed and put together for a purpose, e.g. that they are so formed and adjusted as to produce motion, and that motion so regulated as to point out the hour of the day; that if the different parts had been differently shaped from what they are, or placed after any other manner or in any other order than that in which they are placed, either no motion at all would have been carried on in the machine, or none which would have answered the use that is now served by it. [...] the inference we think is inevitable, that the watch must have had a maker – that there must have existed, at some time and at some place or other, an artificer or artificers who formed it for the purpose which we find it actually to answer, who comprehended its construction and designed its use.

Living organisms, Paley argued, are even more complicated than watches, thus only an intelligent designer could have created them, just as only an intelligent watchmaker can make a watch. According to Paley:

That designer must have been a person. That person is GOD.

Paley's metaphor was already negated in his time by Hume and other contemporary philosophers. And much more emphatically in recent times by Dawkins in his *Blind Watchmaker* (1990, p. 5), who answers back after almost two centuries:

Paley's argument [...] is wrong, gloriously and utterly wrong. The analogy between telescope and eye, between watch and living organism, is false [...] Natural selection, the blind, unconscious, automatic process that Darwin discovered [...] has no purpose in mind [...] it does not plan for the future, it has no vision, no foresight, no sight at all.

Dawkins reminds us (Dawkins, 1990 and 2002) that nature proceeds without an *a priori* purpose, but rather according to the "tinkering" mechanisms advocated by Monod's colleague François Jacob (Jacob, 1982) a few years earlier. See also the above citation by de Duve. This is an important point; we will come back to that in the next chapter, talking more in detail about contingency.

One of the arguments used by ID people pivots on the idea that according to evolutionary science, life has no purpose, and therefore is deprived of moral codes. The ID movement also uses strange pseudoscientific arguments, for example funny and elaborated probabilistic calculations to show that the creation of a living cell from the molecular components by mere chance would be impossible (as if a serious scientist would have ever proposed something so silly). In addition, most of them insist in negating natural selection and evolution as the origin of biodiversity and emphasize alleged weak points and certain aspects of Darwinism. Of course, as in any other theory, there are points in Darwinism that are not yet fully clarified. Darwinism certainly does not explain everything – and the entire structure of Darwinism is actually undergoing an evolution in this period. But, of course, these points are not ignored by Darwinian scholars and are amply discussed in the normal classes and conferences on Darwinism. What ID people do is dwell in a classic negationists' attitude, without providing any alternative hypothesis or theory. In fact, ID has not published a single paper in a referred science journal.

The problem, as already mentioned, is that ID is spreading, carrying with itself ignorance and misconceptions about science.

The success is due to a combination of factors – but basically, all pivots on the fact that preaching ignorance and fear is rather efficient to convince uncritical, simple-minded listeners; and this combined with a philosophy well rooted in fundamentalism, and turned into a political movement with political lobbies, usually associated with wealthy and right-wing-oriented, influential politicians. To fight against this cloud of falsehood should be the duty of any scientist who cares for truth and respect of reason. It is the old battle between the clergy and Galileo, the eternal fight against superstition and blindness.

1.6 SETI and the anthropic principle

Interestingly enough, in addition to ID, the notion of the inevitability of life is present in several other fields of science. In my opinion the anthropic principle (AP), for example, belongs to this category – and I believe that finally, at its roots, is equivalent to a kind of ID. AP can be expressed in different ways but the basic idea is that the universal constants, the geometric parameters, and all things of the universe are so finely tuned in order for life and humankind to appear and develop (Barrow and Tipler, 1986,1988; Davies, 1999; Barrow, 2001; Carr, 2001). It is the *post hoc* argument that since we are so improbable, our presence must signify a purposeful universe.

The anthropic principle can be expressed in more sophisticated forms, but I believe that my simplification given above is not far from the target. In fact, one reads in the primary literature, for example in Paul Davies' book (1999):

If life follows from (primordial) soup with causal dependability, the laws of nature encode a hidden subtext, a cosmic imperative, which tell them: 'Make life!' And, through life, its by-products, mind, knowing, understanding. . .

Contrary to the ID, the adherents to the anthropic principle are scientists, often with an impeccable pedigree. In addition, it is interesting that the anthropic principle finds more supporters among physicists than biologists, who remain in general rather skeptical about this (see, for example, Erwin, 2003).

The view of the astonishing fine-tuning is held, although not always expressed as an adherence to the anthropic principle, by several authors in the field. For example, Freeman Dyson (1985) writes:

As we look out in the universe and identify the many accidents of physics and astronomy that have worked together to our benefit, it almost seems as if the Universe must in some sense have known that we were coming.

We can even add a citation (Shermer, 2003) from Stephen Hawking, a self-defined atheist (although in his book, the word "God" appears on almost every other page):

And why is the universe so close to the dividing line between collapsing again and expanding indefinitely? [...] If the rate of expansion one second after the Big Bang had been less by one part in 10^{10}, the universe would have collapsed after a few million years. If it had been greater by one part in 10^{10}, the universe would have been essentially empty after a few million years. In neither case would it have lasted long enough for life to develop. Thus one either has to appeal to the anthropic principle or find some physical explanation of why the universe is the way it is.

Do the scientists of the anthropic principle present scientific arguments in favor of their view – other than the wonder about the fine-tuning of the universal constants? One scientific argument that is often used by adherents of the anthropic principle is the ubiquity of biological convergence: the fact that the paths of evolution are relatively few (see, for example, Conway-Morris, 2003). However, it has already been argued by Gould that architectural constraints limit adaptive scope and channel evolutionary patterns (see also Erwin, 2003).

There are fierce arguments against the anthropic principle's view. The most general is the argument that the great laws of nature are the way they are because otherwise there would be no life; this is a truism at many levels. For example, things are also the way they are in other parts of the universe, and slight changes in geometric distances would bring about cosmic catastrophes. Yet there is no life there. Aside from the fine-tuning of the constants of the universe, there would be no human life if in our biosphere there were a little more oxygen, or less oxygen; or a somewhat higher temperature, or a lower one; or less humidity, or more humidity, and so on. The same is true in the molecular world: if on Earth there had been only diketo-piperazines instead of amino acids; or if sugars did not have the size they have; or if lipids were three times shorter, then there would have been no life. Of course, life, being life, is a granted mystery. Somebody once said, "If you believe in the existence of life, then you can believe in anything." The point is to accept and wonder about its beauty, but without switching off the inquiring reason.

Also note a general, important point about all this fine-tuning (both astro-physical and chemical): that fine-tuning can only concern the *necessary conditions* for life, and by no way is compelling regarding the *sufficient conditions*.

Concerning AP, it must be added that though it may not be valid at all from the scientific point of view, it is nevertheless a beautiful, fascinating story.

For me – and several others – part of the fascination arises with the link between AP and the modern aspects of cosmology and quantum physics. For example, in the book of the particle physicist Susskind (2005) – as reviewed by Ellis (2005) – the fine-tuning is seen in terms of the "multiverse" – the existence of a huge number of "pocket universes" – and with the "landscape of possibilities," *per sé* a recent discovery in string theory. Moreover, Paul Davies (2007) adds a touch of sophistication when he links the question of the constants of the universe to an even larger context: the old question of Leibniz and Heidegger, "why there is something instead of nothing." Paul Davies likes to see consciousness as a creative force in nature, as a teleological, immanent principle that makes life and consciousness appear as a necessary feature of the universe. Thus, for the universe to be meaningful, it must include us – *us* meant as life and embodied consciousness. Interesting, this view has been present in oriental philosophies – and classic Buddhism in particular – for thousands of years.

Let us now leave AP and ID to consider another scientific movement that in my opinion seems to operate outside the framework of contingency. This is the field of SETI (Search for Extra Terrestrial Intelligence). In this camp, scientists are trying to catch signals from the cosmos. This is a research based on the assumption that alien civilizations may exist and are willing to communicate with us (Huang, 1959; Kuiper and Morris, 1977; Sagan, 1985; Horowitz and Sagan, 1993; Sagan, 1994; Barrow, 2001; Wilson, 2001; and perhaps you will want to reread the article by G. G. Simpson, 1973, on possible alien civilizations).

The assumption of intelligent life elsewhere is based on the assumption that the same or a similar set of conditions for creating intelligent life is operative on that other (unknown) planet. Not only should one then believe in the determinism of life on our planet, but also in a kind of cosmic determinism that leads to the occurrence of life on other planets: determinism squared.[6]

Again, it is far from me to throw a lance against SETI. Personally, I think that this is a great vision, and that visions in science should be encouraged, particularly in an era in which mostly pragmatic and applied or military research projects find support. My point is only to emphasize that this SETI movement is also based on the belief that life is inevitable and widespread.

Generally, people who are firmly grounded in contingency are not too sympathetic to the idea of alien life. I mentioned above the sentence by the "contingentist" Monod, according to which we are alone in the universe. See also the

---

[6] For an interesting discussion on the relationship between SETI and ID (intelligent design), see the article by Robert Camp in *The e-mail Newsletter of Skeptics*, www.skeptic.com, in the *Skeptic Magazine* of February 16, 2006.

consideration made in the previous section, that to have human-like life means, from the evolutionary point of view, the occurrence of several events in succession, each of them very complex and *per sé* unpredictable. It would be a big finding already if we would find somewhere life at the "simple" level of unicellular microorganisms.

1.7 Panspermia – and bringing in C.G. Jung

I believe that conceptually there is a certain affinity between SETI and the idea of a general panspermia, which assumes that life on Earth originated elsewhere in the universe and came to us in the form of some vaguely identified germs of life. This view has appealed already to the ancient Greeks, such as Anaxagoras (500–428 BC), right through to Hermann von Helmhotz and William Thomson Kelvin at the end of the nineteenth century. Then it goes to Svante Arrhenius in the beginning of the twentieth century, ending with Francis Crick (1966, 1968, 1980) and the work of Fred Hoyle and Chandra Wickramasinghe (1999, 2000) in our time (see also Parsons, 1996 and, for example, Britt, 2000). These different versions have different names, such as Arrhenius' radio panspermia, Crick's directed panspermia, ballistic panspermia (meteorites), or modern panspermia from comets (Hoyle and Wickramasinghe, 1999). In the more general and poetic version, the theory of panspermia sees life as a general property that permeates the cosmos and therefore does not need to have an origin (Hoyle and Wickramasinghe, 1999; Britt, 2000).

How far can one go with this idea? Wickramasinghe *et al.* published a paper in the well-respected medical journal *The Lancet* with the theory that SARS has a panspermia origin (see the comments by Ponce de Leon and Lazcano, 2003).

In general, on the issue of contingency versus determinism, the large majority of scientists nowadays are probably on the side of contingency. For most chemists, molecular biologists, and physicists, the notion of contingency is almost trivial. However, it is also true – as we have seen – that a significant part of the scientific population rejects the rationality of contingency and favors an obligatory deterministic view of the origin of life.

How does one explain this basic dichotomy in the same generation of scientists? An easy way to describe the contradiction is to say that scientists, having pushed God out of the front door, let him enter again through the back door. More than God *per sé*, I believe it is the notion of the sacredness of life that has sneaked in the back entrance.

In this regard, Carl Gustav Jung's archetypes of the collective unconscious come to mind. An archetype is the part of the mental structure that is common to all humankind and that, according to Jung and his scholars (von Franz, 1988; Meier, 1992), represents the creative matrix of all conscious and unconscious functions. In their exchange of letters (Meier, 1992), the well-known physicist Wolfgang Pauli and C. G. Jung discuss at length the influence of archetypical mind structures on science.

In our specific case, we would have to invoke a collective unconscious structure (the archetype of the sacredness of life?) that influences the *Weltanschauung* of scientists to maintain the sacred nature of life. This archetype would not appear with the same intensity in all scientists, but would be more manifest in those, for example, who have, or have had, a religious background. Of course, by definition of the unconscious, the beholder is not aware of his mental behavior.

All these "crypto-creationist" movements tend to negate contingency and chance as the constructors of life and humankind, as reiterated by the following extract from Monod (1971):

We would like to think ourselves necessary, inevitable, ordained for all eternity. All religions, all philosophies and even part of science testify to the unwearyingly heroic effort of mankind, desperately denying its own contingency. . .

I will return to the controversy between determinism and contingency in one of the next chapters, with a general, more down-to-earth question: why was a certain type of molecular form selected instead of a different one?

For example, why has the five-membered ring ribose been selected and not, for example, the six-membered pyranose ring? To deal with this question in an experimental way is a constructive way of understanding the nature of life. This is the approach taken by Albert Eschenmoser and his group at the Swiss Federal Institute of Technology (ETH) of Zürich (Bolli *et al.*, 1997a, 1997b). In one of his essays, Eschenmoser reflects on the relation between the cosmic anthropic principle and the fine-tuning of chemical-biological life. He considers the specific case of RNA, and writes (2003):

. . . the strategy may read as follows: Conceive (through chemical reasoning) potentially natural alternatives to the structure of RNA, synthesize such alternatives by chemical methods, and compare them with RNA with respect to those chemical properties that are fundamental to its biological function. (See also Eschenmoser, 1999.)

1.8 Only one start – or many?

This question – whether life started only once in one particular place on Earth, or several times in several places – is also partly related to contingency and determinism. Probably most "determinists" would say that, since life has a very high probability of arising, there is no reason why it should have started only once and only in one magical place. "Only once" is instead a notion appealing to "contingentists": if the conditions to start life were the product of contingency – a particular set of chemicals in particular concentrations at a particular temperature and pressure and pH and so on – it would be almost impossible to have such conditions repeated.

There is, however, an extreme view of the notion of "life starting many times" that does not comply to this scenario. This is the view of C. Schwabe (Schwabe and Warr, 1984; Schwabe, 2001), who starts from the hypothesis that life comes from a distribution of nucleic acids, and that this distribution must have been widespread all over Earth, so that there was not one, nor two or three, but multiple starts. He then goes to the extreme of saying that all species on Earth have an independent origin – a billion independent origins. And, of course, if one assumes such an absolute determinism, then the laws of chemistry and physics would produce the same products at each different start. According to him, then, there was never a time on Earth with only one kind of species, and the development of species was parallel rather than sequential. All this is highly controversial, but worth mentioning as a basis of discussion.

Concluding remarks

We have seen how the simple-minded left panel of Figure 1.2, inspired by the work of Alexander Oparin, is rich in conceptual implications and whose answer is anything but simple. It would be interesting to know what percentage of the world population really accept this picture – that life namely has a natural origin. Certainly, the great majority of scientists do so today, but scientists are only a small minority of people. Thinking of the Christian religiosity in the West, and considering the eastern countries with their religions or mythologies, I wonder whether Figure 1.2 would be the most popular credo if we were to take a global poll. Many of the people who refuse this notion would, however, likely be ready to accept the view that yes, there is a molecular evolution from the inanimate matter to life, but this is the plan of (some) God. Also interesting would be a poll among the scientists who accept Figure 1.2, when we ask: since you accept contingency, are you ready to accept the idea that humankind might have not originated? One would expect a plain "yes" to that, but I know by experience that it is not so simple – here, the archetypes may play an important role.

One argument of the creationists against the scenario of Figure 1.2, and/or Darwinism in general, is that life would not be worth living without the presence of a creator who has given us the moral laws and the aim of existence. I, with many others, believe instead that once we accept the view that the moral laws are self-generated values of the human mind, the human mind being itself the emergence of a complex system of natural processes, then we should arrive at a great respect and honor for life as it is.

We have seen so many ideas on the origin of life, which is to be expected for such an important and difficult area of thought. And it is perhaps important, at this point, to summarize the most important property of life, at least from the point of view of this book and this author – in the sense that this summary will influence the spirit of the whole book. From a molecular point of view, which is the main focus of the prebiotic research on the origin of life, the main characteristic of life on Earth is molecular order. The molecular essence of life is the ordered sequences of proteins, and the ordered sequences of nucleic acids and their mutual interactions. We could say that it is like a language: we have the letters of the alphabet, which are the 20 amino acids and the 4 bases of the nucleic acids, and all these letters must be organized in a precise order – as with an alphabet making words – to have a meaning.

We need, however, one final note on the notion and definition of order, as we use in "ordered structures." Considering, for example, a protein sequence, the sequence does not have a long-range periodicity; there are no elements of symmetry in the chain. We are dealing with "aperiodic order," which is actually the highest form of order: each element (each amino acid) must be in the predetermined position. In nature, the DNA gene, again a case of aperiodic order, dictates this polypeptide order. Aside from the gene causality, however, we can talk about order when there is a mechanism (this can be, for example, the thermodynamic control) that produces multiple copies of the same sequence. In other words, the existence of a polypeptide sequence in many identical copies is a signature of order.

Chapter 2

The hardware

Introduction

In the previous chapter, we have seen the richness of concepts and philosophical implications that the origin of life is bound to arise. The implementation of the origin of life is, however, a chemical process, and it is necessary then to look at the chemistry underlying such a process. In this chapter, we will take a look at the chemical compounds which were there "at disposal" of the emerging life. We will also review some basic reactions taking place among these compounds in prebiotic times. In a way, this will give us a picture of the hardware on which the Oparin's ladder argument might have been developing. Chemistry, however, has its own rules and in the following chapter, we will then see how the constraints of chemistry may influence and possibly hinder the implementation and our own understanding of the process of growth from non-life to life.

2.1 What did we have 4 billion years ago?

Before considering what may happen in a warm little pond, let us have an overview of the inventory of the chemical compounds possibly available on Earth at the onset of the chemical process of the origin of life, say 4 Gya (billion years ago). Generally, we will have to deal with endogenous material, made on Earth; and exogenous material, coming from space. In recent years, bioastronomy (some prefer the term *astrobiology*, others *exobiology*) has, in fact, played a very important role in the field of the origin of life (see Zhao and Bada, 1989; Miller and Bada, 1991; Chyba and Sagan, 1992; Oró, 2002). A very interesting and important point in this regard is the similarity of the molecules from space and the molecules of Earth, which indicates the unity of basic chemistry in the universe.

The amount of space dust falling presently on Earth amounts to c. 40 000 tons per year (much less than in the long past), which, spread over all Earth's surface,

31

Table 2.1 *Major sources (in kg/yr) of prebiotic organic*
compounds in the early Earth (from Oró et al., 2002)

Terrestrial sources	kg/yr
UV photolysis	3×10^8
Electric discharge	3×10^7
Shocks from impacts	4×10^2
Hydrothermal vents	1×10^8
Extraterrestrial sources	
Interplanetary dust particles (IDP)	2×10^4
Comets	1×10^{11}
Total	$\mathbf{10^{11}}$

corresponds to only 8×10^{-9} g cm^{-2} per year (Love and Brownlee, 1993; Miller, 1998; Oró, 2002). On the other hand, accumulation over billions of years can still result in a considerable amount of matter.

Let us consider first Table 2.1, which gives a compact account of the main sources, both terrestrial and extraterrestrial, of prebiotic organic compounds in the primitive Earth. Notice the considerable impact of photolysis (perhaps the main source of spontaneous chemical reactions on Earth) and hydrothermal vents, and, among the extraterrestrial sources, the considerable impact of comets.

It may be interesting to compare these data with another exhaustive inventory of prebiotic material on Earth (4 Gya ago), which was carried out by Chyba and Sagan (1992). They distinguished three sources:

1. Exogenous contribution from meteorites, micrometeorites, interplanetary dust particles (IDP), comets;
2. Endogenous synthesis associated with impact of such bodies;
3. Endogenous production associated with other available energy sources (solar UV, lightning, radioactivity).

The total contribution would correspond to 10^8–10^{11} kg/year, depending upon the degree of atmosphere-reducing conditions.

According to Commeyras *et al.* (2005), the presence of methane in the primitive atmosphere would have oriented endogenous syntheses towards a considerable production of nitriles, especially hydrocyanic acid (HCN). The latter compound could have reached *c.* 10^{10} kg/yr. Accordingly, assuming the volume of the primitive oceans to have been identical to that of the present-day oceans, the primitive oceans could have contained (at a steady state) between *c.*

0.4 10^{-3} g/L of total organic matter in the case of a neutral atmosphere, and 0.4 g/L, in the case of a reductive atmosphere.

One question, which is not yet completely clarified, is the origin of organic nitrogen chemistry, in particular because of the short wavelengths (less than 100 nm) required for the photolysis of N_2 (Selsis, 2000).

Another related question is the production of nitric oxide (NO). Part of this NO, dissolved in the primitive ocean, could have been reduced to NH_3 by Fe^{2+} in the ocean, even in the presence of high levels of cyanide (see Summers and Chang, 1993; Summers and Lerner, 1998). The ammonia thus produced could have made a major contribution to primitive organic synthesis and especially to that of α-amino acids.

Now, let us consider in more detail the chemical structure of these compounds coming from space. Table 2.2 gives an account of the main α-amino acids found in chondrite meteorites.

More than 70 different amino acids (including also non-α amino acids) have been found in chondrite meteorites, mostly in the Murchison and Murray meteorites. This gave a total abundance of 75 μg per gram of meteorite. It is, of course, questionable whether this low rate of exogenous carbon delivery might have been of relevance for the emergence of life on Earth. Asparagine and glutamine were never found in meteorites, but they would have been readily and completely hydrolyzed (into aspartic and glutamic acids, respectively) during either the meteorite aging or extraction process.

Very interesting is the work of Groen and coworkers. They study the aromatic molecules delivered to the young Earth during the heavy bombardment phase in the early history of our solar system and consider that these compounds were likely to be among the most abundant and stable organic compounds available (Groen *et al.*, 2012). To investigate the possible role of aromatic molecules, these authors incorporated different polycyclic aromatic hydrocarbons (PAH) in the membranes of fatty acid vesicles. This was mostly to determine whether PAH could function as a stabilizing agent, similar to the role that cholesterol plays in membranes today. They conclude their study by stating that, in fact, their data represent the first indication of a cholesterol-like stabilizing effect of oxidized PAH derivatives in a simulated prebiotic membrane.

The most striking finding in this field was the chirality of some chondrite amino acids. In particular, six extraterrestrial amino acids showed an *ee* (enantiomeric excess) between 5.0 and 9.2 percent (Cronin and Pizzarello, 1997; Pizzarello and Cronin, 2000). These amino acids were, however,

Table 2.2 *Molecules coming from space. List of most important α-amino acids (from 2 to 7 carbon atoms) and α-amino diacids found in chondrite meteorites. Many others (such as β- or γ-amino acids) are not listed here. (Data kindly put at my disposal by Auguste Commeyras)*

2 carbon atoms	3 carbon atoms	4 carbon atoms	
glycine	alanine	2-amino-butyric acid	N-methyl-alanine
	sarcosine	2-amino-isobutyric acid	N-ethyl-glycine

5 carbon atoms

valine	isovaline	norvaline	proline

6 carbon atoms

leucine	pseudoleucine	cycloleucine	2-amino-2-ethyl butyric acid
isoleucine	2-methyl-norvaline	norleucine	2-amino-2,3-dimethyl butyric acid

7 carbon atoms

2-amino-heptanoic acid	2-amino-2,3-dimethyl valeric acid	2-amino-2,4-dimethyl valeric acid
2-amino-2-ethyl valeric acid	2-amino-3,4-dimethyl valeric acid	2-amino-4,4-dimethyl valeric acid
2-amino-3-ethyl valeric acid	2-amino-3,3-dimethyl valeric acid	2-amino-2-methyl caproic acid
2-amino-3-methyl caproic acid	2-amino-4-methyl caproic acid	2-amino-5-methyl caproic acid
2-amino-2,3,3-trimethyl butyric acid	2-amino-2-ethyl-3-methyl butyric acid	

α-amino diacids

aspartic acid	2-amino-2-methyl succinic acid	2-amino-3-methyl succinic acid
glutamic acid	2-amino-2-methyl glutaric acid	2-amino adipic acid	2-amino pimelic acid

α-methylated derivatives; this argues, together with the D/H and $^{13}C/^{12}C$ isotopic ratio, for their non-biological origin. It is generally assumed that this *ee* might be due not to chemo-selectivity during their synthesis, but rather to enantioselective photo-degradation, probably via circularly polarized UV light.

It has also been found that sugars coming from space are partly chiral (Cooper *et al.*, 2001). The existence of these non-racemic compounds opens up interesting perspectives in the field of homochirality, a very important subject indeed for the origin of life. We will come back to this later on.

In addition to amino acids and other simple compounds, several aliphatic carboxylic acids up to C_8 were detected (Knenvolden *et al.*, 1970), as well as alkyl phosphonates (Yuen and Knenvolden, 1973; Cooper *et al.*, 1992; Pizzarello and Weber, 2004) and heterocyclic pyrimidines (Hayatsu *et al.*, 1975; Stocks and Schwartz, 1982).

Comets are perhaps the most generous source of organic compounds. They are particularly rich in HCN (Miller, 1998), which appears to be very common in space. For this reason, several authors have devoted their attention to it (Oró, 1961; Ferris and coworkers, 1973, 1974, 1978; Matthews, 1975). We will address this further in our discussion on the prebiotic formation of nitrogenous bases of nucleic acids, including an approach to their oligomerization (Ferris and Ertem, 1992, 1993).

Furthermore, membrane-forming compounds may come from space (Lawless and Yuen, 1979; Deamer, 1985; Deamer and Pashley, 1989; Deamer *et al.*, 1994; McCollom *et al.*, 1999). The observation that membranogenic compounds can also be of prebiotic origin is of particular interest, especially in view of the "membrane-first" hypothesis, or more generally, for making the point that membranes and vesicles were present very early in the prebiotic scene (Deamer *et al.*, 1994).

Putting together carbonaceous compounds produced from classic prebiotic chemistry, hypothermal vents, and compounds from space, we reach the conclusion that there was enough material on Earth to start the chemistry of life. Of course, the distribution on Earth of these molecules was not homogenous, and one common bias is to assume that there was one place where the accumulation of material and its distribution was particularly rich and fortunate – the warm little pond of Darwin's memory. As already mentioned, this image has given rise to the metaphor of the "prebiotic soup" – a phrase that has remained in the popular press, but is now rarely used in scientific literature. When it is used, it refers broadly to the set of conditions where the wet organic chemistry of life began.

Conversation with Sandra Pizzarello

Sandra Pizzarello is Emeritus Professor and Research Professor at the Arizona State University (Tempe, AZ 85287–1604 USA), in the Department of Chemistry and Biochemistry.

She is mostly known for her work on astrobiology, and her research of over 30 years has been devoted to the study of the organic material in carbonaceous chondrite meteorites, with emphasis on the molecular, isotopic, and chiral characterization of their soluble compounds. She is the one who discovered with John R. Cronin the presence of enantiomeric excess in meteoritic amino acids (1997, *Science*, **275**, 951–55) and most of her research has focused on the study of the non-racemic amino acids of meteorites, with compound-specific isotopic analyses as well as model syntheses that would mimic their prebiotic catalytic activity and reactions.

She is also president of ISSOL (*International Society for the Studies of the Origins of Life*).

PLL: More and more data are coming from astrobiology, showing that chirality appears to be a cosmic product. Do you think that this is a strong support for the hypothesis that homochirality of our life comes from space? And is this not due to contingency on our own planet?

Sandra Pizzarello: Astrobiology has not produced much in a way of support/proof – people are still checking the possible presence of CPL (circularly polarized light) in UV. However, the study of the products of astronomical processes – meteorites – clearly shows that chirality is present in a-biotic organic compounds, which we may also consider prebiotic. First came the evidence for amino acids, more recently for sugars, see the presentation of G. Cooper at the ISSOL. And it is a fact that chirality may lead to asymmetric catalytic syntheses and their effects may be vast reaching.

PLL: These meteorites, like the famous Murchison meteorite, come from space. Maybe they come not from within our solar system, but from outside of it. Do we have any idea where they come from?

Sandra Pizzarello: Meteorites are, for the most part, fragments of asteroid and asteroid are those small planetesimals that orbit the Sun in great number between Mars and Jupiter. They are the remnants of a planet that never formed, due to various gravitational games of the giant planets. Therefore, they are considered the carriers of pristine early solar materials, as it came to be in the Solar System from collapsing pre-solar materials. In effect, however, many had their own planetary processes, such as heating, water alteration, etc., which contributed to the formation of organic materials, some from pre-solar precursors and some, probably, from scratch.

PLL: Another unrelated question: a few papers indicate that fatty acids – which are the "prebiotic Deamer's surfactants" – may be originated by the Fischer Tropsch reaction, followed by oxygenation. I had the impression that you don't believe too much in this idea. . . is this true?

SANDRA PIZZARELLO: It is not that I don't believe in Deamer but, if citing meteorites, we should remember that their amount of organic compounds usually decreases with chain length – here the length of the fatty acids. Therefore I think that [is] what we would need: a process where compounds are not present in high concentration.

Fisher Tropsch reactions could be a source, for example around volcanoes, an environment preferred by many for the origin of useful molecules if not the origin of life. By the way, let me add that last year I wrote a paper (which I consider rather boring, while apparently it has been broadly read and commented) where I was describing some polyethers in the insoluble part of the carbonaceous chondrite meteorites. Maybe these can be useful. . .

PLL: I remember that some years ago there was some opposition, mostly from the side of "old" members, to integrate ISSOL with astrobiology. One of the arguments was that NASA wanted to incorporate ISSOL and actually take the lead on this research. Now, in the last ISSOL meeting in Nara, about one-half of the contributions were on astrobiology – and some participants thought that this was indeed too much.

SANDRA PIZZARELLO: The *International Society for the Studies of the Origins of Life* (ISSOL) has the distinction to have been founded to present and discuss the scientific theories about the origins of life when they first appeared in the twentieth century. We can argue that, like everything in science, these theories had their basis on previous studies (Francesco Redi and his *Generazione degli Insetti*, Wohler and the synthesis of urea, Pasteur and the end of spontaneous generation and of course Darwin and evolution) but nevertheless the ISSOL was the brain child of Oparin who, with Haldane, proposed a transitional origin of life from non living molecules.

There is also no doubt that the experimental impetus in OL studies came from the fall of a meteorite, Murchison, which led to the first discovery of a-biotic extraterrestrial organic molecules.

That was in 1969 and NASA was fundamental from then on in the funding of these studies through the "Exobiology Program." NASA was actually predominant in its funding. I remember contacting a NSF officer inquiring on the possibility of some funding and I was told, quite rudely I must say, that if I was funded by NASA, they could not help.

Would have I preferred ISSOL kept the old definition from its acronym? Yes but only for esthetic reasons. . . The only concern I have now is that, between the plethora of OL conferences, workshops, and hullabaloos, ISSOL does not deviate from its requirements of scientific honesty and excellence in any of its activities.

So, to the last program and related question. That was redacted on the basis of an equal participation of Bioastronomy and ISSOL in the Origins conferences, which had been created six years ago by wise men.

They are no more due to various contingencies and you may see the ISSOL website for details. The next conference will be ISSOL 2017 and I hope to see there many "origins of lifers" old and new.

PLL: Thanks. Last question: do you see any significant progress, judging from the last two or three ISSOL meetings, on our understanding of the origin of life on Earth?

SANDRA PIZZARELLO: No... extensive progress yes, but still relating to the different venues of chemical and biological evolution, the **crux** of the problem is still there to be discovered...

2.2 Molecules from hydrothermal vents

We have already mentioned that a considerable enrichment of prebiotic moieties may have come from submarine vents and other hydrothermal sources (see, for example, Miller and Bada, 1988; Holm and Anderson, 1998; Stetter, 1998). This kind of investigation started with the 1979 discovery of deep-sea vents with black smokers, which are associated with an extraordinary abundance of the most phylogenetically primitive organisms on Earth. This ecosystem is sulphur based, and is distinct from the more familiar photosynthesis ecosystem that dominates Earth's surface (see also Corliss *et al.*, 1981).

The sulphur-based chemistry brings us to the work of Wächtershäuser, concerning the formation of initial metabolism (1988, 1990a, 1990b, 1992, and 2000). One important metabolic cycle is the one that goes from acetic acid to pyruvate. According to the author, pyruvate is important because it occurs in various metabolic pathways, notably in the reductive citric acid cycle (see also Morowitz *et al.*, 2000). In summary, the pathway goes from the production of acetic acid through metallic-ion catalysis, to the production of three-carbon pyruvic acid by the addition of carbon; and the addition of ammonia to produce amino acids, from which peptides and proteins are possibly derived (see also Lazcano, 2004).

The richness of the research on hydrothermal vents and black smokes, as well as hydrogen sulfite, has given rise to the enthusiasm that this might have been the origin of life. To that, one should remark that in all these vents and smokers, we are still dealing with low molecular weight compounds, and one can have all the low molecular weight compounds of this world and still be unable to make life, not even get close to it. You need ordered macromolecules interacting with themselves, and this is only the necessary, and not the sufficient, condition.

2.3 The chemistry of life. From Oparin to Miller – and beyond

Having indulged in the chemistry of space and of deep oceans, let us go back to Earth to resume the discourse of the origin of life.

As introduced in Chapter 1, prebiotic chemistry started with the theories of Alexander Oparin. As early as 1922, at a meeting of the Moscow Botanical Society, he asserted that there is no fundamental difference between a living organism and non-living matter. The multifaceted combination of manifestations and properties distinctive of life must have arisen as a part of the process of the evolution of matter. Moreover, the simple solutions of organic matter and the arrangement of their atoms into a molecular structure – he said – were governed by the properties of their component atoms. Gradually, the growth and increased complexity of molecules brought new properties into being and a new colloidal-chemical order developed as a successor to more simple relationships between and among organic chemicals.[1]

Oparin's theory was developed in his first book (1924) and expanded upon in the following ones, as briefly mentioned in Chapter 1. In his book of 1936, he suggested that the atmosphere of the young Earth was strongly reducing, containing the raw materials for the evolution of life: methane, ammonia, hydrogen, and water vapor. New properties were determined by the interactions of such, as they bore more complex molecules (emerging properties). He stated that this process brought biological orderliness into prominence.[2]

The first American scientist, who consciously made experimental chemistry based on Oparin's hypothesis (and Haldane's hypothesis[3]), was the graduate

---

[1] Oparin proposed that the "spontaneous generation of life" that had been attacked by Louis Pasteur did, in fact, occur once, but was now impossible because the conditions found on the early Earth had changed, and preexisting organisms would immediately consume any spontaneously generated organism.

[2] Oparin argued that a "primeval soup" of organic molecules would combine in ever more complex ways until they formed coacervate droplets. These droplets would "grow" by fusion with other droplets, and "reproduce" through fission into daughter droplets, and so have a primitive metabolism in which factors that promote "cell integrity" survive, and those that do not become extinct (Oparin, 1953; 1957). Some modern theories of the origin of life still look at Oparin's view and ideas as a starting point.

[3] We have seen also that John B. S. Haldane propounded a similar chemical theory, a few years after Oparin's idea, in a short paper (1929). Independently from Oparin, Haldane postulated that the primitive ocean was a "vast chemical laboratory" containing a mixture of inorganic compounds – like a "hot dilute soup" (distant in dimension scale, but not in meaning, from the "warm little pond" imagined by Darwin in 1871) – in which organic compounds could have formed. Under the solar energy, the reductive atmosphere containing carbon dioxide, ammonia, and water vapor gave rise to a variety of organic compounds, "living or half-living things." The first molecules reacted with one another to produce more complex compounds, and ultimately the cellular components. At some point, a kind of "oily film" was produced that enclosed self-reproducing molecules, thereby becoming the first cell (Haldane, 1929, 1954).

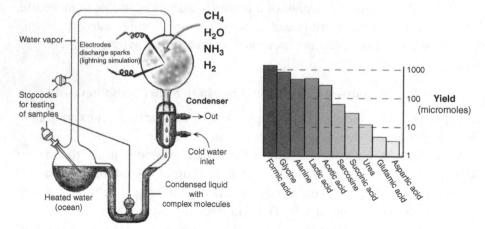

Figure 2.1 The famous Stanley Miller experiment: a strongly reducing atmosphere containing the four gaseous components (CH_4, NH_3, H_2, and H_2O), considered by Oparin the constituents of the initial Earth atmosphere exposed to electric discharge. *On the left*, the experimental scheme. *On the right*, the yield of an experiment. More details in the text.

student Stanley Miller, in the early 1950s. Fascinated by Oparin's idea, he tried and finally succeeded in convincing his already famous mentor, Harold C. Urey, to give him some time to try an experiment. How the young Miller arrived at celebrity is now part of the narrative on the origin of life (see also Bada and Lazcano, 2003 and Lazcano and Bada, 2003).

Miller filled a flask with the four gaseous components assumed by Oparin to be the constituents of prebiotic, reductive atmosphere: hydrogen, ammonia, methane, and water vapor (see Figure 2.1). By passing electrical discharges through this flask to simulate primordial lightening, he was able to witness the formation of several α-amino acids and other substances having biological importance (Miller, 1953).

The experiment was published in 1953, the same year as the discovery of the double helix by Watson and Crick, a memorable year indeed for biochemistry (in the same period, Sanger had shown that proteins are linear chains of amino acids covalently bound to each other).

There is only limited agreement on whether primitive atmospheric conditions were really reductive. However, this is not the key point. The key point is that relatively complex biochemicals, amino acids in particular, can be formed from a mixture of very simple gaseous components in a chemical pathway that can indeed be regarded as prebiotic. As we will discuss in the next chapter and in more detail in Chapter 5, there is now a consensus that prebiotic amino acids are 10 in number.

Miller (1953) explained the formation of the α-amino acids by the Strecker reaction (Equation 2.1), whereby the start is a carbonyl compound – a ketone or an aldehyde – which is condensed with ammonium chloride to yield α-amino nitrile, which is subsequently hydrolyzed into the amino acid. For example, in the original Strecker reaction, acetaldehyde, ammonia, and hydrogen cyanide combined to form after hydrolysis alanine.

$$(2.1)$$

While usage of ammonium salts gives unsubstituted amino acids, primary and secondary amines also successfully give substituted amino acids. The usage of ketones gives α,α-di-substituted amino acids. The Strecker reaction is also considered the most plausible pathway for meteoritic α-amino acid synthesis.

There is an interesting follow up of the original Miller experiments. Jeffrey Bada and collaborators (Parker *et al.*, 2014) noticed that Miller intermittently added the plausible prebiotic condensing reagent cyanamide, but that, for unknown reasons, an analysis of the samples was not reported. These authors analyzed the archived samples for amino acids, dipeptides, and diketo-piperazines; they discovered 12 amino acids, 10 glycine-containing dipeptides, and 3 glycine-containing diketo-piperazines. According to the authors, these results highlight the potential importance of condensing reagents in generating diversity within the prebiotic chemical inventory. The discovery of simple peptides is particularly interesting, and this for two reasons. Firstly, peptides can be used for the condensation into longer polypeptide chains, which may eventually fold into proteins; furthermore, some of these shorter peptides are endowed with catalytic properties – mostly with protease activity – and this may also lead to the biogenesis of longer chain peptides.

Going back to Miller's synthesis in the flask, the question is why α-amino acids have been obtained and not, for example, β-amino acids, nor diketo-piperazines.

This is a very important question for the origin of life on Earth. It has to do with the general question: "Why this and not that?" in the products of nature; and more specifically, it has to do with the fact that many stable, good compounds are excluded from the construction plans of nature, whereas other molecules, which are less stable and less common, have been selected as important bricks of life. We will discuss this general question in Chapter 3, as I believe that this is one of the most interesting questions concerning the biogenesis of the molecules of our life.

About these arguments, let us see the opinion of two distinguish researchers in the field, David Deamer and Eors Szathmáry.

Conversation with David Deamer

David Deamer began his academic career in 1967 at the University of California, Davis, and has been a research professor at the University of Santa Cruz, California, since 1994, where he chairs the Department of Biomolecular Engineering. He is well known for his work on membrane biophysics and for proposing that the base sequence of single nucleic acid molecules can be determined by drawing them through a nanoscopic pore in a membrane.

In his research on the origin of life, Deamer was among the first to show that stable membranous vesicles can assemble from simpler molecules than the phospholipids used by all life today. For instance, fatty acids like oleic acid readily form vesicles. In 1985, he extended this observation to the amphiphilic compounds present in carbonaceous meteorites, showing that they can also assemble into microscopic vesicles. His pioneering efforts inspired later developments in Luisi's and Jack Szostak's group.

PLL: One thing that concerns me about origin of life research is that we haven't seemed to make much progress beyond the time of Stanley Miller. To illustrate this point, I can think of six questions that represent vast gaps in our knowledge:

1. What is a plausible mechanism for the prebiotic synthesis of polymers like nucleic acids or proteins?
2. The first such polymers would have random sequences, so how did the sequences become ordered into structures required for them to function as genes or catalysts?
3. An even more difficult question is how polymers began to replicate?
4. How did the genetic code emerge within systems of interacting polymers that ultimately evolved into ribosomes?
5. What were the first metabolic pathways that could capture energy and nutrients from the environment?
6. And finally, given that systems of replicating catalytic polymers might be encapsulated within membranous compartments to form protocells, how did the protocells divide into daughter cells?

A pessimist might conclude that these questions will never have satisfactory answers, but I know that you are more optimistic. It would be nice for our readers to hear your thoughts.

DAVID DEAMER: I understand your skepticism/pessimism very well and could make a similar argument if I wanted to. I suppose my relative optimism comes from considering the history of DNA. In 1923 it would have been impossible to understand the molecular structure of genes, but 40 years of incremental progress finally allowed Watson and Crick to deduce the double helix. With that single insight, the field of molecular biology has blossomed into the enormous enterprise we see today, in which the entire human genome can be scrutinized by anyone with internet access.

We have also made 40 years of incremental progress in research on the origin of life, but I am not at all disturbed that no one has yet discovered how life can begin. It's a much more difficult problem than DNA structure, isn't it? And so few people are devoting themselves to the problem. Nonetheless, I have the impression that "we are looking through a glass darkly."

PLL: Your introductory notes at the ISSOL conference in Nara were very much appreciated. There, among others, you enumerated the facts that we know now in the field, and which we did not know a few years back. Can you please summarize these facts?

DAVID DEAMER: I can't list everything, of course, but I can provide a few examples that represent incremental steps toward understanding how life can begin.

First off, I think you will agree that we have gone far beyond Miller in understanding how the simple monomers of life can be synthesized in plausible prebiotic simulations. Not just amino acids, but also nucleobases, carbohydrates, and amphiphilic compounds like fatty acids. These compounds are also present in carbonaceous meteorites, so we know they can be synthesized by non-biological reactions in the early solar system and delivered to the Earth.

We also know that there were oceans over 4 billion years ago, and microfossils have been discovered in rocks 3.5 billion years old, so life began in that time interval. We didn't know that 40 years ago.

We have learned about molecular self-assembly. For instance, polymers like proteins fold into the functional structures we call enzymes, and membranes spontaneously self-assemble from simple soap-like compounds and readily encapsulate large molecules to form protocells. That's much easier than we would have guessed 40 years ago.

The next answer is very important. We know that RNA can fold into structures called ribozymes that have catalytic activity and can evolve. A ribozyme can even function as a polymerase, and it's possible that the first life lived in an RNA World, and that DNA and protein enzymes came later. Crick and Orgel were guessing about catalytic RNA 40 years ago, but now we know that it's real, and that a ribozyme even serves as the active catalytic center of ribosomes.

We can now determine the base sequences in DNA and RNA, and from this deduced that life on Earth exists as three domains. One of the domains has extremophiles that thrive at elevated temperatures, in acidic pH ranges and salty water, precisely the conditions that were likely to prevail when life began on the early Earth, or even on Mars.

Now, we have all heard the saying that there is room at the top, and for us that means we can look out at the universe and imagine that life might exist in solar systems around other stars. But Richard Feynman famously said that there is also room at the bottom. He was referring to particle physics, but for us it means nanotechnology, because research on single molecules is now possible. In my lab we routinely watch single DNA molecules being captured by a single polymerase enzyme and processed in real time. The molecular machinery of a living cell works at nanoscopic dimensions, and I think that single molecule techniques have much to tell us about the origin of life.

PLL: This is all well and good, but does not solve the problem of the origin of life. In this field, the dominant idea is still the RNA-world, best exemplified by the work of Jack Szostak and Jerry Joyce. My impression is that researchers working with RNA may think that all problems have been more or less solved thanks to ribozymes and their ability to evolve. Something about this confidence reminds me a bit of fundamentalism. What is your feeling on this?

DAVID DEAMER: Luigi, as active scientists, you and I both know that research, like politics, is the art of the possible. Each of us evolves during a career in science, picking up knowledge and tools that match our particular interest. If we are clever, we then apply those tools to a problem that can be solved. The scientists working with RNA have developed a very sophisticated set of tools and can use them to test significant hypotheses. For instance, who would have guessed that RNA can evolve in a test tube as it goes through multiple generations? Well, it does, and that gives us insight into how polymers can evolve when subjected to selection and amplification. Who would have guessed that a ribozyme would be the active center of a ribosome? In fact, no one guessed that, but it was discovered by Harry Noller and adds weight to the conjecture that RNA came first. These, I think, are the incremental steps, the individual pieces of the puzzle that will finally make sense when we can fit them all together.

PLL: Last question. In the past, in addition to your work on vesicle chemistry and nucleic acid biophysics, you managed to produce a CD using DNA sequences to make music. Can you tell us how it was done, and whether we should expect more?

DAVID DEAMER: I took accordion lessons as a teenager growing up in Ohio, so music has always been part of my life. During the 1980s the first base sequences of DNA began to be published, and when I looked at a sequence I was struck by the fact that it resembled musical notation. The four bases of DNA are adenine, thymine, guanine, and cytosine (ATGC). Three of them happen to be musical notes, and by changing T into the note E, the base sequences in DNA can be played on a piano by anyone. Most sequences just sound like random notes, but occasionally a remarkable melody emerges. For instance, the insulin gene turns into a sprightly jig, and the ALU transposon, representing 10 percent of the human genome, becomes a haunting melodic waltz. An internet search for "DNA music" now turns up dozens of musicians worldwide who are translating nucleic acid sequences into music.

I still enjoy looking for musical rhythms and melodies in DNA. Recently I had a memorable experience in which someone wrote to say that she had a mutated BRCA1 gene and was undergoing treatment for breast cancer. She asked whether I could send her a melody for the correct BRCA1 gene without the mutation so that she could listen to it and pass the time during chemotherapy sessions. In fact, the BRCA1 mutation can be a single wrong nucleotide out of thousands in the gene, so she would not hear any difference. Instead, I used a sequence of 30 bases around one of the reported single point mutations, then corrected the wrong base and turned it into a nice repetitive melody that could be recorded on our synthesizer. It was sent to her as an email attachment, and I hope it helped her cope with the treatment and then recover.

Conversation with Eörs Szathmáry

Eörs Szathmáry is a Hungarian theoretical evolutionary biologist at the late Collegium Budapest Institute for Advanced Study and at the Department of Plant Taxonomy and Ecology of Eötvös Loránd University, Budapest.

His main interest is theoretical evolutionary biology and focuses on the common principles of the major steps in evolution, such as the origin of life, the emergence of cells, the origin of animal socictics, and the appearance of human language. Together with his mentor, John Maynard Smith, he has published two important books which serve as the main references in the field (*The Major Transitions in Evolution*, Freeman, 1995, and *The Origins of Life*, Oxford University Press, 1999). Both books have been translated into other languages (so far, German, French, Japanese, and Hungarian). He serves on the editorial board of several journals (*Journal of Theoretical Biology, Journal of Evolutionary Biology, Origins of Life and Evolution of the Biosphere*, and *Evolutionary Ecology and Evolution of Communication*).

Professor Szathmáry was awarded the New Europe Prize in 1996. The Hungarian Academy of Sciences acknowledged his outstanding scientific contribution with the Academy Prize in 1999. He was invited to prestigious institutions, including the Wissenschaftskolleg zu Berlin and the Collège de France. He is a member of Academia Europaea and the Hungarian Academy of Sciences.

PLL: You have been interested in the general field of the origin of life for many years. Have you seen any major finding or development since Miller's experiments in 1953? What is (are) for you the most important new discoveries relevant for prebiotic chemistry since then?

Eörs Szathmáry: Yes indeed. I think that the experiments by Miller, although "catalytically" important for the field, have not delivered too much; the problem with the reductive atmosphere is just one of the concerns. I think there have been important conceptual and experimental advances. In the first category I would name the science of autocatalytic sets, which has been developed by people at various levels of abstraction. Rosen, Varela, Kauffman, and Gánti belong to these visionaries.

Interestingly, they published their first ideas roughly at the same time. It has taken quite a while for people to take notice, since they have been beguiled by molecular biology and the Miller experiments. Also, the work by Manfred Eigen (again in the same period), especially quasispecies theory and the molecular cooperation problem (but not his favored solution to it) also stands out. Experimentally I value the studies on spontaneous template replication (von Kiedrowski) and some of the work on repro-ducing vesicles (Luisi). Sometimes important experiments have been driven by theory (Wächtershäuser). This could be said for the RNA world and ribozymes, also (Joyce, Szostak). For a later stage of evolution, the work by Yarus on the possible ingredients of the emergence of the genetic code is exciting. And we shall see how far Sutherland will get with the small molecules.

PLL: The most popular and most active research in the field is within the RNA-world. What is your general thinking about that? How do you see the chance of success of this scheme, based on self-replication and evolution of a prebiotic self-replicating RNA molecule?

EÖRS SZATHMÁRY: The RNA world is a very important idea. I am convinced that it DID exist, but not in its naked and conceptually depauperate form. It is unlikely that it all started with a single RNA. I am quite sure there had been collectively autocatalytic systems of different kinds before, with limited evolvability, and RNA may not have been the first informational template replicator either. The RNA world could have flourished only by passive (on rocks) or active (protocells) compartmentation. And it could not have been internally clean, by which I mean that there must have been many small molecules around for metabolism, including amino acids and (unavoidably) peptides. Look at nucleotide synthesis today: some amino acids are needed as raw material.

PLL: In the RNA-world scenario, there is an identification between the criterion of life and the criterion of self-replication/evolution. Do you agree with this identification?

EÖRS SZATHMÁRY: Yes and no. Many evolutionary biologists for obvious reasons share the view you mentioned. They have a point, because it is the Darwinian dynamic that results in adaptation. As a student of Gánti I abide by the view that conceptual sloppiness is detrimental. There is the living world and there are living systems. Both real and some computer (!) viruses do evolve, but they are not living systems. Units of evolution and units of life have a large, but not complete overlap in my view. A mule is not a unit of evolution but it is alive. A virus is evolvable but not alive. According to James Griesemer, if there was an RNA world before metabolizing protocells, we should say that the living world preceded living systems (a bonus for the cited view). These thoughts square nicely with Gánti's concept of absolute and potential life criteria. The latter are necessary for the creation and maintenance of a living world only, not for individual living systems as such. Now, of course, we would like to see unfolding life, so the NASA approach, although conceptually sloppy, is practically useful. For living systems, of course, we are interested in the primitive ones here. Gánti's conceptualization (chemical super system composed of three different auto-catalytic systems, i.e., chemoton = metabolism with template replication with boundary) is not only elegant, but also erects the right goalpost. It is a cornerstone of the emerging field of systems chemistry.

PLL: The notion of compartment is not present in the research related to the RNA-world, except for the work of Jack Szostak. Do you think this is the right approach?

EÖRS SZATHMÁRY: See above. Parasites ruin the RNA world without some form of compartmentation. But in the beginning it may not have been protocells. Rock surfaces and pores are a poor man's compartments.

PLL: What are your present scientific interests?

EÖRS SZATHMÁRY: Ehhhm, my most exciting adventure is to find out whether complex thinking is based on real-time evolution in the brain. Actually, there are analogies here. For example, my favoured computational account of language is the Fluid Construction Grammar (FCG) by Luc Steels. It *is* very chemical in nature, with linguistic molecules and catalysts in the system. More specific to the origin of life, I have the following on the table. First, the evolvability of chemical systems without template replication. Then the role (positive or negative) of sex in the protocell era; also, the coevolution of metabolism with membrane selectivity, and experimental demonstration (with Andrew Griffiths in Paris) of some possible interim stages of early evolution (such as the one portrayed in the stochastic corrector model, which is a bag of genes running metabolism in a reproducing compartment). In addition, there is the grand picture: I have just submitted a paper: "Towards major evolutionary transition theory 2.0."

2.4 Prebiotic nitrogen bases

Following Miller's experiment, several chemists successfully synthesized other compounds of biochemical relevance under prebiotic conditions, thereby demonstrating convincingly that several molecular bricks of life – in addition to those coming from space – might have been endogenously present on prebiotic Earth.

This is the case for the bases of nucleic acids. A possible prebiotic route to adenine has been described long ago (Oró, 1960; Oró and Kimball, 1961, 1962; see also Shapiro, 1995), as shown in Figure 2.2. For details, see also Miller's review (Miller, 1998). Guanine and the additional purines such as hypoxanthine, xanthine, and di-amino-purine could also have been synthesized by variations of the above synthesis (Sanchez *et al.*, 1968).

The prebiotic synthesis of pyrimidines is based on cyano-acetylene, which is obtained in good yields by sparking mixtures of methane and nitrogen; and by the reaction between cyano-acetylene and cyanate (Sanchez *et al.*, 1966; Ferris *et al.*, 1968 – see Miller, 1998). This is shown in the Figure 2.3, taken from Miller (1998), which also shows that cytosine can be converted into uracil.

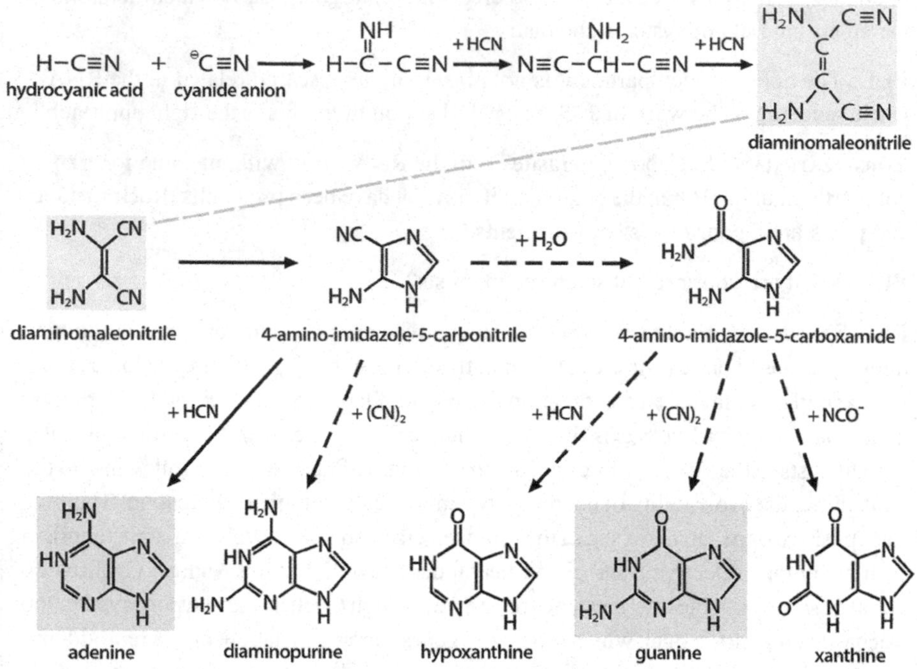

Figure 2.2 Prebiotic synthesis of purines (redrawn, from Oró, 1960).

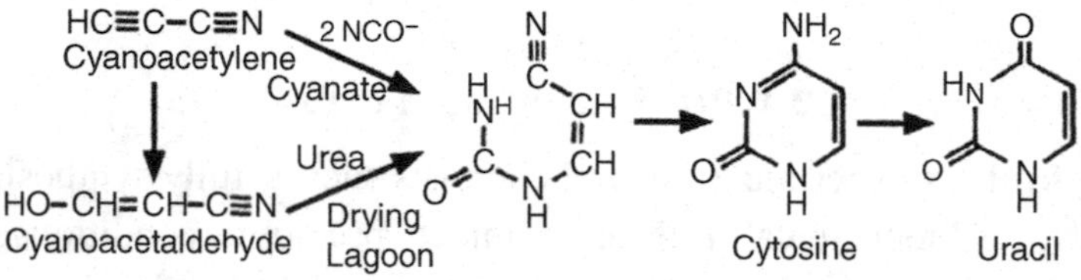

Figure 2.3 The prebiotic synthesis of pyrimidines cytosine and uracil. (Adapted from Miller, 1998.)

It should be kept in mind that these prebiotic syntheses concern only the bases, and not the mononucleotides of the nucleic acids. The mononucleotide consists of three moieties attached to each other – the base, the phosphate, and the sugar. Whether mononucleotides could be the products of molecular pre-biotic evolution has been a question for several years, and only a few years ago, the Sutherland group gave a positive answer to that (Powner *et al.*, 2009 and 2010). The route is described in Figure 2.4, reproducing a scheme from a Sutherland teamwork paper (Powner and Sutherland, 2010). It represents the crowning of intense studies by this group (Buchet and Sutherland, 2006; Sutherland, 2007; Anastasi *et al.*, 2007). Their latest paper is also very beautiful

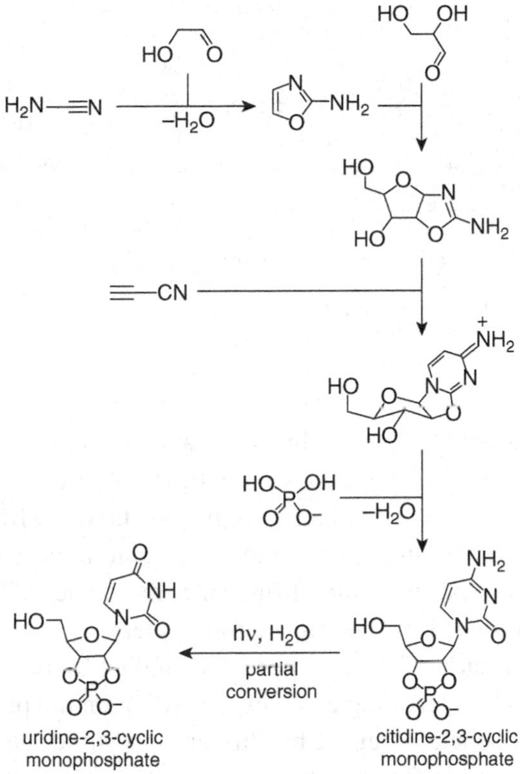

Figure 2.4 The presumed prebiotic pathway of the biogenesis of mononucleotide phosphates. (Adapted, with a few modifications, from Powner and Sutherland, 2011.)

and important (Patel *et al.*, 2015), in which they show that precursors of ribonucleotides, amino acids and lipids can all be derived from hydrogen cyanide and some of its derivatives. The implication is that all the cellular subsystems could have arisen simultaneously through common chemistry, and possibly within the same "warm pond." The key reaction steps are driven by ultraviolet light, use hydrogen sulfide as the reductant, and can be accelerated by Cu(I)–Cu(II) photo redox cycling. We will come back to this paper a couple of times in this book, paying special attention to its relevance for the minimal cell.

2.5 Sugars

Concerning sugars, let us begin with the well-known Formose reaction (see Figure 2.5), an autocatalytic non-enzymatic process that, starting from formaldehyde, proceeds, through glycolaldehyde and glyceraldehyde, to reach four-carbon and five-carbon sugars. This reaction, however, gives a

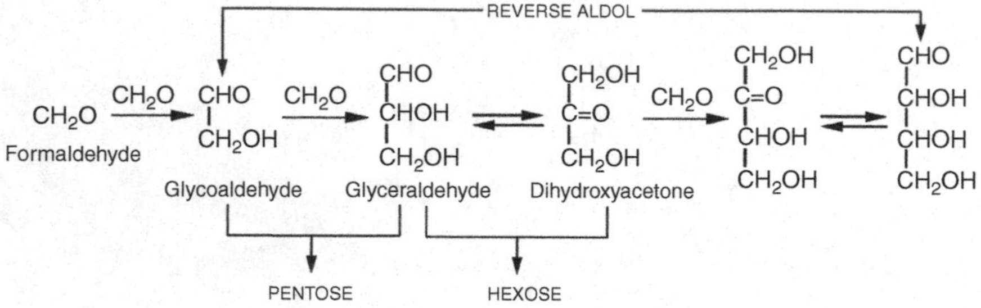

The hardware

Figure 2.5 The Formose reaction.

wild variety of sugars, both straight chain and branched. Ribose occurs in the mixture, but is not particularly abundant (Decker *et al.*, 1982); furthermore, this compound is chemically rather instable (Miller, 1998). The interest in this reaction lies also in the fact that it is a process of self-reproduction: note that we start from one molecule of glycolaldehyde, which is again produced at the end of the cycle.

The prebiotic formation of sugars has been studied by Eschenmoser and his group at ETH Zurich. Interesting is the investigation on the prebiotic formation of ribose-2,4-diphosphate obtained by this group (Müller *et al.*, 1990); some results are summarized in Figure 2.6.

Regarding the prebiotic ribose, see also Shapiro's critical analyses (1984, 1986, and 1988).

2.6 Redox reactions

One should mention here again the work of Wächtershäuser (1988, 1990a, 1990b, 1992, 1997, and 2000). Having in mind the possible source of reductive power in a prebiotic scenario, Wächtershäuser and coworkers gave the following equation:

$$FeS + H_2S \rightarrow FeS_2 + 2e^- + 2H^+ \tag{2.2}$$

They stressed that this reaction of FeS with H_2S has a reducing power to drive the primordial metabolism. They argue that a number of non-spontaneous reactions can be coupled to the previous redox process, including the reaction of CH_3SH, from which thio-esters such as CH_3COSH can be synthesized (Heinen and Lauwers, 1997; Huber and Wächtershäuser, 1997). And from the latter compounds, other interesting and more complex molecules can be obtained, for example, fixation of CO_2 (Nakajima *et al.*, 1975):

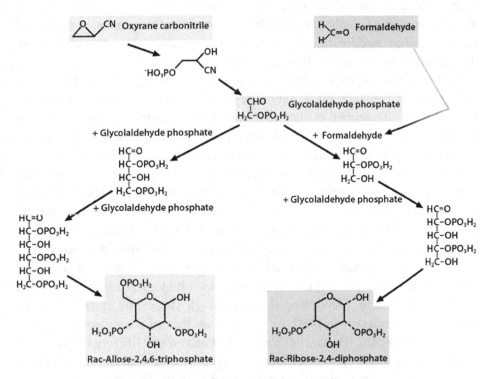

Figure 2.6 The biogenesis of sugars poly-phosphate, in particular the ribose diphosphate, starting from formaldehyde and oxirane carbonitrile, as obtained by Eschenmoser' group. The central role of glycolaldehyde phosphate in this synthesis is shown (redrawn from Müller *et al.*, 1990).

$$CH_3COSH + CO_2 + FeS \rightleftarrows CH_3COCOOH + FeS_2 \qquad (2.3)$$

This work is part of a general concept on the origin of life from Wächtershäuser, and we will come back to that later on.

2.7 The Fischer-Tropsch reaction

This reaction converts carbon monoxide and hydrogen into liquid hydrocarbons:

$$(2n + 1) H_2 + n CO \rightarrow C_nH_{(2n+2)} + n H_2O \qquad (2.4)$$

where n is typically 10–20, mostly as straight-chain alkanes. It was first developed by Franz Fischer and Hans Tropsch in Germany in 1925 and used extensively in Germany during the Second World War as low-sulfur diesel fuel, an alternative to petroleum.

In addition to alkane formation, competing reactions give small amounts of alkenes, as well as alcohols and other oxygenated hydrocarbons. A variety of catalysts can be used for the Fischer-Tropsch process, and the most common are the transition metals, cobalt, iron, and ruthenium.

There are two reasons why this reaction may have biological importance. The first is that it can take place naturally, giving rise to abiotically originated hydrocarbons; the second is that the products of oxidation of alkanes can give origin to fatty acids, which are considered prebiotic membrane-forming agents.

Concerning the first point, one should recall that there was – and still there is – a controversy concerning whether petroleum is really of biological origin (the common view) – or whether instead it has an abiogenic origin. The beginnings of this story dates back to the nineteenth century, when the French chemist Marcellin Berthelot and the Russian chemist Dmitri Mendeleev proposed plainly that hydrocarbons such as natural gas and oil are primordial materials – i.e., they were originally embedded to Earth during the process of planetary accretion and have no intrinsic connection with biological material near the surface of the Earth. Mendeleev also made such statements in 1877, but his theory was revived only in 1950. As we all know, the orthodox view about oil is that it is coming from remnants of buried plant and animal life (organic detritus) hundreds of meters deep, where the action of pressure and temperature for a long time, in geologic scale, would convert kerosene into hydrocarbons.

One should notice that the idea of the abiogenic origin of oil has noteworthy supporters. Here is what Sir Fred Hoyle had to say on the subject in 1982:

The suggestion that petroleum might have arisen from some transformation of squashed fish or biological detritus is surely the silliest notion to have been entertained by substantial numbers of persons over an extended period of time.

And the Nobelist Sir Robert Robinson said in 1947:

Actually it cannot be too strongly emphasized that petroleum does not present the composition picture expected from modified biogenic products, and all the arguments from the constituents of ancient oils fit equally well, or better, with the conception of a primordial hydrocarbon mixture to which bio-products have been added.

There is the question of some biomarkers found in oil, but this is also controversial; for example, Thomas Gold (1979) says that these biomarkers are actually organic contaminants of natural hydrocarbons, as the deep biosphere biomass surpasses all mass of surface biosphere and, of course, primordial hydrocarbons are the food of these deep microbes.

I do not want to enter here in this controversy, which from a quick look into the literature appears to be ongoing. I would simply like to point out the question, which is a big one. For the origin of life, an environment rich in hydrocarbon might be of some interest, as it permits the formation of water in oil compartments, such as reverse micelles and water-in-oil microemulsions, and these, in turn, as we will see in the next section of this book, can be seen as possible intermediates for the construction of the first protocells.

2.8 The N-carboxy-anhydride condensation

While the meteorites that have arrived on Earth are rich in amino acids, they are poor in peptides. Shimoyama and Ogasawara (2002) found the first exogenous peptides in the Murchison and Yamoto-791198 meteorites. They showed, however, that the only peptides present are Gly-Gly and the corresponding diketo-piperazine, with the indication that these compounds were probably formed by dehydration processes. We have also mentioned the presence of some simple peptides in the old archives of Miller, as found by Beda and coworkers (Parker *et al.*, 2014) – also in poor yields.

A formal treatment concerning the possible biogenesis of peptides in prebiotic times is given by the work of the French group of Commeyras in Montpellier. They started from the consideration that the early atmosphere contained nitrous acid (Summers and Chang, 1993), which permitted the synthesis of N-carbamoyl amino acids. Rather than being free amino acids, according to them, these were the products of a multicomponent chemical system operating in the early Earth. The "multi-component system," consisting of prebiotic molecules (hydrogen cyanide, several carbonyl compounds, ammonia, alkyl amines, carbonic anhydride, sodium bicarbonate, borate, and cyanic acid) evolved towards two families of compounds: the α-hydroxy acids and N-carbamoyl amino acids (main product).

The latter compounds are stable in water around neutrality, and they can be converted into N-carboxyanhydride (NCA) amino acids, which can crystallize. As shown in Figure 2.7, NCA amino acids can be converted into amino acids when adding water at a pH less than 4, while at a higher pH they give rise to peptides. And since the primitive oceans had a pH around 6.5, the existence of primitive prebiotic peptides, according to the authors, is very likely (Taillades *et al.*, 1999; Commeyras *et al.*, 2005; Pascal *et al.*, 2005; Collet *et al.*, 2010).

The scheme of Figure 2.7 (redrawn from Commeyras *et al.*, 2005) shows the reaction patterns, which have brought – according to the French authors – carbamoyl amino acids to NCA-activated amino acids and, from these compounds, to peptides.

The pathway going from a NCA-activated amino acid to the corresponding dipeptide is shown in Figure 2.8, in the example of alanine.

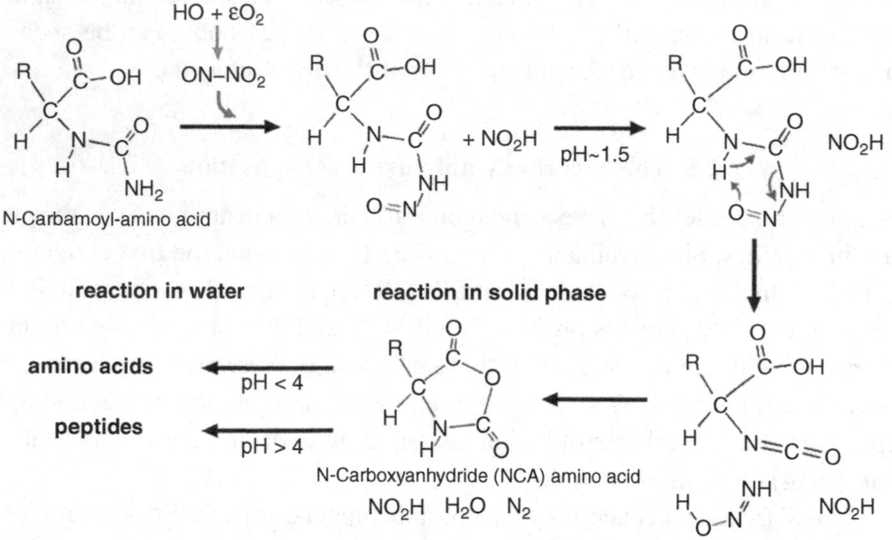

Figure 2.7 The reaction patterns leading, in prebiotic conditions, from N-carbamoyl amino acids to N-carboxyanhydride (NCA) amino acids, which may give rise, in water at pH > 4, to peptide condensation (redrawn from Commeyras *et al.*, 2005).

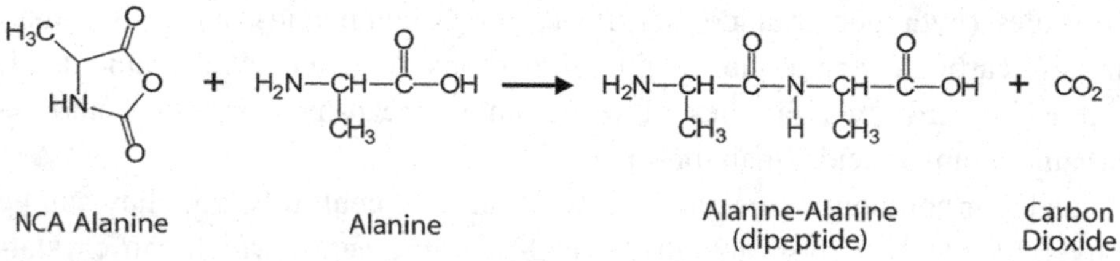

Figure 2.8 The pathway going from a N-carboxyanhydride (NCA)-activated alanine to the condensation of alanine-alanine dipeptide.

As mentioned above, the Commeyras group was able to connect the NCA derivatives of amino acids to the multicomponent system, which was operative under prebiotic conditions.

Aside from the prebiotic conditions, the conversion of NCA to peptides as a condensation method has been in the chemical polymer literature for a while. The term *Leuchs' anhydrides* is often used in this field, as Hermann Leuchs reported on them for the first time in 1906.

There are actually too many publications on the subject to review all of them here. A review by Kircherdorf (1990) describes the formation of polypeptides by ring-opening polymerizations, and reports on new aspects of the polymerization processes, such as the formation of cyclic polypeptides, as well as the formation of polypeptides with various architectures, such as di-block, tri-block, and multi-block sequences, and star-shaped or dendritic structures are also mentioned.

The ring-opening polymerization is also discussed in the review by Cheng (2012), while for the attempts to control the polymerization of the NCA amino acids, one should refer to the older articles of Hirschmann *et al.* (1971) and the several papers by Iwakura and coworkers (1970–1972). Hayakawa, Windsor, and Fox considered, back in 1967, the copolymerization of the Leuchs anhydrides of the 18 amino acids common to protein. On the mechanism, see also Brack (1998).

Now, going back to the work of Commeyras' group, it should be mentioned that, for them, the formation of the NCA amino acids is only one part of a more general scheme in the scenario of the origin of life.

Concluding remarks

In the case of the origin of humankind, we know that Africa is the most likely place for it to have occurred, and it would be nice to be able to identify a location with certainty. There is no way to know for sure, however. The other unanswered question, as we mentioned already in Chapter 1, is whether it all started in a single small spot, or whether instead conditions for starting life were present in a vaster region, for example, in the ocean or a very large lake. The difference may not be so trivial: in the first case, the start could have been due to a kind of frozen accident – a very particular contingency in one small place; the other case would indicate that conditions for life were ripe as a thermodynamic general situation at a global level. The recent work in Sutherland's group may indicate that, at least, all main basic components may have been present in the same pot (Patel *et al.*, 2015). The presence of lipids might suggest that primordial cells or protocells might have been present at the beginning, and this, together with the phenomenon of spontaneous overconcentration (described in Part V of this book) may have posed the conditions to start the first primitive forms of metabolism (Luisi *et al.*, 2010).

However, low molecular weight compounds, even in the presence of suitable membranes, would not have been enough to produce life. The main problem for understanding and possibly reproducing the origin of life lies in the control of molecular order at the level of macromolecules. Proteins and nucleic acids are orderly sequences, and we need to understand how this molecular order originated, and then how the two classes of macromolecules have begun to interact with each other in a causal cyclic order. This point has been already mentioned, however, it is right to emphasize that the problem of the origin of life will not be "cracked down," as the media love to say, until the mystery of the origin of molecular order has been unveiled. The complementary point is the following: that you cannot solve or address this problem starting with already ordered, functional macromolecules, such as a high molecular functional RNA. Using a well-known metaphor, this would be like building a house and starting with the roof.

With this in mind, let us go to the next chapter, which discusses how we can possibly ascend the ramp of bio-complexity.

Chapter 3

Ascending the ramp of complexity

Introduction

Looking again at our "famous" Figure 1.2, we may realize how bold Oparin was to assume that there should be a spontaneous increase of molecular complexity from the simple molecules up to the first cells. Bold because this process should take place in a prebiotic scenario deprived of the "intelligence" of enzymes and of the "memory" of nucleic acids: only the basic natural laws were then in operation.

There are many processes in nature that permit an increase of complexity and are attended by a negative free energy change – namely they are under thermo-dynamic control – "spontaneous" reactions, just to use a simplified term. Think of the bubbles being formed in soap – spherical structures made up of billions of molecules, which spontaneously find their place in such a regular aggregate. Or, think of some spontaneous polymerizations – for example, styrene under the effect of sunlight, or the de-mixing of a water-oil emulsion. These are also examples of self-organization, processes which are thermodynamically favored and constructing at the same time higher order. We will devote an entire chapter to self-organization in Part IV of this book, where we will be discussing the self-organization of surfactant molecules, which produce aggregates such as micelles or vesicles, bilayers, and liposomes.

The increase of size given by self-organization would not be enough to proceed towards life. We need an increase of functionality as well, namely the arising of novel properties. And this is indeed what we will discuss: when smaller parts are brought together – be it surfactant molecules as in the bubble or in membranes, or protein moieties forming tissues – new properties arise (new in the sense that they are not present in the constituent parts). These are the "emergent properties," a very important element to construct life. This is, in fact, the most important departing point from the vision of reductionism and we will deal in detail with this notion – together with self-organization – in Part II of this book.

Thus, going up Oparin's ladder along the pathway of molecular evolution is marked by a natural increase of molecular complexity and the arising of novel properties. This is a very general and vague statement that requires further discussion and examples, which will be done more extensively later on in the book.

3.1 The creativity of contingency

We have observed that the entire biochemical world is based on aggregation and in particular self-organization of smaller parts, proteins forming oligomer structures that are going to form tissues, these forming organs and then organisms . . . and think of the strands of DNA making the duplex, and then the complex formation of histones and then ribosomes, and so on. Seen in this way, life displays a huge variety of structures. One question worth examining is where the initial structures, proteins and/or nucleic acids, come from – in particular, how they have been selected.

Concerning the Stanley Miller experiment, we have already asked why α-amino acids are being formed, and not β-amino acids or some other isomers. The simple-minded answer is that – in general – in a chemical process, the structures that are formed are those which are thermodynamically more stable under the given initial conditions. This answer is linked to some more complex, subtle questions: Why certain compounds "are with us," while others, including molecules that are stable and common, have not been chosen as basic building blocks of life? Why nucleic acids possess ribose and deoxyribose as sugars, and not glucose? Why hemoglobin is composed of four chains? And why is it that the chains of alcohol dehydrogenase and lysozyme were made precisely in that way?

Let us repeat that "our" enzymes are not with us because their structures correspond to thermodynamic energy minima, as in the case of the α-amino acids of Miller. They are with us due to a long series of sequential steps in the chemical evolution, each of which is governed primarily by the contingent set of factors operating at each step.

Contingency is a very important concept in evolution, both prebiotic and biological. As already stated, contingency is the convergence of several factors, which, although independent from one another, act simultaneously in a given point/time and by that action, determine the course of the event.

The main point that we have to accept here is that the succession of a series of biochemical steps in prebiotic evolution is not an obligatory pathway, but is mostly determined by the contingent set of parameters that happen to be operative at each particular step. In a way, then, contingency has to do with the initial conditions, and the initial conditions are the very core of the chemical thermodynamics – this is why I consider contingency a chemical factor.

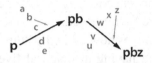

Figure 3.1 The hypothetical contingency pathway of the biogenesis of a protein as fragment condensation of shorter peptides. In a biologically mature situation, the growth might be at the level of nucleic acids strands, but the concept does not change. We start from the penta-peptide p. If "chance" would have been at work, any one of the possible 25 (5x5) peptides 15-residues peptides could have been formed. The contingent set of initial conditions at each step, with the structural determinism of the growing peptide, determines the choice of the reagent; in our example, first **b**, then **z**.

Is contingency equivalent to "chance"? The answer is: no, it isn't. This will be clear with the following example.

Let us imagine the ideal pathway by which a protein is being born during prebiotic evolution – its biogenesis. We can start with a simple penta-peptide **p**, prebiotically originated (this is realistically possible by the N-carboxy anhydride reaction, as we have seen in the preceding chapter – see Commeyras *et al.*, 2005, Taillades *et al.*, 1999). Let us assume now that, in a solution of the "prebiotic broth," **p** finds as reaction partners five different penta-peptides, which we will call **a, b, c, d**, and **e** (see Figure 3.1). With which of them will the reaction take place?

If reaction would be determined by mere chance, one of those five reactants at random would be chosen, or all of them. It will not be so, because of simple chemistry laws: firstly, there will be a choice due to the "structural determinism" of **p**. If, for example, **p** is positively charged, the three negatively charged compounds **b**, or **c**, or **d** (assume so) will be preferred.

Which one of these three? This depends on the relative reactivity of the three peptides under the given environmental conditions – for example, by the pH, or salinity, temperature, and concentration. All this amounts to the simple consideration that the thermodynamics of the reaction will be dictated by the initial conditions. If **b** would then be selected, we will obtain the deca-peptide **pb**, as indicated in Figure 3.1.

If the environmental conditions would have been different, for example a higher pH or, say, a lower temperature, then the partner **c** might have been selected – and the biogenesis story would be different.

This is the notion of contingency in action in the simple case of prebiotic selection. In fact, parameters and factors like concentration, pressure, salinity, and pH are generally independent from each other. This kind of bifurcation can go on: starting now from **pb**, if another set of penta-peptides is given as possible reaction partners (**u, v, w, x**, and **z**, as in Figure 3.1) then, for example, the particular 15-residues peptide **pbz** will be ideally formed, and none of the other ones. Would the set of contingent parameters have been different, then perhaps the product **pbw** would have been formed.

Note in these simple examples how and why "this" can be formed instead of "that." In fact, consider that having at disposal five reaction partners in two successive steps, ideally one might construct 25 possible 15-residues peptides. But by the action of contingency, most of them will not be formed.

Which environmental factors will be present at that particular moment and space of the reaction may be ascribed to chance, and also their relative statistical weight may be due to chance. Chance also determines which peptides – in our case one among **a, b, c, d, e** (in the first reaction) and one among **u, v, w, x**, and **z** (in the second reaction) – will be present in each particular reaction. However, the outcome is not a random event: there is a high degree of determinism in these selection phenomena, due first of all to the structural determinism of the growing peptide, as we have seen in the example, and also due to the chemical structure of the possible partners. Any physical factor, like pH, salinity, pressure, gravity, and concentration, plays its own role – and each of them has, of course, a deterministic origin. Chemistry does not proceed by chance. Structural determinism is a concept dear to Maturana and Varela (1980, 1998), the authors of the theory of autopoiesis (see Part II of this book).

This example of the formation of the longer peptide **pbz** is illustrative of the general concept: the final product has not been formed because of a predetermined plan. This point – the lack of a plan – is very important and we will see that in the next section.

In conclusion, we can say the following about contingency in prebiotic evolution (and by inference to biological evolution, as well):

1. Contingency should not be confused with chance, or random event, as it works based on the structural determinism of the growing entity, and the outcome depends on selection chemistry. What is due to chance is which factors and which reaction partners will be present in that given moment, and their relative statistical weight. This finally corresponds to the definition of the initial conditions of a reaction. The point here is that these initial conditions may be and generally are changing and arbitrarily changing by each step due to the vagaries of the environmental conditions, as well as the very long time stretches possibly involved in prebiotic reactions.
2. Because of the above mechanism, the structure being formed is not the product of a predetermined plan. It comes by itself, but at the same time, there is some degree of determinism in the process. We will come back to this point in the following section.
3. Because of the above, the products of contingency cannot be predicted *a priori*.

4. For the same arguments, we will not be able to know the detailed mechanism of the biogenesis of each "modern" macromolecular product. In particular, for the origin of life, this means that we will not be able to know the pathway by which chymotrypsin or lysozyme have originated.

And it is proper at this point to remember the following quote by Eschenmoser and Kisakürek (1996):

... the aim of an experimental etiological chemistry is not primarily to delineate the pathway along which our (natural) life on earth could have originated, but to provide decisive experimental evidence, through the realization of model systems ("artificial chemical life") that life can arise as a result of the organization of the organic matter.

Let me finish this section by considering that for many people, creationists in particular, the main message of contingency is generally not accepted – in particular, the notion that mankind might not have been arising. This poses problems, possibly by emotional nature, also to scientists.

3.2 The primacy of structure

The mechanism described above, according to which a macromolecular structure is built by successive steps, each of them determined by its own set of contingent parameters, illustrates a very important point undermining all evolution, *prebiotic* and *biological*: that the structure precedes the function – simply because there is no predetermined plan for the function. As we already mentioned, there was not a predetermined plan to make hemoglobin as a four-chain protein as a carrier of oxygen; or to make an alcohol dehydrogenase capable of oxidizing ethanol to acetaldehyde. The function was found later and established on the basis of the chemical (or biological) structure, which was born first.

With the caveat that you should substitute the term "chance" with *contingency*, consider the following excerpt from Monod's *Chance and Necessity* (1971), when he writes:

... the various structures built by nature are the result of chance. There is no aim, no predetermination – only chance assembly processes, and random structures. If one of these structures happens to perform an useful function, it may be codified and preserved. Then *Chance* becomes encoded in DNA, it becomes *Necessity*, i.e. the hard law of genetic invariance.

This idea, that structure precedes function, may sound like the outcome of modern molecular biology and the Darwinian modern synthesis. In fact, it is a very old idea.

Already Aristotle would say (translation from Wikipedia): *nature adapts the organ to the function, and not the function to the organ*; but it is Lucretius, in his *De Rerum Natura*, who would really surprise us. In the fourth book, from the verse 833 on, he writes:

> ... nothing is part of our body
> To be used in a very precise manner.
> It is the use that originates from the organ.
> sed quod natum est id procreat usum
> Nobody has ever seen before having the eyes
> And nobody spoke before having the tongue
> But the tongue existed before the spoken language
> And the ear arrived to us before we heard the sound
> Therefore one should say
> The organs all existed before the function:
> Therefore they did not develop at the aim of giving us a service
> . . .
> The habit to give rest to the body when the body is tired
> Is much older than the invention of the bed
> And everybody was able to drink
> When there were no cups around
>
> . . .
> Other things existed out of a spontaneous creation
> And only later they could be of some use
> Among those things, certainly
> The reason, and the arms.
> We need therefore to reject the false theory
> According to which all this was given
> To serve to us.

This is a text that could have been written a few years ago by Stephen Jay Gould or any modern evolutionary biologist. It is interesting that certain minds over 2,000 years ago could see things that neither our modern creationists nor intelligent design (ID) people can see today. The arguments of these people, who are by profession anti-Darwinist and anti-evolution in general, are known. Those are arguments which are easy to be communicated and, unfortunately, not so anti-intuitive as science often is. They say, for example, that the eyes have been created to permit to see; and the tongue and the ears have been made by God in order to permit man to speak to and hear each other.

Here, we can only reiterate that, whereas the dialogue between science and religion can be and is, as a matter of principle, useful, the confrontation between science and fundamentalism is not, and is actually reminiscent of the century-old Galilean battle between knowledge and obscurantism.

3.3 Thermodynamic and kinetic control

Self-organization, emergence, and contingency are important determinants for going up Oparin's ladder. But there is something more.

When considering the reaction pathway of our peptide **p**, we have to give consideration to the general principle that reactivity can be determined by thermodynamic as well as kinetic factors. Among the factors that bring to reaction our starting peptide **p** (Figure 3.1), we have not expressly considered catalysis. In fact, in the biological reactions of our life, catalysis and in particular enzymatic catalysis plays the most important role. We have mentioned earlier that the α-amino acids under the Stanley Miller conditions were produced under thermodynamic control.

Kinetic control occurs when the reaction product is determined by kinetic factors, in particular because the competitive activation energy to reach the thermodynamically more stable product is too high under the given conditions, whereas the activation energy to arrive at the other product is smaller and can be overwhelmed under the given conditions. Thus, once an enzyme complex is formed in a stereo-chemically selected way, there is only one way to go – and so it is for most of our enzymatic reactions. Kinetic control can also exist in organic chemistry without the intervention of enzymes. A simple representation of the two effects is shown in Figure 3.2.

Going back now to Figure 3.1, the "choice" of the reaction partner of **p** can also be due to kinetic control, and this is true in particular if we are in the presence of a more or less specific catalyst. Enzymes are later comers in the prebiotic evolution pathway, but metal ions, and simple peptides, are likely to be present in our prebiotic scenario.

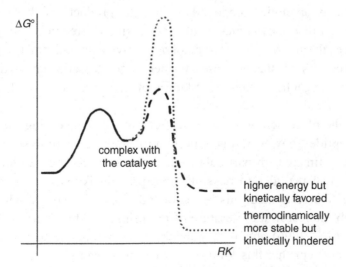

Figure 3.2 A schematic representation of kinetic versus thermodynamic control.

We know, now, that metal ions are very important in several enzymatic reactions, and we also know that metal complexes can catalyze stereospecific reactions of technical importance – a well-known example is the stereospecific polymerization of α-olefins by Ziegler-Natta catalysts (a complex of $TiCl_3$ and aluminum alkyls), which permit the synthesis of isotactic poly-propylene. We accept in the field that metal ions may have played a role in some form of catalysis, and so it is for clay, but it appears difficult to draw on that a general scenario for the origin of life.

On the other hand, several literature data show that simple peptides can display proteolytic functions. Shen's group reports on the prebiotic synthesis of histidyl-histidine. This is interesting in view of the catalytic properties of this molecule (Shen, Lazcano, and Oró, 1990; Shen, Mills, and Oró, 1990). Studies on the prebiotic synthesis of histidyl-histidine were also carried out by Oró and coworkers (1990). Glycine and diglycine, Gly-Gly, were investigated as possible catalysts in the prebiotic evolution of peptides (Plankensteiner *et al.*, 2002).

Of particular interest is the dipeptide Ser-His, described originally by Li and coworkers in Beijing (2001). In fact, this simple compound is capable of catalyzing the hydrolysis of peptide, esters, and nucleotides bonds. These authors also indicate that even simpler peptides, such as Gly-Gly (mentioned before), can have a similar catalytic action.

Based on the principle of microscopic reversibility, Ser-His should also catalyze the synthesis of peptide bonds, and in fact, it has been successfully used in this sense by our group (Gorlero *et al.*, 2009) to synthetize a series of peptides – and later on used to catalyze the synthesis of oligonucleotides (Wieczorek *et al.*, 2013).

This may be important for the origin of life, as peptides as simple as Gly-Gly can be seen as prebiotic compounds – i.e., the products of thermodynamic control, so that thermodynamics would give a kind of free ticket for catalysis. Furthermore, the importance of proteolytic activity should not be underestimated: one can assume that proteases, formed by condensation and hydrolysis of polypeptides, might have been a first family of enzymes, from which the others may have evolved.

It is not the place here to review all these studies concerning the catalytic action of peptides; however, it is important to emphasize that these studies are rather underestimated. In particular, the action of prebiotic peptides for the condensation of peptides to possibly give protein-like structures has never been properly investigated; this is one of those open questions, which somehow find no interest in the literature of the origin of life. There is one paper (Chessari *et al.*, 2006) about the condensation of random co-oligopeptides, a proof of concept that this kind of condensation can give rise to proteins. In fact, one *ex-novo* foldable protein with over 40 residues has been obtained.

In this case, however, the condensation of the co-oligopeptides has been carried out by the Merrifield method, as at that time the action of peptides such as Ser-His was not yet known. (In a later discussion, we will see this reaction in detail.)

We close this section by noting that kinetic control, possibly mediated by catalysts, may be another chemical determinant of the prebiotic evolution, and again we are not able to say, in the case of the biogenesis of chymotrypsin or lysozyme, whether and when this constraint was effective and at which step.

One could see the kinetic control as one of the factors of contingency, as it is in a way part of the initial conditions operating at one particular growth step, but catalysis is conceptually different from contingency, and it is better from the heuristic point of view to see these two effects separate from each other.

3.4 Self-replication – and the concentration threshold

The going-up process in Oparin's ladder is basically based on chemistry, and the basis of chemistry implementation is concentration. This is equivalent to saying that there is always a concentration threshold, below which the reaction does not effectively take place. In chemistry, the measure of the interaction between two or more partners is generally given by an equilibrium constant, or an analogous quantity. Thus, in enzymatic reactions, the concentration of the substrate must reach the K_M value; in a bimolecular reaction, concentrations must be such to overwhelm dissociation forces, and so on. This may sound very trivial, but it is something that is occasionally, if not often, forgotten by theoreticians who use to work without mentioning concentration.

A very common case in this field is the treatment of self-replication.

There are some good reasons why, in the field of the origin of life, the emergence of self-reproduction is of paramount importance. Before the implementation of self-replication, any interesting structure that might have originated in the prebiotic scenario would have decayed due to degenerative processes, and would have disappeared leaving no trace. Instead, with self-reproduction (as soon as the rate of self-reproduction is larger than the rate of decay), an increase in concentration of this structure would be possible.

Moreover, if the self-replicating structure had chemical information and, in addition, had the capability to mutate, then self-reproduction, information, and evolution would occur all by the same system: something already very close to life. It is then clear why the search for self-reproduction mechanisms is the holy grail of the research on the origin of life.

Shreion Lifson (1997) utilized a nice arithmetic to illustrate the power of the autocatalytic self-replication. He took the example of a normal hetero-catalytic process that makes one molecule of **B** from **A** at the rate of one per second. Then, it would require 6×10^{23} s to make one mole of **B**. If instead there is an autocatalytic process by which **B** gives rise to 2**B**, then 2**B** give rise to 4**B**, then 4**B** to 8**B**, and so on, then the process requires only 79 s to make one mole of **B**!

The dramatic power of nonlinear growth was already known in ancient times. You may have come across the story of the Chinese Emperor who – according to one particular version – played chess with a concubine. The smart woman had requested – should she be victorious – only a little bit of rice: one grain of rice in the first square of the chess board, two grains in the second, four in the third, and so on. The emperor nodded scornfully; he lost, and discovered that he did not have enough rice in his kingdom to pay his debt; and the same unproven version continues that this is how a concubine became empress of China and the emperor became a disillusioned monk.

The arithmetic expressed in Lifson's previous calculation, as impressive as it is, may bring the reader to misleading conclusions. For example, one may be induced to think that one single molecule can do the whole trick: arise by chance and begin to self-replicate, giving birth to a vast family of molecular progeny.

Obviously, with one single molecule no real chemistry can be achieved. Here comes the threshold of concentration. In normal wet chemistry, in order to self-replicate, the replicator **A** must bind to another molecule **A**. The need to form the **A-A** complex (**A$_2$**) from two **A** molecules is a severe constraint. First, in order to make an appreciable concentration of this complex, there must be a significant amount of **A**, so as to overcome the effect of diffusion; and the real difficulty arises when spontaneous decay is introduced. If the concentration of **A** is low enough, the population will decline no matter how large the growth rate is. This means, that the concentration of **A** in solution must be high enough to bring about a sizable concentration of **A$_2$**, so that the natural decay diffusion forces do not destroy the dimer itself.

The precise calculation of the critical threshold concentration of **A** is not easy, as we would need a good estimation of the dissociation constant K_d for the dimerization process $2\mathbf{A} \rightleftarrows \mathbf{A_2}$. Assuming for example a $K_d = 1$ μM, with an initial **A** concentration of 10^{-12} M (1 pM), the conversion into the dimer would be around 0.0002% (would be 0.2% if $K_d = 1$ nM). In other words, there would be no significant formation of the dimer below one picomolar concentration of **A**. One picomolar solution of **A** (defined as 10^{-12} M) still contains (since the Avogadro number is 6.02×10^{23}) 6×10^{11} molecules, i.e., 600 billion. And even dealing with one microliter we would have in it 6×10^5 (i.e., 600,000) identical copies of **A**. This means that we would need several hundred thousand identical copies of **A** in order to start replication, even restricting the volume to one microliter. In other words, in order to have an effective self-replication of a hypothetical sr-RNA, we would need a preliminary metabolism to make a useful concentration of such a molecule.[1]

This is one constraint given by chemistry to the notion of self-replication. And it is not the only one. The other concentration problem is with regard to the quantity of basic material involved in a nonlinear growth process of replication.

Suppose you were to start from a solution containing the replicator at 1 µM concentration. The solution will become ca. 1M in 20 nonlinear (2, 4, 8, 16, and so on) replication steps. Imagine what it would take to do that in a small warm pond of, say, 1,000 liters; then, for a polynucleotide of 10,000 Dalton, we would have after the 20 steps, 10 tons of material, an amount really unconceivable, at least with this example. Here, we see again the threshold of concentration at another level, namely given by the fact that such an amount of material would not be realistically available in any terrestrial pond.

With these concentration thresholds, there are severe difficulties in the pre-biotic RNA-world, and we will come back to this point in the next chapter.

Let us move to another quite different level, where concentration is, in fact, a problem. This is at the level of the first cells. In this field, terms such as *overcrowding, super-concentr*ation, and *similar* are often used to indicate the particular concentration of the macromolecules inside the membrane compartment.

The high local concentration in the cell is an accepted notion, and is the prerequisite for their biological activity. If you squeeze the content of 1 kg of *Escherichia coli* in the previous large pond, you will have in that pond all DNA and RNA and enzymes – and no life. In addition, when we do *in vitro* protein expression using a commercial cellular extract and a given plasmid, we know that dilution by only a factor of two-three of the cellular extract inhibits the protein synthesis.

Thus, it will not be enough to find a way to make ordered sequences of DNA, RNA, and proteins (still open questions), but we will have to find a way to concentrate them in a small compartment having dimensions of a few microns.

---

[1] We can make some rough calculations starting from contemporary RNA replication (Szathmáry and Luisi, unpublished). RNA self-replicase does not exist in nature, but considering the actual concentrations of *Qβ* replicase and template-RNA in a single cell, and considering this in terms of the smallest dimension of a bacterium, a minimal concentration of ca. 10 nM can be calculated for RNA in vivo. Let us compare this with *in vitro* experiments.

For *in vitro* kinetics, in typical experiments, nucleotides are provided in 0.2–0.3 mM concentrations (Biebricher *et al.*, 1981), while the enzyme replicase is in the range of 80–140 nM (still about 10^{12} enzyme molecules per reaction sample!). Under such conditions, the duplication time of molecules (about 120 nucleotides long) is about 2.3 hours, and becomes months or years when reagent concentrations are in the picomolar range and below. It may be argued that reaction times of months or years are not prohibitive in a prebiotic scenario; but here again one should not forget the problem of the decay rate of the replicator. Wong has calculated that for a genome consisting of three genes, the replication time must be shorter than 8.6 years (Wong and Xue, 2002). This seems to be a relaxed condition, but in fact, it is not, once the actual concentrations are taken into account as calculated above.

As we will see in the next chapter, this is indeed one major operational problem in the "compartmentalistic" hypothesis of the origin of life. When we come back to this point, we will also see some possible solution to this concentration threshold. For the moment, this is something to keep in mind as another constraint that chemistry imposes to the Oparin pathway.

3.5 Ordered macromolecular sequences

This is one of the greatest stumbling blocks in our uphill movement in Oparin's ladder. Life is governed by enzymes and nucleic acids, which are not simply polymers (i.e., repetition of the same monomer unit along the chain), but are ordered co-oligopeptides or co-oligonucleotides, and are present in many identical copies. We will come back to this point a few times in the book, as I consider the biogenesis of ordered sequences a fundamental item in the origin of life. You can have all the molecular weight prebiotic compounds in any amount you like, but you will not start life unless you have ordered macromolecules. I will insist on that, because it is one of the unanswered questions (how to make them), and because it is a question that is not taken seriously enough in the literature. How many groups are really tackling this question?

There are literature reports on the synthesis of homopolymers – only one amino acid, for example, poly(phenylalanine), which is the most studied case. However, this kind of study on polymerization has little to do with the macromolecular chemistry of ordered copolymerization. In fact, consider that in the first case (only the monomer **A**), you will deal with a chain growth characterized mostly by one kinetic constant, k_{AA}. When you have two monomers, **A** and **B**, you are dealing with four competitive kinetic processes, characterized by k_{AA}, k_{AB}, k_{BB}, k_{BA} (the first two being shown in the following scheme of Figure 3.3); and generally, the outcome will be a function of the kinetic ratios R_1 and R_2, where $R_1 = k_{AA}/k_{AB}$ and $R_2 = k_{BB}/k_{BA}$, and it might be that you do not get any –AAAA– polymeric chain sequence in this case.

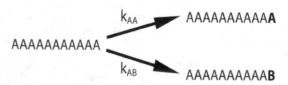

Figure 3.3 The kinetic scheme for a copolymerization of two different co-monomers, **A** and **B**. This is to show that a copolymerization is different from a homopolymerization. Actually, a study in the homopolymerization is not very helpful for studying and understanding copolymerization – a point which is usually not well considered in the literature.

When instead of two co-monomers you have three, four, or more of them, then their kinetic becomes almost impossible to handle.

Imagine the complexity when you have the copolymerization of 20 different monomers. In fact, at the best of my knowledge, there is no copolymerization scheme for a mixture of amino acids.

Later on in this book, we will take up this problem again and see some possible way out. However, the fact remains that, until now, there has been no clear way to obtain ordered sequences by prebiotic means. The reason why the emphasis of "many identical copies" is important became apparent in the previous section, where we spoke about the concentration threshold: with one single copy, you do not do any chemistry.

3.6 The question of homochirality

One of the questions associated with the origin of life on Earth is why our α-amino acids have almost exclusively the same absolute configuration L, and sugars have generally the D configuration. In addition, we should accept the idea that, without such homochirality, life as we know it would not be possible. Take a small protein with 50 residues in an ordered sequence: if its biogenesis would be possible with both L and D forms of the constituting amino acids, this very protein sequence would exist in 2^{50} stereoisomers (*c.* a 10 followed by 15 zeroes). By applying the trick of using only L form, nature reduces this number to just one. Indeed, this is one of the greatest ordering factors in the structures of biological life.

Chemists know too well that the two enantiomeric forms have the same energy and the same properties, and therefore the same probability of synthesis when operating with normal laboratory techniques; the amino acids in Miller's synthesis were, of course, racemic. Then the question arises: why in nature only one form is present, which appeared from the very beginning as a riddle, and of course it represents a problem for going up Oparin's ladder.

Discussion in literature on that matter has been presented in terms of two well-known alternative hypotheses: contingency or determinism. The first term, as we have learned earlier, is implying that homochirality was due to an accidental combination of parameters. The second term would instead imply that there is some basic principle (*ex-lege*, using the expression chosen by Martin Quack and his group in a paper published in 2002) by which one of the two enantiomers is the preferred one.

Until recently, the *ex-lege* mechanisms were extremely complicated, having to do with the symmetry of elementary particles breaking, and the calculated effects were uncertain and extremely small (Quack, 2002; Quack and Stohner, 2003a, 2003b, 2014). In fact, chemists in general would prefer much simpler chemistry explanations, as that described in the experiments by Kondepudi (Kondepudi et al., 1990) or Lahav and collaborators (1999).

However, things took a turn with the discovery by Cronin and Pizzarello (1997) of the enantiomeric excess in amino acids found in meteorites, as was mentioned in Chapter 2. In the following years, more of these observations were made, until the findings that sugars from meteorites have a significant prevalence of the D form (Cooper et al., 2001; see also Bada, 1997).

Thus, it appears that the space of our solar system (as meteorites are, in general, bodies orbiting around the sun) has the capability of producing a certain percentage of *ee* (enantiomeric excess). The consensus today is that this effect, more than being due to asymmetric synthesis, has to do with preferential breakdown of one of the two enantiomers due to circularly polarized light.

Researchers are still looking for sources of circularly polarized light in space, and at this point the mechanisms underlying this chirality production are still unknown. But it is a fact that material arriving on Earth from space has a certain enantiomeric excess – so the hypothesis that there is an *ex-lege* effect is nowadays more concrete. As already stated in Chapter 1, it is a long stretch to go from this point to say that this is the origin of homochirality in our biological life. However, it is certainly a tenable hypothesis.

The enantiomeric excess in meteorites is rather modest, and one should also consider that the earlier compounds were not composed of the 100 percent enantiomeric purity as those used nowadays in our prebiotic experiments, but they were just mixtures of stereoisomers. This is probably not ideal for a clean beginning of chemical organization and reactivity. The order, so necessary for the real origin of life, must have implied some form of amplification of the enantiomeric purity. This is not impossible, even starting from minute enantiomeric excess, as demonstrated by Soai (2008) and his collaborators, but we should be aware of the fact that when we start to play origin of life using 100 percent enantiomeric purity, we are already a step further than the real beginning; therefore, we are already cheating.

Concluding remarks

We have highlighted in this chapter some of the main features that should be taken into account when we try to climb Oparin's ladder, either conceptually or experimentally. Some of these features are real constraints, and we do not know as yet how to disentangle them. In addition, I did not mention other important stumbling blocks, for example the problem of bio-energetics, or the formation of membranes – certainly essential to reach our final goal: life. We will examine some of these issues later on in the book, particularly the question of protocell compartments.

The features we have described in this chapter are well known and rather clear – things like emergent properties or self-organization or homochirality. This does not mean, however, that we understand well their origin. For example, the question of how ordered sequences of macromolecules, chirality, and kinetic control originated is still part of the mystery and fascination that, in general, surrounds the origin of life. Even at the level of emergence, there are subtle questions on how and why properties, which are completely new and unexpected, pop out at a certain level of complexity. We will consider some of these subtle points later on in the book, specifically in the special chapter devoted to emergence (Chapter 9). Here, it is proper to say that, because of this difficulty to go to the very core of the origin of things, science finds its limits.

One general notion we have learned from this third chapter is that the process of going up in Oparin's ladder is not the simple-minded picture that may appear in the left panel of Figure 1.2.

In the next chapter, we will see the hypotheses that have been put forward by the most active researchers in the field of the origin of life to explain the passage from nonlife to life in terms of molecular details. We will have to keep in mind that these mechanisms should take care of the determinants illustrated in this chapter.

Chapter 4

Experimental approaches to the origin of life

. . . l'uovo cadde dal ciel
e come a Dio piacque
l'uovo si ruppe
e la gallina nacque.

Introduction

This chapter is a review of the various proposals for the origin of life, limited to those approaches that have been or can be tested, in one way or another, by experiments. Therefore, I will not consider here the very many theoretical models, which have only a speculative, theoretical standing. This is certainly a shortcoming of this analysis, but on the other hand, the critical enumeration of all that has been proposed in the secondary literature on the subject (apparently, there are more than 30 models) would go beyond the space available in this book.

Rather arbitrarily, then, I will mention only a few of these approaches, beginning with the three most known: *the prebiotic RNA world, the compartmentalistic approach*, and *the enzyme-free metabolism approach*. A similar classification is made by Eschenmoser (Bolli *et al.*, 1997a, 1997b).

We have examined in the previous chapter the chemical constraints to the understanding of the origin of life. They will be put in evidence in each of the various approaches to the origin of life.

4.1 The prebiotic RNA world

Walter Gilbert first used the expression "RNA World" in 1986, in a hypothesis for interpreting in a unitarian scheme the various properties of RNA, including catalysis (Gilbert, 1986). Apparently, this idea had been in the air before, for example, in Carl Woese's book *The Genetic Code* (1967), as well as in two papers by Francis Crick (1968) and Leslie Orgel (1968).

The basic idea is the primacy of RNA to originate all other biopolymers (proteins and DNA) and the main mechanisms up to the origin of life. The simplified scheme is shown in Equation 4.1, with the addition that the active form was a self-replicating RNA (sr-RNA), which in its replication would undergo mutation and corresponding production of catalytic RNA forms (ribozymes), which in turn would catalyze the formation of proteins and DNA.

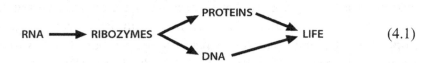

$$\text{RNA} \longrightarrow \text{RIBOZYMES} \underset{\text{DNA}}{\overset{\text{PROTEINS}}{\big\langle}} \longrightarrow \text{LIFE} \tag{4.1}$$

This is the most widespread and accepted scenario for the origin of life. The supportive observations offered in the literature are multifold. First, RNA contains genetic information and catalytic properties in one single structure. Then, RNA preceded DNA in life scenarios on Earth. Also, some viruses use RNA as their genetic material, rather than DNA (Lindahl, 1993); ribozymes catalyze a series of reactions, including polymerizations (Forster and Symons, 1987; Johnston *et al.*, 2001; Zaher and Unrau, 2007), and the ribosome has a ribozyme function, this last point being emphasized by the latest data from Yonath group (see infra). Also the presence and action of viroids (Diener, 1971) is added to the list.

Despite all this, the RNA world remains a hypothesis, and criticism and objections are not missing in the literature, mostly based on the improbability of the spontaneous biogenesis of RNA, and on its chemical fragility. (See, for example, Cairns-Smith, 1982; Shapiro, 1984, 1986, and 2000; Lindahl, 1993; Pääbo, 1993; Anastasi *et al.*, 2007; Kenyon and Mills, 1996; Sutherland *et al.*, 2007; Cech, 2011; Fry, 2011; Yarus, 2011; Bernhardt, 2012). I offered some criticism, based on different arguments, in the previous chapter.

Some authors, with the argument that starting the RNA world is so difficult, suggest a *pre-RNA* world – for example, with simpler nucleic acids (Orgel, 2000a, 2000b), or involving PNA (peptide nucleic acid), as suggested by Nelson and coworkers (Nelson *et al.*, 2000). Other suggestions for "simple" pre-RNA nucleic acids have also included threose nucleic acid (TNA) or glycol nucleic acid (GNA) (Sutherland *et al.*, 2007). Even Robertson and Joyce (2012) voice some concern:

However, the problem of the origin of the RNA World is far from being solved, and it is fruitful to consider the alternative possibility that RNA was preceded by some other replicating, evolving molecule, just as DNA and proteins were preceded by RNA...

There is a very active and very high level of research in the field of the RNA world, and it would be impossible to review all relevant papers. Among the many examples, let me mention the work of Joyce and coworkers (Jaeger, Wright, and Joyce, 1999), and in particular, the work of Joyce with Paul (Paul and Joyce, 2002, 2004; Paul, Springsteen, and Joyce, 2006). See also Teramoto, Imanishi, and Yoshihiro (2000), as well as Eddy (2002). Landweber and Pokrovskaya (1999), using a constant region and a randomized region, were able to detect the formation of *de novo* RNA with ligase activity. See also the work of Szostak and coworkers (Chapman and Szostak, 1995; Adamala and Szostak, 2013; Sheng *et al.*, 2014; Blain *et al.*, 2014). See also the work by Irene Chen and coworkers (Jiménez *et al.*, 2013).

The term *RNA world* is nowadays used in the literature in a rather undifferentiated way, covering most of the research dealing with RNA. This includes also the work, which belongs more properly to synthetic biology, based on the reaction between highly ordered and functional RNA macromolecules. This is why I will use the term *prebiotic RNA world*, to indicate more specifically the work that deals directly with the origin of life.

The original version, the one shown in Equation 4.1, is based on the existence of a prebiotic form of RNA capable of self-replication, a sr-RNA. The idea that a sr-RNA may come out spontaneously from a prebiotic soup was dubbed as "the molecular biologist dream" by Joyce and Orgel several years ago (1986). Despite these early warnings and Orgel's more recent analysis (2004), this "dream" is generally found in between the lines of most modern literature dealing with the origin of life from the viewpoint of the RNA world.

Actually, the new line of research based on the prebiotic RNA world and conducted by contemporary authors, cleverly enough, tends to ignore this point. Modern authors start from extant RNA, bypassing the question of how RNA originated prebiotically – and they then proceed in analyzing the destiny of an evolving RNA.

I have outlined in the previous chapter the main shortcoming of this prebiotic RNA world scheme. That criticism was relative to the concentration threshold, both for the replication and for the mutations into novel ribozymes. I would like to add here a couple of more general critical considerations.

The first is based on the main standpoint of this book, according to which the main question for the origin of life is the origin of molecular order. There is no need to repeat that all mechanisms of life pivot on the orderly sequences of proteins and nucleic acids; and on the dynamic relation between these two families. You cannot simply bypass this central argument, and present solutions about the origin of life starting from already ordered – even functional – macromolecules. It is actually a very simple point. You have to explain first how the high molecular order present in ribozymes came about.

Talking about ribozymes, one should mention here the work of Ada Yonath's group on ribosomes (Yonath, 2010, 2012; Bashan *et al.*, 2010; Belousoff *et al.*, 2010; Fox *et al.*, 2012; Krupkin *et al.*, 2014), based primarily on X-ray diffraction work. They propose that the primordial ribosome was actually a ribozyme, with the proposition that the proto-ribosome is the entity around which life has evolved. These authors also arrive at the proposition that there was a co-evolution between the ribosomes and the proteins to arrive at the genetic code (Krupkin *et al.*, 2014). See also the conversation with Ada Yonath at the end of this section.

Ada Yonath's work is of great relevance for biochemistry at large, and so is the proposition for life evolving from a proto-ribosome. However, concerning the origin of life, the problem for me is the same as for the RNA-world: starting from already highly organized macromolecular material and ignoring the question of how this pristine order came about.

This is perhaps the main point of people looking for a pre-RNA world, as cited above. This is tantamount of saying that you need some kind of mechanism to generate orderly RNA sequences, and in many identical copies (or as an active quasi-species). To me, and to many others, this consideration suggests that some primitive forms of catalytic activity should be at the starting point. For this kind of activity we should look at prebiotic compounds, and in my opinion this might well be present in peptides, even simple dipeptides. In fact, we will see later on that Ser-His is capable of catalyzing the synthesis of peptide, and even of nucleotides bonds. In addition, we will see a proof of concept whereby fragment condensation of relatively simple co-oligopeptides gives rise to a folded, thermodynamically stable protein at a micromolar concentration. The same can, in principle, be proposed for RNA. This idea of a pristine interaction between RNA and catalytic peptides or proteins, is present in the literature under the name *Peptide-RNA world* (see for example, Kunin, 2000).

The other criticism is also general in character, but is again related to the question of macromolecular order. Suppose we accept the scheme proposed above, by which a sr-RNA that evolves into a ribozyme and catalyzes the formation of peptide bonds. And, so what? The problem is not the formation of peptide bonds, *per sé*; several simple dipeptides can do it, even better than the present-day ribozymes. The problem is the orderly sequence of amino acids in a chain. The same can be said for the synthesis of DNA by a properly born special ribozyme. In other words, the original version of the RNA world leaves the question of the genetic code untouched.

I mentioned that one of the general criticisms of the prebiotic RNA-world view lies in the lack of consideration of how to make a pristine family of RNA with a unique sequence in many identical copies. In fact, we do not even know any prebiotic nucleotide polymerization, aside from the question of the stereo-specific linkage. Even assuming that there is such a random and stereospecific polymerization process, how, then, can we select from this huge library a unique RNA family (or at least a quasi-species)? As I have pointed out repeatedly, there is not much in the literature along this line of reasoning.

This kind of selection can be achieved by using modern molecular biology tools, and in this perspective remarkable is the work by Bartel and Szostak (1993), who reported that starting with trillions of random RNA sequences, selecting for ligase activity and then amplifying, after some cycles of ligase activity, the initial RNA gel was resolved into a half dozen RNA bands.

The work of Fabrizio Anella (one of my former students) in his Ph.D. work may show that a kind of sharp selection may be indeed possible (Anella, 2011). He started from a random library of *c.* 10 billion clones, applying to that the Foster selection method (based on the action of a S1 ribonuclease at different temperatures, see De Lucrezia *et al.*, 2006a and 2006b). Thus, he reached the point of having in his hands a reduced population (millions of clones) with a fold that was stable at the action of the S1. From this library, Anella has arbitrarily chosen 24 sequences for further analysis, and in this group, there was a sequence that was represented 8 times (8 out of 24). This means that this particular sequence represented *c.* 30 percent of the entire population after S1 treatment.

Then, it does not appear impossible, by pushing the Foster assay further, to arrive at one single molecular form. Fabrizio Anella could not complete this task – Ph.D. theses are always too short! Nevertheless, his work shows that such a reduction to one single folded RNA form is, in principle, possible.

Remarkable in Anella's work is also the observation that certain macromolecular RNA sequences are present in multiple copies. It is noteworthy because in a purely random polymerization process yielding relatively long chains, the probability of having two identical chains is practically zero. This might then indicate that there is a certain tendency to build chains, which are favored from the energetic point of view. And if we have, in the original pool, sequences that are particularly stable, these will accumulate with time, while all others are being degraded.

In this kind of selection work remains, of course, the problem of how to make all this – the starting (stereoregular) RNA library and the catalytic process – in a prebiotic scenario – when molecular biology had still to be invented.

Conversation with Ada Yonath

Ada E. Yonath is the current director of the Helen and Milton A. Kimmelman Center for Bimolecular Structure and Assembly of the Weizmann Institute of Science. In 2009, she received the Nobel Prize in Chemistry along with Venkatraman Ramakrishnan and Thomas A. Steitz for her studies on the structure and function of the ribosome.

She graduated from the Hebrew University of Jerusalem with a bachelor's degree in chemistry in 1962, and a master's degree in bio-chemistry in 1964. In 1968, she earned a Ph.D. in X-Ray crystallography at the Weizmann Institute of Science. Ada Yonath accepted postdoctoral positions at Carnegie Mellon University (1969) and MIT (1970). In 1970, she established what was, for nearly a decade, the only protein crystallography laboratory in Israel. Then after visiting the Max Planck Institute for Molecular Genetics in Berlin, she was the head of the research unit of the Max Planck Society in Hamburg, in parallel to her faculty position at the Weizmann Institute.

PLL: At the last ISSOL (International Society for the Study of the Origin of Life) meeting in Nara (Japan), several people considered your results on the primordial ribosomes, if we may refer to them this way, one of the few outstanding new pieces of science. By that, you were able to link the origin of life and the origin of protein expression. Would you condense all this in better terms for our readers? What is the message that you like to leave behind with this research?

ADA YONATH: Your description is very good. I can only add that based on fully conserved structural features at the active site of all contemporary ribosomes, we assume that it represents a prebiotic bonding entity, where peptide bonds could be formed.

PLL: I know that you do not like the question of "where r-RNA comes from." We leave it there. But you agree, perhaps, that the complexity of ribosomes must imply a preliminary metabolism capable of affording RNA, its polymers, which then also implies the action of specific enzymes as well. So, there must have been an active world prior to ribosomes. The broth was already boiling...

ADA YONATH: It is not a question of liking the question: "where r-RNA comes from." I just do not know the answer for it and assume that the distinguished scientists that are attempting to reveal the answer to this question will eventually find this out.

Your statement: "there must have been an active world prior to ribosomes" is logical. However, the extent of "activity" and its biological relevance is still to be investigated. Also, please note that our hypothesis does not concern ribosomes as such. We assume that there was a *proto-ribosome*, which is just an "enzyme" (actually a ribozyme) for making bonds.

PLL: Something more personal. Having the Nobel Prize and such a wonderfully active research going on now, do you feel satisfied with yourself as a scientist? Do you feel, as people used to say, "fulfilled"?

ADA YONATH: I am sorry – I cannot answer this question since my intellectual energy is fully devoted to questions related to life processes and life origins, not to myself. . .

PLL: The origin of life for most scientists is a natural process governed by basic chemistry law, without any transcendent intervention. For some others, the life complexity we are experiencing cannot be conceived without the action of God. What is your opinion on this question?

ADA YONATH: As in my answer to third question, I expect that once all "secrets" of life processes and life origins are revealed, we will be able to disclose also the possible contribution of God. . .

PLL: The field we are working in is still dominated by big, fundamental questions about the origin of life, of mankind, of consciousness. . . Is there something special that you may like to say to young people entering this field?

ADA YONATH: This is one of the most exciting interdisciplinary fields to be investigated.

4.2 The ribocell

As an extrapolation of the prebiotic RNA-world scheme, one can think of a cell containing initially only RNA and ribozymes. This would be a RNA cell, or "ribocell." This was described a few years ago (Szostak *et al.*, 2001) and in fact represents, among other things, a nice example of the coming together of the RNA world and the compartment world.

The general idea is illustrated in Figure 4.1. It consists of a vesicle containing two ribozymes, one (**Rib-2**) capable of catalyzing the synthesis of the membrane component; the other (**Rib-1**) being an RNA replicase that is capable of replicating itself, and reproducing **Rib-2**, as well. In this way, there is a concerted shell-and-core replication, and there is therefore a basic metabolism, self-reproduction, and – since the replication mechanism is based on RNA replication –evolvability.

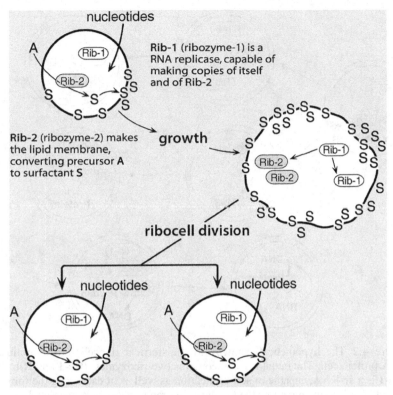

Figure 4.1 The simplest "ribocell," consisting of a lipid membrane containing two ribozymes, **Rib-1** and **Rib-2**, whereby **Rib-1** is a **sr-RNA**, capable of replicating itself as well as **Rib-2**. **Rib-2**, on its part, builds the membrane, converting the precursor **A** into the surfactant **S**. Since the entire system is based on a **sr-RNA**, it is capable of evolution. (Adapted from Szostak *et al.*, 2001, and Luisi *et al.*, 2002.)

How realistic is all this? One may recall at this point the previously mentioned self-replicating ribozyme reported by Paul and Joyce (2002); however, as beautiful as this work is, we are still far from an RNA polymerase that catalyzes the synthesis of itself as well as the synthesis of another ribozyme.

Thus, Figure 4.1 is still a theoretical construction, and this scheme implies a series of assumptions. For example, it is assumed that both the precursor **A** and all mononucleotides, present in large excess outside, can permeate into the cell. It is also assumed that the cell divides by itself mechanically giving rise to identical daughters, whereas a statistical distribution of the macromolecules in the resulting cells should be expected (some having no **Rib-1** and/or **Rib-2**; and only some having both of these ribosomes). The RNA cells

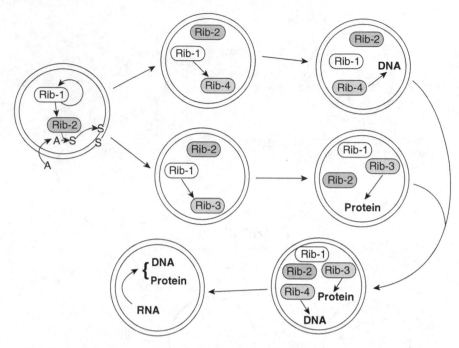

Figure 4.2 The hypothetical pathway to transform a ribocell into a minimal DNA/protein cell. The initial ribocell contains two ribozymes, **Rib-1** and **Rib -2**: **Rib-1** is a **sr-RNA**, capable of self-replication as well as of catalyzing the formation of **Rib-2**, a ribozyme involved in the membrane synthesis. As depicted in this hypothetical scheme, **Rib-2** catalyzes the reaction that produces the surfactant **S** from the precursor **A**. During its self-replication, **Rib-1** may evolve into new ribozymes, as **Rib-3** and **Rib-4**. The first one is capable of catalyzing the peptide bond formation, and into another ribozyme, **Rib-4**, capable of catalyzing the formation of DNA. In the idealized figure, it is assumed that the two mutations take place in two distinct compartments, which eventually merge together giving the final DNA/protein cell. (Modified from Luisi *et al.*, 2002.)

eventually have to evolve into protein/DNA cells. And this is a long and certainly not easy pathway. However, this is the beauty of the prebiotic RNA world: that at least on paper, a possible pathway leading to DNA and proteins can be conceived. One ideal pathway showing the transition from the RNA to the DNA cell is illustrated in Figure 4.2.

The constructs of Figures 4.1 and 4.2 are still ideal, hypothetical systems, nevertheless very interesting in some respects: they show that, at least in principle, cellular life can be played with a very limited number of RNA genes. Those who believe in the RNA world may add that this basic simplicity emphasizes the importance of RNA in the early stages of life.

Regarding these arguments, an interesting point of view may be the opinion of a distinguished researcher in the field, Gerald Joyce.

Conversation with Gerald Joyce

Gerald Joyce is a Professor in the Department of Chemistry and an Investigator of the Skaggs Institute for Chemical Biology at The Scripps Research Institute in La Jolla, California. He received a B.A. from the University of Chicago and both an M.D. and Ph.D. from the University of California, San Diego.

His research involves the test-tube evolution of nucleic acid enzymes, with relevance to the origins of life on Earth and with applications in medical diagnostics and therapeutics.

PLL: Several years ago you and Leslie Orgel wrote a paper, ironizing about the molecular biologists' dream, with a self-replicating RNA popping out from the prebiotic broth and evolving so as to give rise to ribozymes capable of catalyzing all the rest. The research of the last 20 years has shown that the self-replicating RNA is hard to produce, and the dream is even harder to forget. How do you see this situation?

Gerald Joyce: Actually, I see the trend as moving in the opposite direction, with a RNA-first origins now seeming more likely than it did 20 years ago. At that time the two biggest obstacles appeared to be prebiotic "clutter" and enantiomeric cross-inhibition. The former is not yet overcome, but the creative new approach of Powner and Sutherland to the prebiotic synthesis of RNA has shown a potential path forward. Also, there are now several examples of chemical processes that break chiral symmetry, most notably the work of Blackmond and colleagues. The chemists have not been sleeping!

PLL: At the 2014 ISSOL meeting in Nara, we had again our lifelong discussion about a definition of life, more or less in the same terms that we had in a discussion with Francisco Varela in California over 20 years ago. His autopoiesis views were so sharply criticized that he finally left the room, slamming the door. The definition of life he was proposing, from the theory of autopoiesis, reads: "Life is a self-maintaining system due to regeneration of the components from within a boundary of its own making." I still believe this is the most universal description at the level of each single individual, as there is no living system that does not comply with this. But for some reason you did not like this at that time. Is it the same now?

Gerald Joyce: Yes. I regard the capacity for Darwinian evolution as the key distinguishing feature of life, subsuming self-replication and self-maintenance. Living systems record their adaptive history as genetic information, something that is not the case for non-living systems.

PLL: This goes to one major point of the RNA-world view, according to which there is an identification between the criterion of life and the criterion of self-replication/evolution. Do you really think that these two things (life and self-replication) are one?

GERALD JOYCE: No. Self-replication is necessary but not sufficient for life. Self-replication must be combined with the capacity for Darwinian evolution to provide the basis for life.

PLL: You have done brilliant experiments with synthetic biology of RNA. Quite in general, where do you think that synthetic biology will be heading?

GERALD JOYCE: There are many different aspects of what is now called "synthetic biology," ranging from metabolic engineering to genetic reprogramming to the construction of synthetic organisms and potentially fully synthetic life forms. I expect there will be continued, rapid progress in all of these areas.

PLL: Do you see the arising of life on Earth as an obligatory, quite deterministic process – as the late Christian de Duve was advocating – or can you also say that the occurrence of life on Earth might not have to be?

GERALD JOYCE: I would call this the "cell biologist's dream." De Duve argued that the origin of life on Earth, even if a very low probability event, would have been inevitable because there were a vast number of opportunities for such a rare event to occur. I do not know if this is the case because I do not know how rare that "low probability event" might be. It may be human nature to hope that it is not so rare as to make us alone in the galaxy, but the relevant data are lacking.

4.3 The compartmentalistic approach

We have shown above the hypothesis of a ribocell, which combines the notions of the RNA world with the notion of membranes and compartments, something that is usually ignored in the classic RNA world.

We will consider now the classic "compartmentalistic" approach. The main conceptual framework here is that everything should have started from within a closed spherical boundary. The main argument in favor of this view is that all life on Earth is cellular, and only cellular – i.e., all based on closed compartments. A thought experiment in support of this view is the following: take all cells of this world and squeeze their content into the vast ocean; you will have all the RNA and DNA in the world – and no life. The result of this experiment is, in fact, all possible genetic material at a great dilution, without boundary. In a cell, life's functions are guaranteed by the flux of material and information through the boundary of the compartments, and by a high concentration of the components, which in turn permits the dynamic interactions of metabolism.

Given that such an asset is so important for life, the "compartmentalists" argue, in order to consider the origin of life we have to start from a primitive semi-permeable closed boundary.

This kind of approach, although with different wording and emphasis, has been in the literature for some time. Actually, the first books on the origin of life emphasize the importance of a cell-like compartment as the prime act of the pathway that eventually leads to the self-reproducing cells. Oparin (1924, 1953) was the first one, and he emphasized coacervates as the primitive compartments. These aggregates were not stable and poorly reproducible, however, and studies with coacervates were forgotten. In a paper with the suggestive title "Oparin's reactions revisited," Walde, Goto, and coworkers (1994) have repeated some of the original work of Oparin by entrapping the same enzymes and reactions into vesicles.

Morowitz (1992) and Dyson (1985) all discuss the case of "membrane first," the concept that the origin of life must be seen in closed boundaries. Several other authors have worked with this background idea, most notably David Deamer (Deamer, 1985 and 1998; Pohorille and Deamer, 2002) and, of course, my own group. All this has given rise to the field of minimal cells, which will be discussed in the last part of this book (Part V) as a chapter within synthetic biology, including detailed references. We will also see in detail how this work is based, operationally, on the possibility of incorporating enzymes, nucleic acids, and other biochemicals inside these compartments, so as to make possible models for primitive biochemical cells. This scenario is reinforced by the fact that cell-like compartments – e.g., vesicles – form spontaneously with molecules of imputed prebiotic origin, as indicated by the work of Deamer and collaborators (1989, 1994).

In this section, then, we will limit ourselves to a few general considerations about the compartmentalistic approach.

One idea, well known since the beginning of prebiotic chemistry, is to think to dehydration processes in a pool containing diluted components. As also recently suggested by Damer and Deamer (2015), due to dehydration, concentration can increase, and if in that pool there is a membrane, components may be entrapped – and repeating many times the hydration/dehydration cycles, the higher critical concentration can be achieved. One has to assume here that the repeated hydration does not bring back to a higher dilution. If entire vesicles are already in the pool, we are back to the mentioned problem of permeability.

One of the basic working ideas of compartmentalism for the origin of life is that the spontaneous formation of vesicles may have encapsulated some simple

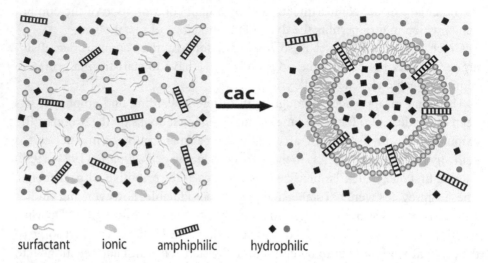

surfactant   ionic   amphiphilic   hydrophilic

Figure 4.3 A schematic representation of the formation and closure of a vesicle at the critical aggregate concentration (**cac**), with encapsulation of the solutes in the bulk solution. Water-soluble solutes can be localized in the aqueous interior of the vesicle, the amphiphilic in the hydrophobic bilayer, and the ionic ones on the charged surface of the vesicle.

peptide catalysts together with other molecules, and that a primitive protocell metabolism may have started in this way. A simple schematization of this is shown in Figure 4.3, which summarizes one of the basic ideas in this part of science, namely that a realistic scenario of the emergence of life can be based on a gradual transition from random mixtures of simple organic molecules to spatially ordered assemblies displaying primitive forms of cellular compartmentation, self-reproduction, and catalysis.

The next step in this hypothetical scenario would be something more complex and detailed (Figure 4.4). We see here that ideally DNA and other biopolymers can be incorporated into vesicles, or liposomes (liposomes are vesicles made out of phospholipids), giving rise eventually to a first metabolism and then to the self-reproduction of the protocell.

However, this is all still a hypothesis, and furthermore in this scenario, we start from already formed macromolecules, and the question of how DNA, RNA, enzymes, and the primitive metabolism really started remains unanswered. You see here one of the basic problems of this approach: it cannot tell us how the ordered sequences of DNA and enzymes have come about.

The other limitation is in the critical threshold concentration needed in a compartment to start life. We will see later on that this last point, at least, can be considered solved thanks to recent findings in literature.

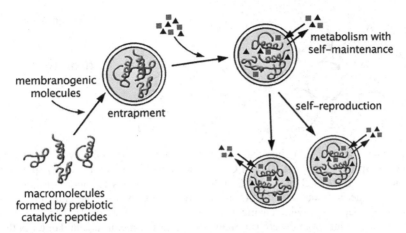

Figure 4.4 Schematic representation of the compartmentalized prebiotic reactions leading to the first cells capable of self-reproduction. As explained in the text, this is still a hypothesis, although certain single steps have been demonstrated.

4.4 Primordial cells without DNA?

One interesting question about primordial cells, or protocells, is whether some form of compartmentalized metabolism might have been present prior to DNA and ribosomal protein synthesis. In this respect, let us consider Figure 4.5.

The picture shows a protocell displaying a minimal metabolism, consisting of only three components, whereby one of the reactions, as byproduct, yields **S**, the membrane-forming surfactant. This metabolism, if **S** accumulates inside the membrane structure, may lead to cell growth and duplication.

The main point is whether one such a structure could be conceived as a product of chemistry, and possibly made in the laboratory.

The meaning of a self-replicating structure due to an internal metabolism, free of DNA, is clear enough: it would be a beginning of a metabolic life prior to – or independently from – nucleic acids. Aside from the big question of the origin of life, this construct would be a magnificent self-replicating system, which simulates the cell behavior – but purely on a chemical basis. As already mentioned, this would be in keeping with the various hypotheses of "metabolism first," advocated by several researchers, and which we will illustrate later on in this chapter. The problem is that no lab until now has been capable of doing it. It remains one of the projects for the near future.

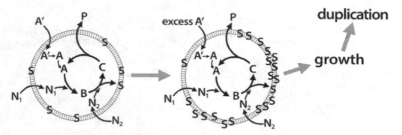

Figure 4.5 Scheme of a chemical protocell. A simple metabolic cycle, based on three components only (**A, B**, and **C**) is shown. The idealized metabolic path is: **A** produces **B**, then **B** produces **C**, which in turn (thanks to the incoming N_2) produces the compound **A**. In the intracellular environment, the entering compound **A'** becomes **A**, which is incorporated in the metabolic cycle. The nutrients N_1 and N_2, abundant in the environment, may pass the membrane and enter in the cycle, in reactions **A** → **B** and **B** → **C**, respectively. Moreover, the reaction **B** → **C** affords as product the surfactant **S**. The following reaction, **C** → **A**, leads to a byproduct **P**, eliminated by release in the medium. The excess of **A'** in the medium can result in a significantly higher production of the surfactant **S**, which can accumulate in the lipid membrane, leading to the protocell growth and – eventually – to its duplication. The reproduction mechanism will be described in more detail later on in the book, in a discussion about self-reproduction of vesicles.

4.5 The phenomenon of spontaneous overcrowding

The basic working idea of compartmentalism is that the spontaneous formation of vesicles may have encapsulated some simple peptide catalysts together with other molecules, and that a primitive protocell metabolism may have started in this way, as schematized in Figure 4.4.

And here, as already mentioned, an acute problem of concentration threshold arises. Assuming that the starting scenario is the ocean, or a vast lagoon, it is inconceivable that the concentration of all important biopolymers and/or cofactors would be above the micro-molar range; actually, one should expect that dilution would be extreme and would prevent the formation of macromolecular complexes and their interactions. And even if one assumes that in some particular local spot the concentration of biopolymers may be in micro- or milli-molar range (e.g., after the tide retires), then the question arises, how all this material can be locally incorporated into vesicles, also considering the well-known lack of permeability of vesicles and liposomes. This is, in fact, one of the main riddles that shakes the basis of the compartmentalistic, approach.

However, this riddle may have recently found a solution. We will give here only a short synopsis of the finding, as the whole subject of overcrowding is presented in some detail in Part V of this book.

The finding is the following: when vesicles forming surfactants are added to a solution containing biopolymers, the formation and closing of these vesicles *in situ* acts as an "attractor," which determines the spontaneous accumulation of macromolecular solute inside the compartment at a concentration that can be more than one order of magnitude larger than the bulk concentration. Only a small percentage of vesicles display such a property, while the majority of the vesicles remain "empty" – so that the situation can be depicted in first approximation as a situation of "all-or-nothing." A typical experiment is shown in Figure 4.6. At this preliminary point, it is important to recall that the authors suggested (Luisi *et al.*, 2010; see also de Souza *et al.*, 2011, and Stano *et al.*, 2011) that this spontaneous overcrowding – with prebiotic membrane forming surfactants – might be considered as a primordial mechanism that allowed the first forms of metabolism.

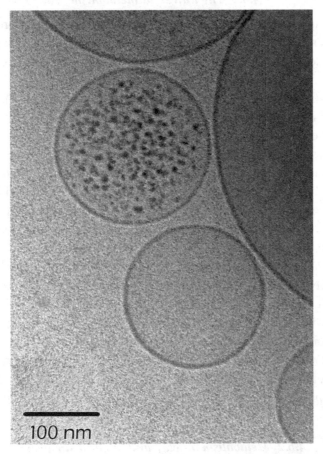

100 nm

Figure 4.6 The overcrowding of a protein (ferritin) inside phospholipid vesicles. This is a kind of all-or-nothing encapsulation mechanism. (Luisi *et al.*, 2010 – see also de Souza *et al.*, 2011, Stano *et al.*, 2011, and de Souza *et al.*, 2012.)

For the compartmentalistic approach, there remains another main stumbling block: the lack of information about the biogenesis of the biopolymers. Again, this is the question of the biogenesis of orderly macromolecular sequences. We will see in the next chapter some attempts to overcome this obstacle.

4.6 The "prebiotic metabolism" approach

The origin of life based on a prebiotic metabolism appears to be the aim of a few distinct research groups, who are inquiring into the possibility of prebiotic metabolic pathways prior to enzymes. There is a rather complex kaleidoscope of authors and views, and I present in the following paragraphs only some of the main trains of thought.

4.6.1 The universal metabolism

In a rather original approach, H. J. Morowitz and co-workers (Morowitz *et al.*, 1991, 1995, and 2000) examine the chemistry of a model system that starts with carbon dioxide and reductants and uses redox couples as the energy source. They start with the large database of organic molecules, *Beilstein on-line*, at www .beilstein.de (see also www.indiana.edu/~cheminfo/cciim34.html), to investigate the reaction networks that might emerge. On the basis of certain assumptions, from the 4.5 million entries in *Beilstein*, they notice the emergence of only 153 molecules, which happen to be the ones that contain all 11 members of the reductive citric-acid cycle. According to the authors, these calculations suggest that the metabolism corresponds to a universal pathway chart and is therefore central to the origin of life. Enzymes would have come only later and accelerated the cycles, eventually taking over; the bottom line is the idea that there was a metabolism prior to the origin of catalytic macromolecules.

It must be said that this view is not universally accepted; for example, Orgel has argued forcibly against the *Beilstein* approach and generally against the metabolic cycles (Orgel, 2000b). His point is that theories involving the organization of complex, small-molecule metabolic cycles, such as the reductive citric-acid cycle on mineral surfaces, have to make unreasonable assumptions about the catalytic properties of minerals and the ability of minerals to organize sequences of disparate reactions. Another conclusion is that data in the *Beilstein Handbook of Organic Chemistry* can be interpreted in a different and more plausible way.

4.6.2 Metabolism on clay and mineral surfaces

When considering clay and minerals, one should necessarily mention the pioneer work by Cairns-Smith (Cairns-Smith, 1977, 1978, 1982, 1990; Cairns-

Smith & Walker, 1974; Cairns-Smith *et al.*, 1992; see also Bujdak *et al.*, 1994). This author, developing an earlier idea suggested by Bernal, showed that clay mineral surfaces could adsorb organic compounds, building up sufficient concentrations of these, as well as acting as templates for polymerization in an aqueous environment. The "primeval soup" would thus become a concentration of organic compounds not in solution but absorbed on mineral surfaces. The original intuition was probably that the replicator-catalyzing agents were actually crystals to be found everywhere in the clay that lay around primitive Earth. Crystals are structurally much simpler than any biologically relevant organic molecule, and they appear to grow and reproduce by breaking into smaller seeds that can grow further. The next step is to propose that these very primitive "organisms" started by incorporating peptides found in the environment, which may open the road to a gradual increase of complexity – up to the genetic takeover.

All this sounds more like a metaphor and this is probably why the idea was so catchy – it draws on visual evidence of simple and well-known inorganic mechanisms. However, from a metaphor to a chemical reality there is a wide gap: how are proteins and nucleic acids built? And, how is the lipid metabolic compartment built? Actually, from the experimental point of view, there is very little that substantiates the take-over scenario. This view of Cairns-Smith has taken a good niche in the literature of the origin of life, being constantly refreshed and reconsidered (Cairns-Smith, 1978, 1982, 2008). Of course, the chemistry on clay and mineral surfaces as primitive forms of matrices for primitive metabolism eventually has to evolve into the world of membranes and lipids.

Cairns-Smith's original approach may sound different from Morowitz's approach mentioned earlier; but in fact, it also postulates a kind of metabolism without and prior to enzymes.

4.6.3 The beauty of pyrite

Along similar lines unfolds the work of G. Wächtershäuser (Huber & Wächtershäuser, 1997; Wächtershäuser, 1990a, 1990b, 1992, 1997, 2000). We have mentioned this author in Chapter 2, talking about redox reactions. More in general, the basic idea is the proposal of an autotrophic emergence of life, based around a reductive Krebs cycle (enzyme-free of course), which is "running in reverse" (Lazcano, 2004). Synthesis and polymerization of organic compounds would take place on the surface of pyrite under extremely reduced volcanic settings resembling those of deep-sea hydrothermal vents. Life then began with the reduction of CO_2 and N_2, coupled with the reducing power of pyrite formation, and (of course) all this became another "world" – the "iron-sulfur-world."

It is worthwhile to remark that the metabolism is a surface metabolism, based on a two-dimensional order, and based on negatively charged constituents on a positively charged mineral surface. Actually, Wächtershäuser sees this as an interesting part of a broader philosophical view (Huber and Wächtershäuser, 1997). At this regard, since this chapter is devoted to the various views on the origin of life, it may be relevant to catch the difference between this approach and the "prebiotic soup" approach (Wächtershäuser, 2000):

It is occasionally suggested that experiments within the iron-sulfur world theory demonstrate merely yet another source of organics for the prebiotic broth. This is a misconception. The new finding drives this point home. Pyruvate is too unstable to ever be considered as a slowly accumulating component in a prebiotic broth. The prebiotic broth theory and the iron-sulfur world theory are incompatible. The prebiotic broth experiments are parallel experiments that are producing an increasingly greater medley of potential broth ingredients. Therefore, the maxim of the prebiotic broth theory is "order out of chaos." In contrast, the iron-sulfur world experiments are serial, aimed at long reaction cascades and catalytic feedback (metabolism) from the start. The maxim of the iron-sulfur world theory should therefore be "order out of order out of order."

Several authors in the field have not spared their criticism on these views, e.g., Stanley Miller, Christian de Duve (de Duve and Miller, 1991), and Leslie Orgel. One argument by Orgel (2003) is that this "pyrite metabolism" might reflect not a pyrite-dependent primordial metabolism but the unique, deterministic way in which a given chemical process can take place. In addition, Orgel (2000b), as already mentioned, repeats his general point: that the theories advocating the emergence of complex, self-organized biochemical cycles in the absence of genetic material are hindered, not only by the lack of empirical evidence, but also by a number of unreasonable assumptions about the properties of minerals and other catalysts required spontaneously to organize such sets of chemical reactions.

Despite these criticisms, Wächtershäuser's ideas have enjoyed considerable popularity. (See, for example, the book by W. Day, 2002, in the list of books on the origin of life in Chapter 1.)

4.6.4 Other metabolic approaches

To the list of enzyme-free metabolism one should add the work by August Commeyras and his coworkers in Montpellier (Taillades *et al.*, 1999; Plasson *et al.*, 2002). We have already mentioned these authors when we described the N-carboxy anhydride condensation of amino acids. This mechanism is part of a larger molecular engine mechanism, which could have taken place on primitive beaches in the Hadean age. This mechanism consists of a primary pump, which relies on a reaction cycle made up of several successive steps, fed by amino acids, and fuelled by NO_x species. The proposed mechanism assumes a buffered ocean, emerged land, and a nitrosating atmosphere. The French authors argue that this primary pump might have been the prebiotic mechanism that gave rise to oligopeptide sequences, which in turn started the reaction that led to the origin of life.

This primary pump shared its energy with phosphoric acid, so that – according to these authors – NCA-activated amino acids gave rise to aminoacyl-phosphate, able to react with nucleotide mono phosphates (NMPs), originating NDPs, as well as to react with NDPs to give NTPs, in particular ATP (Biron and Pascal, 2004; see also Commeyras *et al.*, 2005).

Moreover, the primary pump induces the peptides to assume a particular globular structure (the so-called Dendrigraft, see Figure 4.7). This structure permitted – in the Hadean scenario depicted by the authors – an ordered packing of peptides [Figure 4.7 (**a**)] and orients their positive N-terminal groups ($-NH_3^+$) towards the surface of the globular structure [Figure 4.7 (**b**)]. This attracted – according to the authors – the negative groups of several nucleotide triphosphate (NTP, and ATP in particular) on its surface [Figure 4.7 (**c**)], letting them concentrate around the globular peptide structure. This saved them from hydrolysis. So, argue the authors, all these processes gave permanent activation of nucleotide monophosphates into nucleotide triphosphate (NTP, and ATP in particular). This may have led to nucleic acids, whereby the scenario is considered a co-evolutionary one.

For such a logical sequence to operate effectively, the French authors argue that the primary pump must have a mechanical component. This may be constituted by the tides, which periodically dilute or concentrate solutes (Commeyras *et al.*, 2005). Citing them:

At high tide, the water inundates the land surface with a batch of diluted material. Upon withdrawal, evaporation concentrates the N-carbamoyl amino acids to the crystallization point. Thus, evaporation could have been the first natural technique to concentrate the starting material of the primary pump.

The details are rather complex; please refer to the original literature by Commeyras and coworkers for further explanation.

Still concerned with a prebiotic metabolism without the help of enzymes is also de Duve's "thioester world" (1991). And in this respect, Shapiro's ideas (1986) should also be mentioned.

Of course, this list is not exhaustive; theories on the origin of life abound in the literature, and as I mentioned before, here I want to present a basic *spaccato* without giving a general and detailed review. In addition, as I have mentioned in the introduction, the theoretical work falls too short regarding all of this. However, I would like to cite at least one influential author such as Stuart Kauffman. In his well-known article (Kauffman, 1986), he connects the origin

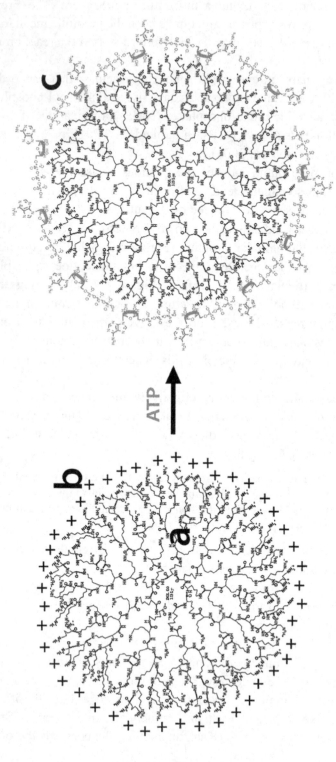

Figure 4.7 Dendrigraft poly-L-lysines (**a**) and its ordered and oriented structure, with positively charged N-terminals at the globular body's surface (**b**). The interaction of dendrigraft with several ATP molecules is also shown (**c**). See the text for more details or refer to the original work (Commeyras *et al.*, 2005, from which, with some modifications, the dendrigraft picture was taken).

of life to the spontaneous rise of a catalytic set of peptides (see the guest essay by this author in Chapter 9). The main suggestion is the possibility of an emergent self-organizing collective property, in particular the emergence of self-replicating protein systems in prebiotic evolution. Similar principles may apply to the emergence of a primitively connected metabolism. The question, whether and to what respect this view connects with the experimental findings cited in this chapter, is still open.

Finally, I would like to mention Doron Lancet, a pioneer of the so-called "lipid world" scenario. So, let us take a look at his view on the subject of prebiotic evolution and related ideas.

Conversation with Doron Lancet

Doron Lancet has been acting as Professor at the Department of Molecular Genetics, Weizmann Institute of Science in Israel. As director of Crown Human Genome Center at Weizmann and Israel's National Center for Genomics, he pioneered genome research in Israel.

He has been actively involved in teaching bioinformatics and in initiating a Systems Biology effort at the Weizmann Institute. A pioneer of olfactory research, he studied the genomics and population genetics of human olfaction, as well as the evolution of the olfactory receptor superfamily. In the field of prebiotic evolution, Doron Lancet pioneered a Lipid World scenario, in which compositional information is replicated with mutation, and hence may jump-start Darwinian evolution without DNA or RNA.

PLL: Since Miller's time, what is (are) the experiment(s), which you consider the most significant for our understanding of the origin of life?

DORON LANCET: My avid view is that the Miller-Urey experiment made a great contribution to understanding how small chemical building blocks, such as amino acids and fatty acids, got generated abiotically from yet simpler atomic combination, i.e., inorganic gases. In perspective, the significance of this experiment, with all due respect, is limited: there are other sources of similar carbon compounds, chiefly infalling comets, meteorites, and interstellar dust particles. In my view, important experiments should portray a path from such small building blocks to complex molecular assemblies, not necessarily involving long covalent polymers, and likely somehow involving amphiphilic lipids. Under some conditions these could manifest a rudimentary capacity for information storage and propagation, possibly empowered by mutually catalytic networks as described by Stuart Kauffman and Freeman Dyson. These experiments are only beginning to emerge, and some names relevant in this context are Lee Cronin from Scotland and Tetsuya Yomo from Japan.

PLL: Orgel and Joyce wrote several years ago a paper referring to the molecular biologist's dream: a self-replicating RNA coming out from the prebiotic soup and making by successive evolution/mutations all necessary ribozymes to make life start. For several people this is still the most likely scenario for the origin of life. Do you belong to this group?

DORON LANCET: I definitely do not. The worst enemy of prebiotic RNA is the likely random chemistry nature of prebiotic organic mixtures. If molecules similar to RNA formed abiotically, they would appear along with numerous isomers. Same goes for sugars. Under conditions where sugar would be phosphorylated, there will be a mixture of molecules different in phosphor-site counts and positions. Getting a linear homopolymers of just one monomeric configuration would be next to impossible. There are other hurdles, mentioned by many, such as RNA's instability and the difficulty of abiotically forming polynucleotides sufficiently long to stand a chance of being self-replicating. There is also a semi-philosophical problem: why should we at all assume that life began with anything that resembles present-day convoluted molecular replication? A good analogy is what might happen long after the technological singularity. Under some circumstances, all organic life might disappear, leaving behind complex, cognitive silicon-based creatures. Imagine them trying to trace back how they had emerged on earth. Imagine the elaborate theory and experimentation regarding the spontaneous emergence of computer chips. Using RNA as a target for scrutinizing life's origins could be just as futile.

PLL: You have been active within the "party" of the compartmentalists, emphasizing the relevance of liposomes in the early start. This kind of view is not very popular in the field; membranes and closed compartments do not receive much coverage in international meetings and reviews. Do you have an explanation for that?

DORON LANCET: A possible explanation rests in the misnomer "compartmentalism." The school I belong to asserts that the same molecules (lipids, and generally amphiphiles) that form today's compartments in living cells are also capable of doing much more in a world devoid of "smart" biopolymers. Their being able to form compartments may be less relevant to life's origin. In fact, the computer-simulated model we have devised pertains much more to micelles than to vesicles. A vesicle's inner aqueous compartment plays no role in the models. What does is the focus on lipid-like molecules because they spontaneously accrete in water to form discrete microscopic assemblies that can have idiosyncratic composition. Here of course the random chemistry nature of a prebiotic soup suddenly becomes an advantage. Only highly heterogeneous chemical mixtures can yield highly diverse assembly types. Without assembly diversity there is no meaning to information content and to its propagation by homeostatic growth and fission of micelle-like entities. Compositional information-based entities can show behavior akin to selection and evolution. Because of such misunderstanding, our lectures are often placed in a "compartments" session, where they do not belong. And such misinterpretation is the reason why it is difficult for assembly-first proponents to become more popular. The crowd sticks to the misconception that compartments are "foolish" barriers, and are less inclined to follow the subtleties of compositional "lipid World" essentially a specific instance of the "metabolism-first" scenario.

Concluding remarks

We have seen the major approaches to the origin of life, those that are more discussed in the literature; and some less known approaches. Most of these will be discussed again later on in the book, with due consideration of the experimental procedures. Here I have presented mostly the basic ideas. These are all hypotheses, in the sense that none of these views can claim to have solved the problem of the origin of life – and we have seen in fact that each of these approaches has its serious limits. These different hypotheses are also displaying different philosophies. The fact that we have so many different views indicates the obvious: that we do not have one that is ascertained and convincing for the majority of researchers. Partly, these views are seen in the literature as opposing each other; I have experienced fierce fights in the past between those in the RNA world and those advocating autopoiesis/compartmentalism – a controversy that did not change much over the years. Another one is Wächtershäuser's polemic against those who still hold to the "prebiotic soup."

However, it should be obvious that these views are not mutually exclusive; rather, they are often complementary. In particular, concerning the dichotomy between the prebiotic RNA world and compartmentalism, we pointed out first the work by Szostak *et al.* (2001), on the convergence between the prebiotic RNA-world and the compartmentalistic approach. One should also consider that most of the criticism against the RNA world would be highly attenuated, once agreed that we need a prerequisite pre-RNA world, and one possibly based on amino acids. Then, the dichotomy would disappear in favor of a unified vision – as the compartmentalistic hypothesis would be the basis for the formation of RNA.

The prebiotic RNA world remains the most popular view on the origin of life, the one which is blindly accepted by most undergraduate and graduate students in the life sciences. I have mentioned a couple of reasons for this popularity. First, the fact that in this generation we live in the "century of the gene," in which most of the life sciences are nucleic acid centered; then, the success and the beauty of the research on RNA at large (mostly synthetic biology), and the fact that there are no alternative views on the origin of life that are really convincing.

Even with the compromise, that a cellular metabolism must precede the RNA world, there remains the old basic dilemma: the biogenesis of DNA and of the first enzymes. For that, we need perhaps a fresh view on the origin of life, and this is the primary focus of Chapter 5 – the idea that we should return to basics in order to tackle the biogenesis of the biopolymers of life.

Chapter 5

Origin of life from ground zero

Introduction

We have seen in the previous chapter the most common approaches to the origin of life, and despite the conceptual and operational differences, there is something in common which is quite interesting: they do not solve or even tackle the biogenesis of the basic macromolecular components, like DNA, RNA, and proteins. Several of these approaches start programmatically from a scenario in which the ordered sequences are supposed to be already extant. As said in a common metaphor, this is likened to building a house and starting with the roof.

It is necessary to work on the foundation. One may recall in this regard the work of old master Ponnamperuma, who in a 1976 review (Hulsof and Ponnamperuma) makes the point that all biopolymers are formally made out of the condensation of the corresponding monomers. Those reactions, that are unfavorable in aqueous solution, can proceed in the presence of condensing agents. Moreover, the authors made a series of important observations, for example that one of the most efficient condensing agent is polyphosphate.

This chapter will focus on the biogenesis of the primordial macromolecular components based on the prebiotic compounds at "our" disposal 4 Gya – a scenario I like to call the "ground zero approach." This will include not only general approaches, but also single experiments, which have the characteristic of guiding us towards the biogenesis of proteins or nucleic acids as ordered structures. In a way, this is a continuation of the previous chapter, which was devoted to the hypotheses on the origin of life, and actually we have to deal with some slight repetitions. These are partly unavoidable, and they are partly called to reify certain points that I consider very important.

5.1 Prebiotic amino acids and peptides

What do we have at ground zero? We have considered this question already, but let us focus now on amino acids and peptides.

There is consensus on 10 prebiotic amino acids, namely the α-amino acids that are spontaneously formed under the Stanley Miller conditions, or similar ones. They are: Ala, Gly, Asp, Val, Leu, Glu, Ser, Ile, Pro, and Thr. Notice that in the 10 primordial amino acids, there are no aromatic ones.

The corresponding one-letter codes are: **A, G, D, V, L, E, S, I, P, T**, and, for the sake of completeness, let me add that **W** is tryptophan, **K** is lysine, **R** is arginine, **F** is phenylalanine, **C** is cysteine, **M** is methionine, **H** is histidine, **Y** is tyrosine, **N** is asparagine, and **Q** is glutamine.

Prebiotic peptides were present in the first of Miller's experiments. As already mentioned, a group of researchers, which includes Jeffrey Bada and Fernadez (Parker *et al.*, 2014) pointed out recently that Miller sparked a gas mixture of CH_3, NH_3, and H_2O, while intermittently adding the plausible prebiotic condensing reagent cyanamide. For unknown reasons, an analysis of the samples was not reported. These authors now analyzed the archived samples, finding a dozen amino acids, 10 glycine-containing dipeptides, and 3 glycine-containing diketo-piperazines.

Important for us, here, is the detection of peptides, something that usually is not linked to the original Miller experiments. These peptides are the results of condensing agents, such as cyanamide, which according to this group of authors are of potential importance in generating diversity within the prebiotic chemical inventory.

Kobayashi and coworkers (Kobayashi *et al.*, 1990, 1998) also obtain 10 different amino acids by irradiating a gaseous mixture of CO, CO_2, N_2 and H_2O with high-energy particles. Peter van der Gulik and coworkers (2009) have studied the relative abundance of these primordial amino acids under different sets of prebiotic experiments, including meteorites. They also conjecture the presence of early functional peptides, which supposedly were short (3 to 8 amino acids long), and made from Gly, Ala, Val, and aspartic acid, which are the most efficiently produced amino acids under prebiotic conditions. They further assume that some traces of these prebiotic peptides still exist, in the form of active sites in present-day proteins.

A ground zero approach is assumed in the work of Ikehara's group (Ikehara, 2002, 2005, and 2009; Ikehara et al., 2002). The starting assumption is that water-soluble globular [GADV]-proteins could be formed with high probability by random polymerization of [GADV]-amino acids, even in the absence of any genetic function. These authors also furnish experimental observations indicating that [GADV]-peptides have protease activity for peptide bond formation. Moreover, [GADV]-proteins could be *pseudo-replicated*: pseudo-replication is a process where proteins having the same constituent set of amino acids (composition), which possess similar but different structures, are generated by a random process without resorting to any exact duplication. This is before the creation of the first genes, which according to Ikehara and coworkers are of the $(GNC)_n$ type. They state that the genetic code originated from a GNC code, which was implemented as singlets. This means that the accumulation of oligonucleotides triggered the generation of the GNC primeval genetic code through stereospecific complex formation among four [GADV]-amino acids and four corresponding GNC-containing oligonucleotides. In fact, according to this thesis, the GNC code comprises four codons and four [GADV]-amino acids.

One general, interesting point stemming from this work is that organized genes, as codon sequences, cannot be created by random polymerization of nucleotides. For this reason, they say, a gene cannot generate prior to the crucial protein (enzyme) synthesis. In contrast, appearance of [GADV]-proteins with catalytic functions preceded the appearance of the genetic system.

In addition, concerning the primordial amino acids and the corresponding primitive proteins, let us examine the work of Longo, Lee, and Blaber (2013). They tackle an essential unsolved question regarding this prebiotic set, whether it contains sufficient chemical information to permit a stable protein folding. To investigate these questions, two "primitive" versions of an extant protein fold (the β-trefoil) were produced in a reduced alphabet size of 12 or 13 amino acids and a percentage of prebiotic amino acids approaching 80 percent. The results suggest that the prebiotic amino acids do comprise a foldable set within the halophile environment.

Aside from that, let us recall the already mentioned works of the French school of Commeyras *et al.* (2005) and Taillades *et al.* (1999), namely the condensation reaction of the N-carboxy anhydrides, giving rise to co-oligopeptides. Concerning this NCA method, there are several studies on homo-polymerization, starting with the old studies by Ballard and coworkers (see, for example, Ballard and Bamford, 1956), and only a few on co-polymerization (see, for example, Habraken and coworkers, 2010). The previously cited Commeyras team, because of their "prebiotic pump hypothesis," have investigated the conditions for the elongation of polypeptides by the NCA method (Plasson *et al.*, 2002; Commeyras *et al.*, 2002, 2005).

Most of the studies carried out so far for making high molecular weight polypeptides refer to homo-polymerization. Let me repeat here a general point, coming from the field of polymer science, and already mentioned in Chapter 3 (section 3.5): The information gathered on homo-polymerizations is of little help regarding co-polymerization.

5.2 Peptides with catalytic power

We were mentioning that simple peptides, of possible prebiotic origin, possess catalytic power. Glycine and di-glycine have been postulated as possible catalytic factors in the prebiotic evolution (Plankensteiner *et al.*, 2002). Erickson and Kennedy (1980) reported the catalysis of peptide bonds due to the action of a dimer His-His in a fluctuating clay environment, adding that this might be a model for a primitive prebiotic synthesis by a "proto-enzyme." The catalytic activity of prebiotic His-His is reported also by other authors (Shen *et al.*, 1990a). Histidine is not in the list of the 10 more plausible prebiotic amino acids, but its synthesis in simulated prebiotic conditions has been reported (Shen *et al.*, 1990c).

Generally, the catalytic power reported for these simple peptides is limited to the hydrolysis of the peptide bond – a protease activity. However, this means, according to the principle of microscopic reversibility, that they should be capable of catalyzing the condensation of peptide bonds. A second, related, very important point is the notion of kinetic control, as discussed previously. In fact, if these catalytic peptides are prebiotic compounds formed under thermodynamic control, we would reach the conclusion that a "free ticket" to catalysis is being offered by thermodynamics.

One of the most interesting catalytic peptides is Ser-His, provided with catalytic functions that are extraordinary in view of the simplicity of this structure (Li *et al.*, 2000). The Chinese group, in fact, reported that this peptide is capable of hydrolyzing the peptide bond, the ester bond, and even the nucleotide bond. They also pointed out that simpler prebiotic peptides such as Gly-Gly have similar catalytic power, although not to such an extent.

Based on microscopic reversibility, this peptide has been utilized to synthetize a series of simple dipeptides and tri-peptides (Gorlero *et al.*, 2009). Table 5.1 gives an account of this work.

Notice that, when condensing an acyl derivative of an amino acid with a free amino compound, the yields may reach 30 percent, but note also the condensation of PNA derivatives, although with lower yields. No systematic work has been carried out in order to optimize the yields – trying, for example, with different pH, temperature, and solvent.

This work is reminiscent of the reverse protease reaction – peptide synthesis instead of cleavage, which is well known in the literature. In fact, proteases have been used extensively for peptide coupling (Jakubke *et al.*, 1985, 1996; Jakubke, 1987, 1995; Luisi *et al.*, 1977b). It has also been shown that even small proteins can be synthesized by block-wise enzymatic coupling (see also Kullmann, 1987; and, for some more recent developments, Celovsky and Bordusa, 2000).

Several methods have been conceived to carry out this kind of reaction and possibly increase the yields. For example, the reaction may take place by freezing the aqueous reaction mixtures to reduce water concentration (Hansler and Jakubke, 1996). Alternatively, one can carry out syntheses in organic solvent-free media containing a minimum water content (Halling *et al.*, 1995;

Table 5.1 *Yields of peptides and PNAs, as catalysed by Ser-His and (Gly)$_n$*

Entry #	Catalyst	Conditions	Acyl donor	Free amine	Coupling product	Yield (%)
1	Ser-His	a	Ac-Phe-OEt	H-Leu-NH$_2$	Ac-Phe-Leu-NH$_2$	15.3
2	Ser-His	a	Ac-Phe-OEt	H-Phe-NH$_2$	Ac-Phe-Phe-NH$_2$	35.2
3	Ser-His	a	Ac-Phe-OEt	H-Leu-Phe-NH$_2$	Ac-Phe-Leu-Phe-NH$_2$	25.4
4	Ser-His	a	Z-Ala-Phe-OMe	H-Leu-NH$_2$	Z-Ala-Phe-Leu-NH$_2$	4.0
5	Ser-His	a	Z-Ala-Phe-OMe	H-Phe-NH$_2$	Z-Ala-Phe-Phe-NH$_2$	0.5
6	Ser-His	a	Z-Ala-Phe-OMe	H-Leu-Phe-NH$_2$	Z-Ala-Phe-Leu-Phe-NH$_2$	4.3
7	Ser-His	b	PNA monomer	PNA monomer	PNA dimer	4.5^f
8	Ser-His	b	PNA monomer	PNA monomer	PNA trimer	5.6^f
9	Ser-His	b	PNA monomer	PNA monomer	PNA tetramer	9.1^f
10	Ser-His	c	H-Phe-OEt	H-Phe-OEt	H-Phe-Phe-OEt	0.5–3
11	Ser-His	d	H-Phe-OEt +H-Trp-OEt	H-Phe-OEt +H-Trp-OEt	H-(Phe)(Trp)-OEt	0.1
12	Gly	e	Ac-Phe-OEt	H-Leu-NH$_2$	Ac-Phe-Leu-NH$_2$	15.9
13	Gly-Gly	e	Ac-Phe-OEt	H-Leu-NH$_2$	Ac-Phe-Leu-NH$_2$	15.2
14	Gly-Gly-Gly	e	Ac-Phe-OEt	H-Leu-NH$_2$	Ac-Phe-Leu-NH$_2$	14.2

Conditions

a: 50 mM reactants (each), 4 mM Ser-His, 24 h, 25 °C;
b: 10 mM PNA monomer; 5.5 mM Ser-His, 35 h, 25 °C;
c: 270 mM H-Phe-OEt; 22 mM Ser-His, 14 days, 4 °C;
d: 50 mM reactants (both), 16 mM Ser-His, 7 days, 60 °C;
e: 50 mM reactants (each), 4 mM Gly (or Gly-Gly; or Gly-Gly-Gly), 15 days, 4 °C.

Note
f: Calculated from the theoretical 100% yield of that product, as the only product in the mixture.

Eichhorn *et al.*, 1997). One can also use a clay support (Bujdak *et al.*, 1995; Zamarev *et al.*, 1997; Rode *et al.*, 1999). The most common procedure, however, is by making the product insoluble, so that there is phase separation that shifts the equilibrium towards the product (Luisi *et al.*, 1977b, Anderson and Luisi, 1979; Lüthi and Luisi, 1984). The scenario of dry and wet cycles, invoked by several authors in the prebiotic field, may be another way of bringing forth the stepwise elongation (see, for example, Saetia *et al.*, 1993, Commeyras *et al.*, 2005; Damer and Deamer, 2015).

Simple peptides as catalysts, possibly products of prebiotic evolution, are of course interesting for the origin of life. The question is whether and to what extent they may be able to produce long polypeptide chains, which fold into proteins. We will see a potential example of that later on in this chapter.

We mentioned that Ser-His can also cleave the nucleotide phosphodiester bond (Li *et al.*, 2000), and therefore can be expected that this simple peptide also catalyzes the condensation of such type of bond. Wieczorek *et al.* (2013) have achieved the synthesis of the nucleotide bonds using this idea. Their preliminary study is mostly limited to the chemical mechanism; apparently, this oligomerization catalysis occurs by a transamination mechanism. In their work only the synthesis of dimers is presented, all with about 10 percent yield, in particular 2'-5'guanine dimer; 3'-5'guanine dimer; SH-capped 2'-5'guanine dimer; SH-capped 3'-5'guanine dimer; and small amounts of guanine trimer (at least the canonical 3'-5'pGpGpG, presumably other trimers are there, too). There was overall 10 percent yield of oligomers after 100 hours of incubation in standard conditions [guanosyl imidazolate (ImpG, 1,5 mM), MES buffer (5 mM, pH 4.5) and Ser-His (8 mM), T = −18.4 °C].

Quite interesting is Figure 5.1, which shows that the simple dipeptide obeys a Michaelis-Menten kinetics: a typical saturation curve and the maximal plateau velocity is directly proportional to the "enzyme" concentration.

Yields are low and mostly only dimers have been produced. However, this work shows that a simple dipeptide may be capable of promoting the condensation of the nucleotide moieties. This is important, as it is suggestive of an alternative scenario in which nucleic acids, including our old RNA, may be the products of peptide catalysis. This idea would be antithetic to that propagated by the RNA-world, and actually goes back to the discourse of the pre-RNA world, with peptide-based mechanisms capable of forming nucleic acids. Clearly more investigation is needed focusing on the "ground zero" peptide catalysis, and we will come back to this in the research proposals at the end of this chapter.

Concerning the prebiotic synthesis of oligonucleotides, it has been reported that hexa-nucleotides containing guanine and cytosine can be condensed using water-soluble carbodiimide (Kawamura and Kamoto, 2000); moreover, there

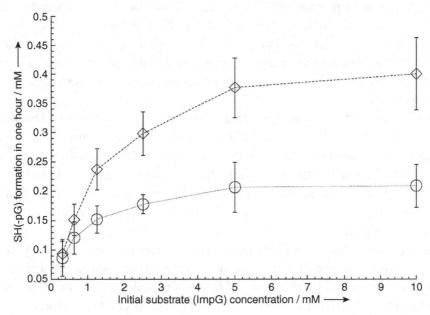

Figure 5.1 Catalysis by Ser-His in the oligomerization of nucleotides. On the ordinate, the concentration of the intermediate of the reaction – see the original paper for more details. (Modified from Wieczorek *et al.*, 2013, with permission of the author.)

are attempts in this direction from the old dehydration-condensation work of Ponnamperuma and Peterson (1965).

5.3 Proteins with a reduced alphabet of amino acids

We have seen in Section 5.1 a few investigations showing that the initial amino acids were only 10 and generally low in number. And we have illustrated the approach by Longo and co-workers (2013), who prepared in the lab a protein containing mostly the 10 prebiotic amino acids.

Before that, a series of investigators have synthesized proteins with a "reduced alphabet" of amino acids. We will mention here some of these studies. One should say from the start that the idea behind these studies was not always the origin of life, and therefore in some case not only prebiotic amino acids and/or prebiotic procedures were used. However, the idea of making proteins with a reduced spectrum of amino acids is, of course, very relevant for the origin of life, because they are important to the general question: whether foldable, possibly functional proteins can be obtained with macromolecular structures much simpler than the present-day proteins.

The problem of proteins with a reduced alphabet has been studied theoretically, as well; for example by Wolynes (1997), Wang and Wang (1999), Murphy *et al.* (2000), and Fan and Wang (2003). In this discussion, we will limit ourselves only to the experimental approaches. For this section, I am indebted to my student Nathalie Caretta, and for more details, see her thesis work (Caretta, 2005).

Let me start with the bold approach undertaken in the mid-1990s by Davidson and Sauer (1994), and later by Davidson *et al.* (1995). They expressed a library of polypeptides with 60–100 residues from random genes codifying only for glutamine (Q), leucine (L), and arginine (R). The aim was to see whether some kind of structure could be obtained in these relatively simple co-oligopeptides. In the investigated cases, the polypeptides (generally poorly soluble) showed a high helical content, as expected.

Riddle and colleagues (1997) undertook a different approach. They constructed variants of the SH3 domain[1] by using a phage display and a reduced alphabet of amino acids. They showed that complex β-structure proteins could also be obtained with an alphabet of 5 amino acids (IKEAG), and studied the velocity of folding of several variants.

With the work of Doi and collaborators, in Yanagawa's group (Doi *et al.*, 2005), we come closer to considerations on the origin of life, as they constructed a library of polypeptides, more than 100 residues long, using an alphabet composed of amino acids considered prebiotic, such as valine (V), alanine (A), aspartate (D), glutamate (E), and glycine (G). Eight VADEG proteins were randomly chosen out of the library; they resulted soluble, but were deprived of interesting structural properties. The solubility is a particularly interesting feature, considering also the previous work by Prijambada *et al.* (1996), who reported that only 20 percent of the artificial proteins formed by all 20 amino acids are soluble. The results of Doi and colleagues with the VADEG proteins appear then in agreement with the idea that the modern set of amino acids may be an evolution from an older, reduced alphabet.

Let us proceed now to an alphabet constituted by seven amino acids, in particular with the work by Schafmeister and colleagues (1997). Their approach

[1] The SRC Homology 3 Domain (or SH3 domain) is a small protein domain of about 60 amino acid residues first identified as a conserved sequence in the viral adaptor protein v-Crk and the non-catalytic parts of enzymes such as phospholipase and several cytoplasmic tyrosine kinases such as Abl and Src. It has also been identified in several other protein families such as: PI3 Kinase, Ras GTPase-activating protein, CDC24, and CDC25.

SH3 domains are found in proteins of signaling pathways regulating the cytoskeleton, the Ras protein, the Src kinase, and many others. They also regulate the activity state of adaptor proteins and other tyrosine kinases and are thought to increase the substrate specificity of some tyrosine kinases by binding far away from the active site of the kinase. Approximately 300 SH3 domains are found in proteins encoded in the human genome.

The classical SH3 domain is usually found in proteins that interact with other proteins and mediate assembly of specific protein complexes, typically via binding to proline-rich peptides in their respective binding partner. Classical SH3 domains are restricted in humans to intracellular proteins, although the small human MIA family of extracellular proteins also contain a domain with an SH3-like fold.

is based on a *de novo* design, in particular on a structure formed by four anti-parallel helices (four helix bundle) – a structure present in several functional proteins in nature. The artificial protein, designed as DHP_1, was formed by connecting four helices with glycine loops, and a protein with 108 residues containing Ser (S), Glu (E), Leu (L), Lys (K), Gln (Q), Ala (A), and Gly (G) was obtained. In conclusion, these authors demonstrated that the four-helix bundle, a structure very diffuse in nature, could be made with only seven amino acid residues.

Silverman and coworkers (2001) took a similar approach, but they focused primarily on another protein structure that is very important in nature, the $(\beta/\alpha)_8$ barrel fold. This is present in 10 percent of the enzymatic proteins and is composed of eight helical loop-strands linked to each other. Focusing on yeast trioso-phosphate isomerase (TIM) as the example, the authors performed point mutagenesis of each residue, whereby the protein library was such to maintain the ratio between polar and hydrophobic residues. This is indeed a very complex study, which cannot be fully reviewed here. One of the authors' conclusions is, however, relevant for the point of this section: they showed that 142 of the 182 residues of the $(\beta/\alpha)_8$ barrel could be reduced to an alphabet of seven residues, namely FVLAKEQ. Again, an indication that the 20 different amino acids are, in principle, not necessary for structure and functionality.

The idea of simplifying an extant protein and maintaining the function, by imposing a reduced alphabet, was used by Akanuma and coworkers (2002) with *E. coli* orotate phosphoribosil transferase (OPRT). Here, we are on an alphabet of nine amino acids. The entire sequence of this 213 residues enzyme is simplified with 22 mutagenesis cycles, which introduce the reduced alphabet. At the end of the process, 88 percent of the residues belonged to the reduced alphabet of nine amino acids (the whole protein was, however, constituted by 13 different amino acid residues, with Cys, His, Ile, Met, Asn, Gln and Trp completely missing).

Recently, also another complex protein, chorismate mutase (CM), has undergone a process of simplification of the amino acid alphabet. Firstly, Walter *et al.* (2001) worked on it, and later on the same group was able to prepare this protein using only nine amino acid residues (Walter *et al.*, 2005), which resulted in an enzymatically active product. Prebiotic considerations were, however, not at the basis of this work.

There is another approach to obtain proteins with a lower number of amino acids, and this is based on the *binary code*, meaning that this is with the combinatorial composition of polar and apolar residues. One can, in fact, recognize that there are two main themes in the structure of globular proteins; one is the tendency to have hydrophobic side chains that avoid contact with the aqueous solvent, the other is the tendency to form periodic structures, like α-helices and β- sheets (see, for example, Beasley and Hecht, 1997). This situation can be arrived at with combinatorial methods by using periodic binary schemes of polar and apolar amino acid residues. The work by Hecht *et al.*, 2004; Kamtekar *et al.*, 1993; Rojas *et al.*, 1997; Wei *et al.*, 2003; and Wei and Hecht, 2004, goes in this direction and will not be reviewed here for space reasons.

As a general point, then, this large bulk of work demonstrates that indeed our proteins could have been made with a lower number of different amino acids, maintaining more or less the structure and the function. This is germane to the general question we have asked in this chapter: whether the prebiotic series of amino acids could have been enough to build the first enzymes and functional proteins. This is also relevant for the question asked in the first chapter: why this and not that? Why 20 amino acids in nature, and not 13 or 12?

But let us conclude this brief section with two important statements coming from all these studies:

1. Starting from the 10 prebiotic amino acids, short prebiotic peptides can be obtained.
2. Proteins with a much simpler structure, even based only on the primordial amino acids, are very likely to have been present in prebiotic time.

A more general kind of conclusion is that, in order to have proteins in prebiotic times, one does not have to wait for ribozymes and their eternal evolution as demanded by the prebiotic RNA-world.

5.4 How to make proteins by prebiotic means?

Let us go back to the problem of the biogenesis of specific macromolecular sequences. The core of the problem lies in the fact that the synthesis of such co-polymeric sequences is not under thermodynamic control. How can we then conceive their formation under prebiotic conditions, i.e., in a time where only spontaneous reactions were possible? The only help may have come from catalytic active prebiotic peptides.

There is one proposal in the literature, which in fact permitted the synthesis of a *de novo* protein, using a procedure that can be considered prebiotic, at least as a proof of concept (Chessari *et al.*, 2006).

The main idea is to operate with fragment condensation starting from a small library of prebiotically formed polypeptides, as those obtained, for example, from the condensation of NCA-anhydrides. Once we have this library of randomly produced decapeptides, fragment-condensation procedures might link the oligopeptides to one another so as to build longer chains.

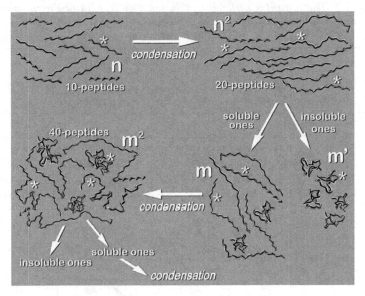

Figure 5.2 Successive steps of fragment condensation of random co-oligopeptides, with the selection principle of eliminating the water-insoluble products. (Adapted from Chessari et al., 2006.) See text for discussion.

Let us see, however, as a matter of proof of concept, what we can achieve by using a scheme of fragment condensation. For this purpose, let us refer to Figure 5.2, taken from the work of Chessari *et al.* (2006). The figure shows a small library of oligopeptides, whereby one (or more) of these products, indicated in the figure with an asterisk, is/are endowed with peptidase activity (alternatively, one can assume a peptide catalyst as an external agent).

These **n** peptides then, under the reaction conditions, can undergo fragment condensation with each other. As shown in Figure 5.2, a family of **n** decapeptides can give rise, ideally, to n^2 20-peptides. Now, let us introduce a kind of selection/evolution criterion. For example, let us eliminate from this library (n^2 peptides) all those that are water-insoluble (say, **m'** peptides). We are left – see Figure 5.2 – with **m** 20-mer soluble peptides, which by condensation with themselves give rise, ideally, to m^2 40-mer peptides. Again, we can eliminate from this library all the insoluble ones, ending up with a much more limited number. Again, by continuing this process of condensation and elimination, one tendentiously will obtain only one, or a very small number, of long polypeptides.

We can concretize all this with some figures obtained from real experiments, as illustrated in Table 5.2. Let us namely take the initial Chessari ensemble of four random decapeptides, the family indicated with **A**, which have fragment-condensed with four other random decapeptides of the **B** family. In this way,

Table 5.2 *The peptides utilized in Chessari* et al. *(2006b) experiments*

A		B	
A_1: YSKFVKSNAQ		B_1: YCADGFWKLQ	
A_2: WARCFLYHQT		B_2: QSWREIMYHS	
A_3: MPRGCDWIED		B_3: PFMHDTNELV	
A_4: PITLMEGHVN		B_4: INKPTCAGVR	
$(A \cdot B)S$		C	
A_1B_2: YSKFVKSNAQ QSWREIMYHS		C_1: QYDDEKSIVWSRLKKFDFWS	
A_2B_2: WARCFLYHQT QSWREIMYHS		C_2: LFVYNAHGTGHCYSREMQPV	
A_3B_2: MPRGCDWIED QSWREIMYHS		C_3: MIFICGWDPHELTGNQEAMP	
A_2B_3: WARCFLYHQT PFMHDTNELV		C_4: QYVCNTAIRWKHPRNLTCMA	

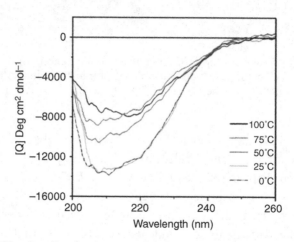

Figure 5.3 Circular dicroic changes of 2.5 µM DDEE-$C_1 \cdot A_2B_2$ in phosphate buffer; T 0–100 °C. (Adapted from Chessari *et al.*, 2006.)

sixteen 20-peptides were obtained; however, only four were soluble, and therefore viable for a further step of fragment condensation. It was, in fact, found that A_1B_2, A_2B_2 and A_3B_2 were completely soluble in aqueous 100 mM TRIS buffer in the 5.2–8.6 pH range; A_2B_3 was totally soluble whereas A_1B_3 and A_3B_3 were insoluble.

The subset $(A \cdot B)_S$ that fulfilled the mentioned criterion of water solubility, were then subjected to chain elongation by condensation with a further small set of 20-mer co-oligopeptide sequences, the family **C**, giving rise to the new library $C \cdot (A \cdot B)_S$ consisting of 16 40-residues-long peptides. None of the latter was soluble in aqueous buffer, but two of them, $C_1 \cdot A_1B_2$ and $C_1 \cdot A_2B_2$, turned out to be soluble in 6 M Guanidine Chloride. The addition of a polar N-terminal extension to them (DDEE) resulted in the 44-residues sequences DDEE-$C_1 \cdot A_1$ B_2 and DDEE-$C_1 \cdot A_2 B_2$. Of these two samples, only the latter was soluble in water and was studied in more detail.

The preceding graph (Figure 5.3) shows the circular dichroic spectrum of DDEE-$C_1 \cdot A_2 B_2$. It is a foldable protein, with some percentage of helical structure, as can be seen by the signal in the 230 nm region; its structure is thermoreversible. This polypeptide is not present in the protein database; it is therefore a *de novo* protein, or, as we call them, a "never born" protein (Chiarabelli *et al.*, 2006a, 2006b).

We have already mentioned the operational shortcoming of this work: fragment condensation has not been accomplished with prebiotic methods, but with the Merrifield machine.

Prebiotic methods for fragment condensation are not really known. Kent's synthesis comes close to the target (1999) but is restricted to particular chemical structures and cannot be considered a generalized prebiotic method. We have seen in this chapter that shorter peptides may have some catalytic activity, particularly with the example of Ser-His. This catalyst was not known at the time of Chessari's PhD work (2002–2004) and represents, in fact, one of the proposals for future work.

However, the proof of concept is there and very clear. It shows that by operating with fragment condensation of shorter polypeptides, one can arrive at foldable proteins. The procedure can in principle be repeated to obtain a family of proteins, each present in solution at micromolar concentration.

The question, at this point, is whether one can do the same with RNA. By this, I mean prebiotic fragment condensation of co-oligonucleotides in order to obtain at the end a single macromolecule at micromolar concentration. This would correspond to the idea of having a folded RNA in billions of identical copies, responding to one of the main objections posed to the prebiotic RNA-world, and meeting the prospect of a pre-RNA world. We have discussed this point earlier, based on the work of Anella (2011), and we will come back to this idea in the research proposals at the end of this chapter.

5.5 About prebiotic vesicles

Since all life is cellular, the pathway to the origin of life must therefore eventually end up in a membrane compartment; one can ask the question of whether some form of compartment might have been present at ground zero, as well.

It is accepted that amphiphilic, membrane-forming compounds may come from space (Lawless and Yuen, 1979; Deamer, 1985; Deamer and Pashley, 1989; Deamer *et al.*, 1994; McCollom *et al.*, 1999). And very important in this regard

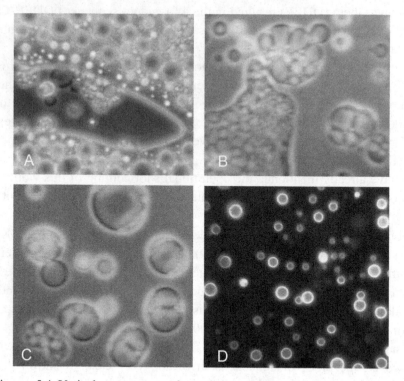

Figure 5.4 Vesicular structures formed by organic compounds from the Murchison carbonaceous chondrite. (Micrographies provided by courtesy of David Deamer, including figure legend). *David Deamer writes*: **A** (160×, original magnification) shows the original dried extract just after a dilute alkaline buffer is added (10 mM sodium carbonate as I recall) and you can see the water penetrating from the left with two vesicles beginning to form. **B** (160×, original magnification) shows the extract 10 minutes later. It has swollen with water and vesicles are beginning to bud off. **C** (400×, original magnification) shows the vesicles at a higher magnification and **D** (160×, original magnification) is a fluorescence micrograph of the vesicles. They are highly fluorescent when excited with near UV illumination because traces of polycyclic aromatic compounds like pyrene and fluoranthene are also in the mixture. The actual amphiphilic compounds are the mixed monocarboxylic acids in the meteorite ranging from 8 to 13 carbons in length.

is the early observation by Deamer and Pashley (1989) that the chloroform-soluble fraction of organic compounds from the Murchison carbonaceous chondrite yields vesicular structures, most likely due to fatty acids. This is shown in Figure 5.4.

Although the fatty acids discovered in meteorites are generally with a rather short chain, Deamer and Pashley's finding suggests that fatty acids are probably prebiotic compounds also on Earth, possibly deriving from the Fischer-Tropsch synthesis, followed by oxidation.

The observation that membranogenic compounds can also be of prebiotic origin is of particular interest, especially in view of the "membrane-first" hypothesis, or more generally, for making the point that membranes and vesicles were present very early in the prebiotic scene (Deamer *et al.*, 1994). Phospholipids, which are ubiquitous in biological membranes and which self-assemble in aqueous environments into stable lipid bilayers and vesicles, are the best candidates for prebiotic membrane components. The prebiotic synthesis of phospholipids has been reported (Hargreaves *et al.*, 1977), and see in this regard the recent work from Sutherland's group (Patel *et al.*, 2015).

5.6 Proposals of research projects from ground zero

This last section is based on two points. The first one, which I have mentioned already, is that both the compartmentalistic approach and the prebiotic RNA world start from a scenario where biopolymers are already there. Still a worthwhile research enterprise, but the foundations are missing. The second point is more personal: I have retired from active experimental research on the origin of life, and in the last years of activity I had accumulated some ideas on the *ground zero approach*, which for obvious practical reasons I could not actively pursue anymore. Here, then, I will present some of these ideas, in the form of research proposals, with the aim of perhaps stimulating some of the younger colleagues to action. There will be more research proposals later on, once we have talked about vesicles and the corresponding synthetic biology.

But before doing so, let me consider a less serious experiment – apparently, less serious, which for many would be perhaps the most basic, and simplest, ground zero prebiotic experiment . . . in your bathtub. You fill it with water under strict sterilized conditions, and add all 20 amino acids, all four nucleotides, both in DNA and RNA forms, mineral salts and phosphates, carbonates, phospholipids, and fatty acids, and whatever low molecular weight compound you may like. And you may even add some ATP – although this is slightly unfair in a prebiotic condition – and some people may like to add mineral and/or clay. The bathtub has a thermostat to change the solution temperature. And it has a little automatic outlet to analyze – say, every month – the products of possible reactions. Stirring, if you like to do so.

The simple question is: will you be able to witness the formation of life, or at least some products that lead to bacterial life? If not, why?

This experiment is good for all those who think that life was an easy event four billion years ago.

And let us go now to more concrete experiments.

5.6.1 The catalytic properties of Ser-His: ideal project
for bioorganic chemists

I have mentioned before the extraordinary catalytic properties of this dipeptide – extraordinary in view of the structural simplicity of the compound, which can hydrolyze and synthetize the ester bond, the peptide bond, the PNA, and finally, the nucleotide bond (Li *et al.*, 2000, Gorlero *et al.*, 2009, Wieczorek *et al.*, 2013). Despite all this, the peptide has not been studied as it should have. Two basic, very challenging research directions are:

1. The specificity of the catalysis. Is the catalytic power of Ser-His restricted to aromatic peptide bonds – as investigated until now – or perhaps valid for all kinds of amino acid residues? The absence of specificity in a primordial catalyst would also be very interesting, and to check this should not be difficult or long. Very interesting, more in general, is the relation between the chemistry structure and catalysis. Note, by the way, that His-Ser is not catalytically active.
2. In this sense, one could then enlarge the specificity by working on the chemical structure, possibly aided by computer design, adding sequentially a longer tail, left and/or right – so to arrive at a small macromolecule, which can be considered a real specific enzyme. The same could be done with the other simple peptides, which show catalytic activity, although not so marked, such as Gly-Gly.

The added tails can be seen as the pathway of prebiotic evolution, brought in by contingency, and the whole project would then correspond to the prebiotic synthesis of an enzyme, perhaps initially limited to proteases.

5.6.2 Ser-His as the catalyst for the nucleotide bond

We have mentioned earlier the work by Wieczorek *et al.* (2013) on the use of Ser-His to catalyze the condensation of the nucleotide bond. This work was actually focused on the mechanism, and only the synthesis of nucleotide dimers, with traces of trimers, was reported. It would now be time to study the optimization of this catalyst for the synthesis of longer oligomers, following the guidelines indicated above for peptides.

One should recall that the conditions for the coupling in that paper are rather peculiar, with the use of the eutectic between water and ice, as under those conditions, the activity of water is lower, and the stability of the starting materials is higher. The work of optimization may be tedious, but the importance of this work would be great. In fact, the notion that oligonucleotides can be formed by the action of peptide catalysts is of obvious relevance for the entire field of the origin of life.

This project can be linked to the previous one obviously, in the sense of chemically enlarging Ser-His to achieve a greater specificity for the nucleotide bond.

5.6.3 The most stable random polypeptide sequences by the NCA method

As mentioned in the text, valuable studies on the NCA condensation of amino acids to form short polypeptides have been conducted by Commeyras, Taillades, and colleagues (2002). This is a reaction considered prebiotic. The reaction has not been studied systematically, probably because it is a tiresome enterprise. However, again, it is something worth doing. The main aim would be to establish whether and to what extent certain co-oligopeptides are preferentially formed. This would be synthesis under thermodynamic control. This work may reconnect with the work by Ikehara, cited above, and other authors, who claim that GADV peptides were present in prebiotic time. It is still not known what is the maximal length of peptides that can be formed.

I would begin with a mixture of three NCA-amino acids at a time, even starting from the GADV series, but using the 10 original prebiotic amino acids. And I would concentrate the work on the longer co-oligopetides being formed, discarding dipeptides and even tripeptides. The analysis of the products, most of which will be insoluble in water, will be tedious but certainly not impossible. Reproducibility will be an interesting subject.

All this should ideally lead to a series of co-oligopeptides that are preferentially formed under thermodynamic control, and therefore preferentially present in prebiotic times. And these compounds can be the basis for the next project: the prebiotic synthesis of proteins.

5.6.4 Protein biogenesis by fragment condensation of prebiotic peptides

The Chessari et al. (2006) work can be generalized in a kinetic scheme which involves two rate constants involved in the fragment condensation (assuming initially only two k_{on} and k_{off} – and considering that hydrolysis of the peptide bond is generally a very slow process – see also Radzicka and Wolfenden (1996) as well as Commeyras et al. 2005); and introducing a strict evolution/ selection rule (e.g., eliminate each product having, say, more than 40% hydrophobic residues). In this first simplified scheme, only the primary sequence is considered (AAVGLG, etc.) – you do not need the relation between primary structure and folding, or solubility. The main target is to develop a simple theoretical model that, although based on a catalyzed combinatorial fragment condensation of co-oligopeptides in solution, yields finally only one chain. The imaginary catalyst can be one of the initial random peptides, or can be added externally (e.g. Ser-His). Once such a theoretical model works (yielding only one chain), you can apply it to a different series of initial peptides – and if you repeat this scheme z times, each time with a different set up of initial co-oligopeptides, you will finally have z folded *ex novo* proteins.

The second phase of the work may involve a 3D framework, using for example the Rosetta method, with a closer, more realistic look at the folding. And you can then see whether the structure and the folding correspond to some known enzymatic activity (towards nucleic acids?).

5.6.5 The notion of "proteases first"

The point made above, that in principle we can have competition between condensation and hydrolysis of the peptide bond, may elicit a scenario of protein evolution, with a library of polypeptides with protease activity (some of them) that are slowly cutting themselves, and making new species continuously by peptide fragment condensation. If the hydrolysis rate is slower (as Commeyras *et al.*, 2005 suggest), then the process leads to the accumulation of foldable peptides. In consequence of partial hydrolysis and a lengthening of fragments, the peptides formed are ideally forced to evolve.

In other words, this scenario may suggest the pristine formation of a family of proteases, which, evolving, may give rise to other enzymes.

The operational framework would be based on the work of Chessari *et al.* (2006). The critical point would be to regulate the addition rate of peptide educts to have a steady moderate increase of the average length of the produced protein family. Another very important point is whether this kind of evolution might be directed towards a specific activity – e.g., RNA polymerase – by adding to the whole mixture or part of it certain binding templates or complexes. This part of the project is, of course, very speculative, but it would be worthwhile to try to implement it, as this would correspond to an alternative view of the origin of life, a kind of "proteases first" hypothesis.

Concluding remarks

The idea of this chapter was to show some alternative ways to look at the origin of life, focusing on the ground zero approach, namely starting from the basic primordial molecules, amino acids in particular. One of the main emphases was on the biogenesis of the first macromolecular sequences. This view is complementary, more than being in contrast, to the commonly accepted views – the prebiotic RNA word and the classic compartmentalistic approach – as these views start from already extant macromolecular species, whether DNA and/or

RNA. The main question is how to make ordered sequences of those, and in many, identical copies.

Going back to the general question of the biogenesis of macromolecular sequences, the "testable model" that I have presented seems to be a good framework on which to base future work. However, when looking at these things in terms of the overall scenario of the origin of life and cells, there arises a problem. The problem is that this mechanism is based on contingency, and contingency, by definition, has no memory.

Once a given enzyme has been originated in this way, there is no way to remake it again and again. This objection can be partly softened by the following consideration: if one protein or a family of proteins is produced because that series of reactions is under thermodynamic control, then this protein or family of proteins can be continuously remade by providing the same initial educts. The other, but more tenuous consideration, is that the "proteases first" scenario may produce, by evolution, polymerases that can start the biogenesis of nucleic acid, which may eventually lead to the first forms of genetic code.

Moreover, I would like the reader, at this juncture, to take a look at the Appendix of this book, where he or she will find a collection of the *Open Questions about the Origin of Life* (OQOL). It is a wide report on the most intriguing opinions and targets in the field, as they came out in four recent international meetings focused on this subject and, often, are closely related to the arguments discussed in the previous pages. Moreover, these questions/ arguments are presented as they are in the minds of the researchers in the present time. Regarding the questions in the Appendix, several of these have repeated themselves over the years. Although the extensive discussions held during the OQOL meetings do not mean that the researchers have found answers to these recurring items – as they are still unsolved questions – we believe that all those debates have contributed to both a greater awareness of what is missing and, hopefully, an enlightening perspective on the goals of further studies.

In more general terms, the efforts to establish the basic initial principles of prebiotic chemistry – as reported in this chapter – convey one simple truth: that we are still well away from full comprehension of the mechanisms that led to the origin of life, and that therefore, much work is still necessary at the ground zero level.

Questions for the reader

1. We mentioned in Chapter 1 the "NASA definition of life," centered on Darwinian evolution as the main criterion of life. Let us then perform a simple experiment: you observe a colony of bacteria under the microscope

for a few days, or months. You see them in motion and being busy, and in this time interval there is no observable Darwinian reproduction or evolution. Would you then conclude that these bacteria are not alive? Or, rather, that the above definition does not make sense as a criterion of life?

2. More generally: we have talked in Chapter 1 about the difference between the criterion of life and the criterion of evolution (see also the conversations with Eors Szathmáry and David Deamer). And recall that for the first two billion years on the Earth, we had only unicellular organisms, which did not seem particularly eager to evolve. Isn't this kind of consideration enough to illustrate the importance of a clear discrimination between the criterion of life and the criterion of evolution and biodiversity?

3. It is commonly accepted that self-replication of macromolecular sequences is one of the fundamental mechanisms for life. Question (to which you can answer first in terms of intuition): what came first, macromolecular self-replication systems, or should rather be first the arising of a mature cellular or proto-cellular metabolism? And aside from your intuition: do you have any significant data supporting the one or the other view?

4. You have extracted the nucleus from an oocyte. You have a nucleus and the rest of the cell. Neither of these two components – isolated and independent – is alive. Now you insert again the nucleus into the cell – or, in this respect – into a different enucleated cell (as in the cloning experiments). You get a living cell. Can this be taken as a demonstration for the emergence of life, namely that life can be generated out of non-living components?

5. Do you really (in your inner core) accept the view that life on Earth originated from inanimate matter without any contribution from transcendent power?

6. Are you at peace with the idea that mankind might not have originated; and with the idea that we may be alone in the universe? Your brain may say yes: is this also your gut instinct?

7. Section 4.1 develops a series of sharp criticisms to the prebiotic RNA-world (note the term *prebiotic*). Do you agree with this criticism, or what would you argue, scientifically, against this?

8. Finally, a question that can become a research proposal. See the work by Fabrizio Anella at the end of Section 4.1. Do you agree with the statement, made in that section, that by this procedure it should be possible to arrive at a single family of RNA in multiple copies? How would you continue his work to achieve this result?

Part II

What is life?
The bio-logics of cellular life

Chapter 6

Autopoiesis – the invariant property

Introduction

We have discussed in Part I various aspects of the origin of life, and some of the corresponding historical and philosophical aspects that generally attend this question. We will consider now the question, "What is life?" This is epistemically different from the question of the origin of life. In fact, this last query may start from the observation of already existing life, and its multiple forms on Earth – from microorganisms to algae, mushrooms, and plants to fish, birds, and mammals – and ask: "What do all these different things have in common, so that we can say for all of them, that they are living? What is the common denominator?" And, in keeping with the discussion in Chapter 1 of Part I, we want an answer at the level of the single individual. We cannot answer such a question with "self-reproduction." Take all our mature neuron cells, or the cells of the heart muscle (cardiomiocytes), which do not self-reproduce. But they are alive, as they are engaged for all our life in metabolism – a form of homeostasis. And take all animals who do not reproduce because they are either too young or too old – are they not living? And to say that the dog running on the road is alive here and now, because he is the product of his parents' conjunction two years earlier – is really not a sensible thing to say. Nor is it sensible to say that all organisms are alive because they derive from the reproduction activity of parent organisms. Self-reproduction is a *property*, one of its consequences. It is a characteristic of all living things, but it says nothing about the question "why is an organism alive?" We need a criterion of life that answers this question at the level of a single organism, here and now, regardless of its progenitors history. Such a criterion should illustrate the common denominator to all living things, despite their extreme diversity, and clarify the difference between living and not living; and the theory of autopoiesis does that.

Each living system, as an open system, has to interact with its own environment, and we will learn in the next chapter the important notion of cognition, which, still in the theory of autopoiesis, tells us the meaning of this interaction. We will see that this notion is rich in philosophical implications, and not of an easy nature.

6.1 The visit of the Green Man

In order to address these questions on the definition of life first *at the individual level*, let us use a well-known metaphor, utilized already by Oparin (1953) and by Monod (1971) in a different context. This will be useful to introduce autopoiesis in a clear way.

It goes like this: suppose that an intelligent creature from a very distant solar system – a Green Man – comes to visit the Earth in order to understand what life is on our planet. He has received from his superiors a long list of terrestrial things about which they are in doubt as to whether these things are alive or not. Our Green Man is supposed to descend on Earth and clarify the situation. Let us see what happens.

He encounters an intelligent but scientifically naive farmer to whom he asks the question, showing the long list. The farmer very rapidly divides the items of the Green Man into two lists (Table 6.1), a list of living and one of non-living things. The Green Man is surprised by the rapidity by which the farmer has done such a discrimination, and asks him how he did it: he wishes to know the quality which characterizes all living things (left-hand side), and which is not present in items of the right-hand side.

When the farmer, pointing at the mule, says "movement" and "growth," the Green Man nods reservedly, as the tree or the coral in the same living list do not move about, nor show any appreciable sign of growth in a reasonably long

Table 6.1 *The Game of the Two Lists*

LIST OF THE LIVING	LIST OF THE NON-LIVING
the fly	the radio
the tree	the automobile
the mule	the robot
the baby	the crystal
the mushroom	the moon
the coral	the computer

QUESTION:	What discriminates the living from the non-living?
IN OTHER WORDS:	What is the quality (or qualities) which is present in all members of the "living" list and which is not – and cannot be – present in any of the elements of the "non-living" list?

observation time; conversely, a small piece of paper moves in the wind and the moon moves and grows periodically. When the farmer then gives "reaction to stimuli" as an alternative criterion, the Green Man again nods unconvinced, as the mushroom and the tree seem insensitive to a needle; and on the other hand the computer and the radio easily become ineffective upon interference with a stick.

"Living things" – adds the farmer who begins to get irritated – "are uptaking food and by doing so are able to perform their functions by consequent production of energy. Energy is transformed into action." But the Green Man indicates the car and the robot, which are able to move about by doing precisely that – converting energy into action. "Reproduction!" cries the farmer. "All items in this list are able to reproduce themselves!"

"So it is for chicken and men, but it is not so for the mule, who is unable to reproduce," scorns the Green Man. "Nor is this so for all your babies, or old people." "Furthermore," he says, "to reproduce animals takes at least two of a kind, and I want to know about a criterion for life at the level of a single specimen."

The farmer gets more and more angry, but then he arrives at a kind of enlightenment. He looks at a tree and realizes that it loses leaves and fruits in winter, but generates them again in the spring – from the inside. Similarly, it is with his beard and the hair of animals: you cut it and it grows back again – and it comes due to an activity from inside of the body!

The farmer also knows from his personal experience that when his pig is sick and cannot eat, its limbs and organs become smaller; however, as soon as he starts to eat again, limbs and organs grow again. Again, this growth comes from inside his body. He concludes – and tells the Green Man – that in all elements of the "living" list there are internal processes that continuously destroy and rebuild from the inside the structure itself. Living organisms are then characterized by an activity that regenerates their own components!

The Green Man, this time, nods positively. The farmer has finally articulated the quality that discriminates the living from the non-living! The robot, computer, radio, moon, and so on, are not able to regenerate themselves from the inside. If a part of the radio breaks, the radio itself is not going to build it again. However, all items on the left-hand side of the table do have this quality: they utilize external energy to maintain their own structure, and have the ability to regenerate it from within the structure itself. This seems to be the property of life that one is looking for.

The Green Man now draws a figure on the ground (Figure 6.1). In this figure, which represents something that is open to the medium outside, S represents a component of the living system, which is being transformed into a product P;

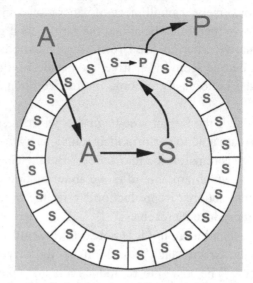

Figure 6.1 The simple-minded drawing of the Green Man: his concept of life on Earth.

the system is able to regenerate S by transforming the entering food A into S again. Actually, the Green Man is rather happy about all this. So, accordingly, he and the farmer make up the following "operational" definition of life: *a system can be said to be living if it is able to transform external matter/energy into an internal process of self-maintenance and production of its own components.*

The farmer nods, and I doubt whether he understands. However, the two have arrived at a "definition" of life by using macroscopic, common-sense observations. Such a simple definition might have been derived by laymen of a couple of centuries ago; you do not need molecular biology or cellular biology for that; however, as it is easy to see, it is also valid for the description of cellular life. It was derived for a single specimen but is also valid for a general case, and is therefore valid both for coded life and for non-coded life.

Figure 6.2 corresponds to a schematization of the behavior of a cell, and we will see – shortly, when discussing autopoiesis – how all of this can be presented in a more rigorous manner. Actually, as will be shown later on, the "definition" of life given above corresponds to the description of the autopoietic unit, the minimal form of life. And we will see, talking more in detail about autopoiesis, that the previous definition can be somewhat improved and made more general. For that, we need to introduce the notion of "minimal and sufficient" when applied to the definition of life.

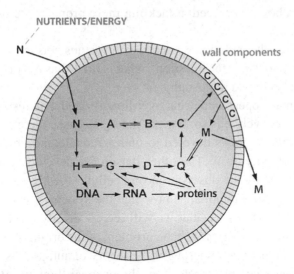

Figure 6.2 Schematization of the work of a cell as an open system. One important feature is the boundary, which is created by the internal network of reactions (a boundary of its own making). The network of reactions brings about a large series of transformations; however, under homeostatic conditions all material that disappears is generated again by the internal machinery and by transformation of incoming nutrients/energy. Thus, the cell (and, by inference, life) can be seen as a factory engaged in self-maintenance.

6.2 Introducing autopoiesis

We saw in the previous section how the Green Man and the farmer arrived at a descriptive definition of life that distinguished the living from non-living. The farmer was ignorant of biology, otherwise he would have answered from the very beginning that all living things are made up of cells – that this is the most discriminating factor. However, the Green Man would have then asked: "What is a cell and why do you call a cell living?"

In fact, the life of a cell is the starting point for the development of the ideas of autopoiesis (from the Greek *auto*, or self, and *poiesis*, producing) developed by Maturana and Varela (Varela *et al.*, 1974; Maturana and Varela, 1980, 1998). The aim of this chapter is to review the notion of autopoiesis and to present it in the context of present-day research in the life sciences, as part of the systems view of life (Capra and Luisi, 2014).

Autopoiesis deals with the question "what is life?" and attempts to isolate, above and beyond the diversity of all living organisms, a common denominator that allows for discrimination between the living and the non-living. Autopoiesis is not directly concerned with origin of life; rather, it is an analysis of the living as it is – here and now, as the authors say. We will see, however, that once the question "what is life?" is clarified by the theory of autopoiesis, we will receive an indication on how to proceed in tackling more properly the question of the origin of life.

We will also see that autopoiesis deals with various epistemic concepts, and this is necessarily so, since the question "what is life?" has not only a dimension in biology, but also one in philosophy.

This chapter may appear somewhat overdimensioned with respect to others in this book, and effectively it is so. It is so, mostly because there is no account of autopoiesis in college books devoted to undergraduate and graduate students, and I wanted to finally do it. The other reason for developing autopoiesis in detail was prompted by the observation that cellular theories, such as autopoiesis, are acquiring more importance now that the systems view is spreading in the science of life; and in particular we will see, as already mentioned, that this view will be useful also to approach the problem of the origin of life. And I have to say at this point that my rendering of the theory of autopoiesis and cognition may depart in some lateral aspects from the original thinking of the Santiago School. I will indicate punctually where and if this happens.

6.3 Short historical background on autopoiesis

Autopoiesis is embedded in a particular cultural background and therefore some historical information may be useful. When, as a student, Francisco Varela met Humberto Maturana in the 1960s at the University of Chile in Santiago, Maturana was already internationally known for his work on visual perception in frogs. This was the basis for the later work with Varela, opposing representationalism in perception (see Maturana and Varela, 1998). The biological basis of cognition had always been an important item on Maturana's agenda (Maturana *et al.*, 1960) and this, too, was destined to influence significantly the interests and the later work of Varela.

The term *autopoiesis* was used for the first time by the Santiago authors in 1971; by the end of that year, Maturana and Varela had prepared a very long manuscript entitled *Autopoiesis: The Organization of Living Systems*. As Varela describes (Varela, 2000), the manuscript was not well received. It was rejected by the most important journals, and colleagues' responses were lukewarm. Finally, a paper on autopoiesis was published in English (Varela *et al.*, 1974). The notion of autopoiesis was very slow to receive recognition. By the mid 1970s, however, some international meetings had used the term in their programs; books on autopoiesis appeared (Zeleny, 1977) and noteworthy biologists such as Lynn Margulis accepted autopoiesis as an integral part of the description of life (Margulis and Sagan, 1995). The notion of autopoiesis had a strong impact in the social sciences where the term *social autopoiesis* was coined (Luhmann, 1984; Mingers, 1992; 1995; 1997). In the 1990s, experimental chemical systems, on the basis of a previously elaborated theoretical scheme (Luisi and Varela, 1990), were developed in Zürich focusing on autopoiesis (Bachmann *et al.*, 1992; Walde *et al.*, 1994b; Luisi, 1996, and references therein).

Still, it cannot be said that the notion of autopoiesis is now familiar in mainstream science. The reason for this will be discussed later on, but it can be anticipated that this is partly due to the fact that autopoiesis is not centered on DNA, RNA, and replication, and makes only minimal use of the term *information*. Furthermore, the fact that it has been used extensively in the social sciences, and not always in a very rigorous way, has given to some the impression that it might even be tainted by a new-age flavor.

6.4 Basic autopoiesis

The autopoietic analysis is based on cellular life, the main argument for this being simply that there are no other forms of life on Earth. We all know that even the simplest cells are extremely complex, encompassing hundreds of genes and other macromolecules. However, beyond this complexity, the question of what a cell really does lends itself to a relatively simple answer. We have seen in Figure 6.2 the simplified representation of a biological cell. What is the corresponding rendering in terms of the simplest possible molecular autopoietic system?

For an answer, consider Figure 6.3.

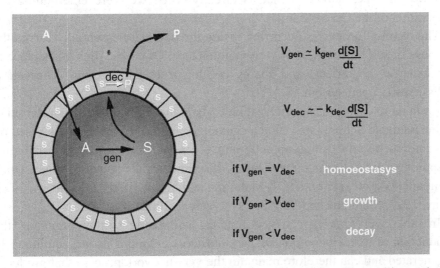

$$V_{gen} \simeq k_{gen} \frac{d[S]}{dt}$$

$$V_{dec} \simeq - k_{dec} \frac{d[S]}{dt}$$

if $V_{gen} = V_{dec}$ homoeostasys

if $V_{gen} > V_{dec}$ growth

if $V_{gen} < V_{dec}$ decay

Figure 6.3 The minimal autopoietic cell, characterized by only two internal reactions, and displaying the three modes of a cell's existence (homeostasis, growth/reproduction, and decay/death) depending upon the relative values of the two velocity constants.

To get an answer in total agreement with the thinking of Maturana, we can read a strong statement in the conversation with him in the next chapter, when he says: "... *discrete molecular autopoietic systems are living systems, and living system are discrete molecular autopoietic systems*," whereby the emphasis, as he adds right away, is on "molecular." This is an important point, as autopoiesis is then situated in the realm of chemistry. We will have to come back to this point in the next chapter. Now, let me offer some personal interpretation of these words, asking the question: how can I give a rendering of this concept for a cellular entity? Since my early time in autopoiesis, I have used the representation shown in Figure 6.3, which is not generally used in the primary literature.

The first thing one observes is the boundary, a semi-permeable, spherical, closed membrane that discriminates the cell from the medium. The term *semi-permeable* means that certain substances (nutrients and other chemicals) are able to penetrate in the interior, whereas most other chemicals cannot. This is a kind of chemical selection and chemical recognition – a notion that will be later on linked to the term *cognition*.

The notion of boundary is, in fact, one central concept in the theory of autopoiesis. Inside the boundary of a cell, many reactions and correspondingly many chemical transformations occur. However, despite all these chemical processes, the cell always maintains its own identity during its homeostasis period. This is because the cell (under steady-state conditions and/or homeo-stasis) regenerates within its own boundary all those chemicals that are being destroyed or transformed, such as ATP, glucose, amino acids, proteins, and so on. We have heard these concepts already when we were entertaining the illustration of the Green Man.

The chain of processes occurring inside the boundary essentially serves the purpose of self-sustenance, or auto-maintenance. Of course, this takes place at the expense of nutrients and energy coming from the medium. The cell is a dissipative, open system.

From these simple, basic observations, Maturana and Varela (often referred to as the Santiago School) arrived at a characterization of living systems based on the autopoietic unit. An autopoietic unit is an open system that is capable of sustaining itself due to an inner network of reactions that regenerate the system's components (Varela *et al.*, 1974; Maturana and Varela, 1980; Luisi *et al.*, 1996; Luisi, 1997; Maturana and Varela, 1998; Varela, 2000).

In other words, it can be said that an autopoietic system organizes the production of its own components, so that these components are continuously regenerated and can therefore maintain the very network process that produces them.

It is perhaps pertinent at this point to cite a recent definition by Maturana himself (Poerksen, 2004):

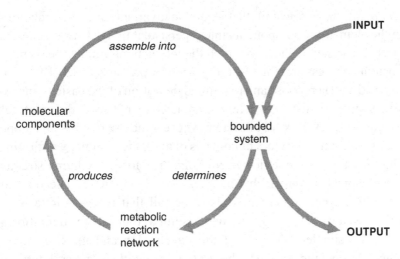

Figure 6.4 The cyclic logic of cellular life. Autopoiesis is the most general pattern of minimal life. It does not specify the actual structures and their processes. The cell, an autopoietic unit, is an organized, bounded system that determines a network of reactions, which in turn produces molecular components that assemble into the organized bounded system that determines the network of reactions that produces... and so on.

When you regard a living system you always find a network of processes or molecules that interact in such a way as to produce the very network that produced them and that determine its boundary. Such a network I call autopoietic. Whenever you encounter a network whose operations eventually produce itself as a result, you are facing an autopoietic system. It produces itself. The system is open to the input of matter but closed with regard to the dynamics of the relations that generate it.

In this way, autopoiesis is capable of capturing the mechanism that generates the identity of the living. A graphic representation of these concepts reveals a circular logic, as shown in Figure 6.4.

The components organize themselves (auto-organization) in a bounded system, which produces the components that in turn produce the system, and so on. In this way, the blueprint of life obeys a circular logic without an identified beginning or end, as pointed out by Maturana in the previous citation. Although the system is open from the physical point of view, from an epistemological perspective it has a logical operational closure (Maturana and Varela, 1998; Varela, 2000). This characterizes the system as an autonomous identity that can be defined as auto-referential. It produces its own rules of existence and therefore has a particular type of bio-logical coherence. These internal rules of the system are what make cellular life.

This is a descriptive definition of the minimal living system, which does not focus on the structure of the cell's components. It focuses, instead, on the *organization* of life's basic unit, that is, on the invariant functional relations that connect the elementary components into a persistent unity. This also means that autopoiesis is a systemic view, in which the interactions among the components that constitute the essence of the thing are the primary focus. By so doing, Maturana and Varela propose an operational definition of the basic living system that is very general, for it is able to characterize not only different cellular systems, but rather every kind of system whose structure realizes an autopoietic organization. Here lies the main hypothesis of autopoietic biology, affirming that the definition of the autopoietic system holds true for every living structure: if a physicochemical system is living, then it is autopoietic. The question of whether the reciprocal sentence is also true (all that is autopoietic is living) will be discussed later. Here, it is perhaps useful to translate the definition given above into the simpler language of the Green Man and the farmer. By not mentioning autopoietic systems, but rather using their general term *living system*, they would have said:

A living system is a system which is capable of self-maintenance through a network of regenerative processes from within the boundary of its own making.

As already mentioned, to be a little more scientific, one should qualify the term *system* as an "open system" that receives energy primarily from the outside (otherwise we would deal with a perpetuum mobile) as a molecular open system.

6.5 Criteria of autopoiesis

The most general property of an autopoietic system is the capability to generate its own components via a network process that is internal to its boundary. The boundary of the system must be "of its own making," a product of the process of component production. Whether a given system is capable of making its own boundary is the most discriminating criterion by which we recognize an autopoietic system.

We mentioned before the statement by Maturana – from his conversation in the next chapter – whereby autopoietic systems are molecular systems. This can be already used as an important initial clarification for the question of whether a system is autopoietic. More in general, the question of the criteria of autopoiesis is formalized at length, but not always clearly, in the primary literature on autopoiesis. Varela, in his latest book (2000), has simplified these criteria into three basic ones, which can be expressed as shown in Table 6.2. The intent of these criteria is to offer a means by which to verify whether the system has a self-produced boundary; that the system encompasses reactions that regenerate the components of the system; and that this network of reactions is produced by the system itself.

Table 6.2 *The criteria of autopoiesis*

Three criteria of autopoiesis	
1 Self-boundary:	Does the system have a boundary of its own making?
2 Self-maintenance:	Is the system capable of maintaining its own identity via dynamic processes, i.e. those components that are being used up are made anew by the system itself?
3 Self-generation:	Does this happen throughout a network of reactions that are generated by the system itself?

Applications of these criteria permit one to discriminate between the living and the non-living. Only if you answer *yes* to all three queries is the system autopoietic and therefore living.

Thus, a virus is not an autopoietic system, as it does not produce the protein coat of its boundary or its nucleic acids (the host cell does this, and it is living). A computer virus is also not autopoietic, as it needs a hardwired computer system to be expressed, and this is not produced by the virus itself. A growing crystal is not autopoietic, as the components are not generated from an internalized network system. Nor is a candle flame, which is actually a dissipative system and does not regenerate the hydrocarbons (wax) that are being burned, and so on.

Recalling the game of the two lists offered by the farmer to the Green Man, the baby, the fish, and the tree are all autopoietic systems. The living is a factory that remakes itself from within. This is the common denominator of the living, regardless of whether we are looking at a microorganism or an elephant. It must be said at this point that in its original form, the theory of autopoiesis was limited to cellular life, and Varela was for a long time opposed to generalizing it. In fact, it took quite a while before he was able to accept publication of the idea (Luisi *et al.*, 1996) that the criteria for autopoiesis could be applied to all higher forms of life, man included.

At this point, we can ask an important question: is autopoiesis equivalent to life? Maturana and Varela explicitly wrote (Maturana and Varela, 1980, p. 82): "...autopoiesis is necessary and sufficient to characterize the organization of living systems." Gail Fleischaker, in an often-cited review (Fleischaker, 1988), writes that whatever is living must be autopoietic, and that conversely, whatever is autopoietic must be living. And Maturana, in his conversation, says the same, as observed above. For some, things are a little more complex than that. Again, this is a point where the current research on autopoiesis may depart somehow from this statement of equivalence (Bitbol and Luisi, 2004; Bourgine and Stewart, 2004), and we will come back to this point in the next chapter, after consideration of the complementary notion of cognition.

Let me mention, now, that I had the great privilege to become acquainted with the two authors Humberto Maturana and Francisco Varela. Humberto granted me a very important dialogue in the conversation that appears in the next chapter, dedicated to cognition. It is an important document. Francisco was a close personal friend, and I came to know the members of his family well, including his son Xavier, daughter Leita, and his wife and great companion and friend Amy Cohen. The following is a conversation with Amy, mostly about Francisco.

Conversation with Amy Cohen Varela

Amy Cohen Varela completed undergraduate and graduate studies in comparative literature at Brown and Columbia Universities, respectively. She moved to Paris in 1981, where she studied clinical psychology and psychoanalysis at the University of Paris-Censier.

She has worked as a clinical psychologist at Sainte-Anne Hospital in Paris where she also did research on psychosis and literary and artistic creation. At the Institut des Hautes Études en Psychanalyse her study has concentrated on the articulation of psychoanalysis and the natural sciences.

She was Francisco Varela's friend and wife from 1987 until his death in 2001.

She is currently chair of the board of Mind and Life Europe.

PLL: You have a clear oversight of the selling and spreading of the books by Francisco all over the world. Do you feel that there is an increase of interest for his work – or a standstill?

Amy Cohen: Today book sales are not the only and may not be the best indicator of the dissemination of ideas. But in spite of (or perhaps because of!) the presence on the Internet of most of Francisco's books in one format or another, many of these books have maintained a steady sales rate over an extraordinarily long period of time. *The Tree of Knowledge* still sells steadily around the world, as does *The Embodied Mind*, of which a second edition is in the works for this year. Another steady seller is

Ethical Know-How: Action, Wisdom and Cognition, perhaps because of the synthetic nature of the presentation. *El Fenómeno de la Vida* has already had two editions in Spanish. It is a precious anthology because Francisco himself chose the texts, ordered them, and wrote short introductions for each. Since this collection contains articles and essays that were, for the most part, translated from the English, it has not appeared in an English version. There *is* a need for a comprehensive anthology of F's works in English, in particular because of the diversity of subjects on which he wrote and the dispersion of his writing in a very heterogeneous array of books, journals, and other publications. An anthology is currently in preparation to be published in France (in French).

[I suppose, Luigi, that one could talk about the development of enaction and neuro-phenomenology by different specific people and schools . . . but this would be very long and I'm not *au courant* enough other than Di Paulo *et al.* and some of the neuro-phenomenology group. . .]

PLL: You have written some papers with Francisco. Could you very briefly summarize what they are about?

AMY COHEN: I was fortunate to have the chance to have a sustained dialogue with Francisco over a decade and a half about all that we both were thinking and writing about. Much of our interchange was, in itself, a practice: that of a shared passion and concern for bridging or finding passages between irreducible domains in order to create new spaces from which novel perspectives might emerge. We evoked this in an article called "Facing up to the embarrassment," in which we discussed possible resonances between the enaction framework and the psychoanalytic process, focusing on the functions of imagination and interpretation. Our discussion of the encounter between two traditionally irreconcilable fields, science and psychoanalysis, was also an occasion to reflect on how meaning is generated between disparate fields not by attempting to create a "grand synthesis" but by a gradual process of change through mutual interpretation. Thus, the rapprochement of the subjective and the objective, the experiential and the experimental, of lived life and biological life (as is the case in neurophenomenology), is necessarily a slowly unfolding dialogue. In another framework, The Mind and Life dialogues were originally designed by Francisco as a setting for this mutually respectful transformative work between Buddhism and science.

PLL: What is your understanding of neurophenomenology, and in which sense you consider this something new and important?

AMY COHEN: As Francisco himself qualifies it, neurophenomenology is a methodology, one that addresses the question of how to give the experiential field its rightful place in our accounts of nature, since, as he said, "lived experience is where we start from and where all must link back to, like a guiding thread." By confronting first and third person data (intersubjectively validated invariants and objective invariants), the neurophenomenological research program aims at nothing less than redefining science by re-conceiving these two traditionally distinct categories of objects as being intimately entwined, or, to use one of Francisco's favorite metaphors, in a perpetual dance. The broader conceptual framework in which this methodology fits is that of enaction, in which the organism-self/world relationship is one of co-determination; mind and world make up an interdependent continuity, say the authors of *The Embodied Mind*, "there is nothing extra on the side of mind or on the side of world to know or to be known further" (Varela, Thompson and Rosch, 1991, p. 225). This perspective is radically anti-reductionist and non-dualistic from its roots in biological identity (autopoiesis) to its method (neurophenomenology). Francisco presents neurophenomenology as "a methodological *remedy* for the hard problem" of consciousness. For those of us, such as myself, who are committed to theoretical and clinical work in psychology, psychotherapy, and psychoanalysis, domains in which primacy is given to lived experience, the enactive perspective is a remedy for a certain deadlock with natural science, providing an opening for dialogue. Even more substantially, this non-reductive position points the way to an embodied ethics that is consistent with that of the psychodynamic approach to human experience and change.

PLL: I liked particularly the last book by Francisco: *El Fenómeno de la Vida*, which I have read in Spanish. Is this being translated into English?
[See above]

PLL: You lived a long life with Francisco: in what way was Buddhism affecting his daily life?

AMY COHEN: I'm tempted to say that Buddhism didn't *affect* his daily life, rather it was woven so deeply into the texture of his thought and days that it is impractical to attempt to disengage particular effects. Of course, one can see the shadow of Buddhist concepts like *sunyata* – groundlessness or vacuity – behind the whole body of his scientific and philosophical work and at each step of its expression. How might concepts like this, and the practice of them, have manifested in his everyday existence? Here I might point to his lack of, and extreme dislike for, dogmatism, and any position that smacked of it. He felt that any politics, be it large-scale or interpersonal, must be based on human awareness and experience as dynamical processes and that one should not depart from these by constructing solid edifices and then entrenching oneself in them. This made him, as you might imagine [and you, Luigi, know!], both a kind and a demanding interlocutor. He had a powerful sense of his personal engagement and responsibility in the events of his life, big and small, which was coupled with an acute awareness of the contingency, the shifting impermanence of both self and world. His consciousness of the paradoxical nature of life – of its self-affirming power and its intrinsic precariousness – grew stronger throughout his long illness, which he took as an opportunity to practice. In his practice he found a resting place apart from the discomfort and anxiety of sickness and its paraphernalia – hospital sojourns, exams, operations, medication. The presence of mind and serenity with which he approached his death is no doubt the most eloquent expression of this practice.

6.6 Zooming into the core of autopoiesis

We have spoken in the previous section about the self-maintenance of the cell, or in general about the autopoietic unity. What is it precisely that is maintained? The answer is: the overall cell organization. Here is where the notion of organization acquires its full value. The organization is the *invariant* of the dynamics of the biological systems; it is the unitarian complex of relations that form the identity of the living. The continuous production and conservation of the basic functional relations permit the living to acquire and regenerate its individuality. All this implies that the attention is not on the components, but rather on the relations that these components have to satisfy to constitute the integer system (see also Damiano, 2006).

The notion of invariance of the relations is what makes autopoiesis a scientifically rigorous description of biological systems, able to distinguish the invariant aspects of the living dynamics (the organizational relations) from the variable ones and to link them to each other. The structure, the concrete unity of the components, is the *variable element*; it varies from cell to cell, and it also varies during development, but these changes do not affect the invariance of the biological organization.

In this link between structure and organization, the theory of autopoiesis is indebted to Piaget and his cognitive biology (Piaget, 1967). This also includes the notion that organization and structure cannot be considered independent one from the other.

The invariance of the relationships among the processes that make the components, as already said, is the reflection of a particular characteristic of the biological self-organization: all rules to produce itself (autopoiesis) are within the boundary, as expressed by the notion of *organizational closure* – a concept also introduced by Piaget (1967). (Nowadays, the term *operational closure* is being more commonly used.) It is a closed domain of relations, which defines a space in which a concrete system can be realized (Maturana and Varela, 1980). It is such a closure that gives the cellular system the capability to specify from within all operations of production, in a closed chain of linked dynamic operations. In other words, all relations of production are coordinated in a system that can be described as a homeostatic system, which maintains the integrity of its own identity through the production of its own components, also under conditions of perturbation.

We see here that this way of characterizing the stabilizing dynamics of the cellular unit speaks a different language with respect to the classic molecular biology model, which is centered on the genetic program (Damiano, 2006). The notions of organization and structure, with the specifying notion of operational closure, define and characterize the biological autonomy of the living systems.

Of course, the notion of operational closure does not mean that the cellular system or the autopoietic system, in general, is an isolated one. As already mentioned, the cellular system, and any living being, is an open system, open from the thermodynamic point of view: it receives energy and/or nutrients from the outside medium. The relation between autopoiesis, autonomy, and self-referentiality is treated in the specialized literature (see, for example, Marks-Tarlow *et al.*, 2001; and Weber, 2002).

There is an apparent contradiction between the biological autonomy and this dependence from the outside medium; and another apparent dichotomy is between structure and organization invariance. In fact, those are complementarities and not dichotomies, and they form the very essence of the living. Living systems transform matter inside themselves in such a way that the product of these transformations is their own organization (Maturana and Varela, 1980). In this way, there is an indissoluble link between the organization of the living and the outside medium.

This link with the outside is indeed another essential facet of the theory of autopoiesis. The main point here is that the *outside* must be recognized from the inside, it must positively interact with the autopoietic organization of the living itself. This gives rise to a series of notions, which will be sketched briefly in the next section: firstly, the notion of *structural coupling*, which, describing the biological interaction with the environment, introduces the new idea of *cognition*, which in turn involves the notion of information and eventually of evolution.

From all the above, it is apparent that autopoiesis belongs epistemologically to systems theory, according to which it is the organization of the components that characterizes the quality of the system. Thus, the life of a cell is a global property, and cannot be ascribed to any single component.

The living cell as an autonomous system can be seen as a "self without localization" (Varela *et al.*, 1991; Varela, 2000), as there is no single component, or single reaction, that alone is responsible for life.

Autopoiesis is still an active area of research. In Santiago de Chile, there is Matríztica Institute, founded by Ximena Dávila Yáñez and Humberto Maturana. Several research places are in Europe and Canada (see, for example, the Side Box by Luisa Damiano, below). I wish to mention also the activities of SAM (*Scuola Autopietica del Mediterraneo*), directed by Rossella Mascolo in Sardinia (autopoiesi.sam@gmail.com). As already mentioned, the notion of autopoiesis is linked to a series of epistemic concepts, and the conversation below with the philosopher Evan Thompson, Francisco Varela's friend and collaborator, will show that in full display.

Side Box 6.1

Autopoiesis: three research directions for future developments

Luisa Damiano

Luisa Damiano (Ph.D., Complex Systems Theories and Methods) is associate researcher in philosophy of science at the Research Center for Complex Systems (CERCO) at the University of Bergamo (Italy), where she established the Epistemology of the Sciences of the Artificial Research Group (ESARG).

My current research work on autopoietic biology relies on one of its most discussed and, in my view, promising operations: leading the two central questions of cognitive biology – i.e., *"What is life?"* and *"What is cognition?"* – to converge in one theoretical solution. Maturana and Varela conceptualized life and cognition as expressions of the distinctive form of autonomy characterizing biological systems. They defined this property as *autopoiesis*, and identified it as the capability of these systems to exercise on themselves an activity of *self-production* through an internal process of permanent (re-)constitution of their elemental components. In line with the emergentist approach developed by the early studies on biological autonomy, the two researchers referred this property not to single components, but to the organization of living systems, that is, to the functional correlation integrating the components in the dynamic units that these systems constitute. This approach defined the main theoretical issue addressed by autopoietic biology: characterizing the organization of biological systems, that is, hypothesizing a form of organization able to generate and maintain their *autopoiesis* – their self-production. Maturana and Varela offered a rigorous solution to this issue at the level of the minimal living system, providing the description of an organizational mechanism supporting the self-productive dynamics of the cell – its topological self-distinction included. This theoretical result, expressed in the notion of *autopoietic organization*, conveys the Santiago School's most innovative contributions to the disciplinary areas related to cognitive biology – results on which my research work draws.

To life sciences, Maturana and Varela proposed this notion as a "synthetic" definition of biological systems, which individuates them not through a list of properties, but by specifying their mechanism of self-individuation (Damiano, 2009; Damiano and Luisi, 2010). Since the late 1970s, this theoretical proposal – largely neglected by traditional biology – has been attracting the attention of the increasing number of research lines exploring biological self-organization. Since the 1990s, due to the collaboration with Francisco Varela that led Pier Luigi Luisi to develop the Chemical Autopoiesis approach, the concept of autopoietic organization has become widely known in the field of synthetic biology, where it is currently considered able to drive cutting-edge research towards the artificial synthesis of minimal cells.

To cognitive sciences, the School of Santiago offered this notion as the theoretical ground of a biologically strong alternative to computationalism – i.e., "classical" cognitive science. The concept of autopoietic organization challenges the computationalist view of natural cognition as a syntactically determined activity of physical symbols manipulation exercised by a biological analogue of the digital computer. The notion introduced by Maturana and Varela proposes to think of biological systems as autonomous agents whose cognitive activity consists not in extrinsic processes of computational problem-solving, but in maintaining themselves – i.e., maintaining the continuity of their processes of self-production – by continuously re-establishing a relationship of dynamical coupling with their environmental niche (Damiano, 2009, 2012; Bich and Damiano, 2012a). Initially, mainstream research did not welcome this proposal, criticizing the attempt of scientifically approaching human cognition by assuming as paradigmatic cognitive processes not human, but cellular interactive dynamics. Since the 1990s, this attitude towards autopoiesis has significantly changed, mainly due to the increasing awareness, within the cognitive sciences community, of the incapability of computationalism to account for the central role of the body in cognitive processes. The emergence of the *embodied cognitive science* – the main competitor of computationalism – stimulated positive evaluations of the Santiago School's effort of grounding human cognition in the deepest processes of self-production of the biological body. On this basis, autopoietic biology has been progressively recognized not only as one of the main theoretical sources of the current mainstream embodied approach to the study of cognition, but also as one of the *avant-garde* theories of its most radical research line – *radical embodiment*. Moreover, autopoietic biology can be recognized as one of the most influential sources of inspiration of *embodied AI*, the emergent research line in AI that declines the classical focus on computer programs expressing high-level human cognitive competences, and re-centers AI inquiries on biologically inspired complete agents – robots – performing interactive tasks.

To both the areas of life sciences and cognitive sciences, the notion of autopoietic organization provided also a third possibility of innovation, related not to their theories of reference, but to their ways of conceiving and practicing science. I refer to a set of epistemological elements that the Santiago School elaborated as developments of the autopoietic theory of cognition, and that can be seen as the basic structures of a new epistemology for science. As shown elsewhere, the latter consists in a constructivist epistemology that finds its roots in the production of the pioneer research on self-organization, and significantly diverges from the representationalist epistemology at the basis of classical science (Damiano, 2009, 2012). It offers to life and cognitive sciences the possibility of realizing a complete paradigm shift: developing not only new theoretical views of life and cognition, but also a new epistemological framework, able to generate and support, consistently with these views, new practices for the scientific investigation of living and cognitive processes.

The first of the three research axes I am developing in relation to autopoiesis draws on this last contribution by Maturana and Varela, and is directed towards the epistemological grounding of the methodological scheme usually called "synthetic methodology." Originally introduced in the late 1940s, it has been rapidly expanding in the last decades as the methodology through which the "sciences of the artificial" intend to positively contribute to the scientific study of life and cognition. This methodological approach – often summarized in the principle "understanding by building" (Pfeifer and Scheier, 1999) – promotes the use of implementations in "software" (computer simulations), "hardware" (robots), and "wetware" (chemical systems) as scientific models of biological and cognitive processes. As argued in various texts (Damiano and Cañamero, 2010; Damiano, Hiolle, and Cañamero, 2011), this "constructive approach" challenges the traditional representationalist scientific epistemology, and assigns to the contemporary philosophy of science an urgent task: defining the possibilities, the limits, and the ways of the synthetic modeling of life and cognition, and establishing its relevance for biological, behavioral, cognitive, and social sciences. Part of my current research work tackles this quest, with the aim of building, on the basis of the autopoietic and other constructivist epistemologies, an epistemological framework capable of guiding the development of the synthetic methodology, and generating answers for the open questions that can impede its establishment as a valid methodological scheme: Can systems built with different materials, and endowed with different embodiments, be considered effective models of natural living and cognitive systems? What are the status and the value of the insights coming from the exploration of synthetic models of life and cognition? What are the criteria that can be established to warrant a productive transmission of knowledge from the sciences of the artificial to the sciences of the natural?

My second research axis related to autopoietic biology deals with the synthetic study of cognition, and is directed towards developing a research program able to productively apply synthetic biology techniques to the scientific modeling of cognitive processes. This extension of the synthetic exploration of cognition is based on recent critical analyses of embodied AI, pointing out its incapability of producing artificial agents able to behave in a context-sensible and "more life-like manner" (Froese, 2007). This limitation of embodied AI is often characterized as its incapability of fully solving, through its present-day approaches, "the problem of meaning in AI": the challenge – insurmountable for classical AI – of building artificial agents able to generate in real-time, for external events, meanings that are pertinent to their maintenance, as well as to the accomplishment of their tasks, in ever-changing environmental conditions. This failure has been frequently ascribed to the ignorance of "some organizing principle of biological systems" (Brooks, 1997), and, more precisely, to limits in understanding and artificially re-creating biological autonomy (Di Paolo, 2003; Froese, 2007; Ziemke, 2007; Froese and Ziemke, 2009). Usually these critical remarks explicitly refer to autopoietic biology as one of their theoretical grounds and rely on the autopoietic constructivist thesis, according to which, in living systems, the activity of meaning generation is an expression of the activity of self-production. On this basis, these criticisms are generally related to positive attempts of defining new design principles for AI, aiming at providing robots with aspects of biological autonomy able to generate life-like cognitive behaviors.

The research program I am setting up with Pasquale Stano and colleagues develops a similar approach, and is directed towards creating artificial forms of autonomous cognition not within the usual AI domains (i.e., computer science and robotics), but within synthetic biology. Our research line, theoretically and technically grounded in Pier Luigi Luisi's Chemical Autopoiesis approach, is not oriented to fully implement biological autopoiesis. Instead, it intends to create, in chemical systems, minimal forms of chemical self-production that can be recognized as intrinsically linked to the capability of generating, for external events, meanings that are pertinent to the maintenance of the coupling with the environmental niche. This work finds its main theoretical source in the "generic model of autonomous system" (Damiano, 2009) that Francisco Varela introduced (Varela, 1979) on the basis of the Santiago School's theoretical and epistemological production, and later re-proposed as the model of the "emergent self" (Varela, 1995). The aim of our research program is not only to deepen current scientific knowledge on the origins of autopoiesis and related cognitive processes. Our main goal is to synthetically study minimal autonomous cognition and its developments, and to establish, between synthetic biology and AI, a bi-directional transmission of knowledge able to maximize the synthetic exploration of autonomous cognition, and to expand the implementation of artificial autonomous agents (Damiano and Stano, 2015) from mechanical to chemical and hybrid (mechanical-chemical) systems.

The Varelian "generic model of autonomous system" is critical also for the development of the third axis of my current research work on autopoiesis. This explorative line, which I am elaborating with Leonardo Bich, takes on the debated issue of the autopoietic characterization of social systems (Bich and Damiano, 2012b). Our work is based on the critical rejection of the approaches that merely transposes the concept of autopoietic organization from the biological to the social domain (Luhmann, 1986). By taking into account more accurate attempts of developing the Santiago School's theory of organization at the "meta-biological" level (e.g., Hejl, 1984), we are drawing on organizational principles originally defined by autopoietic biology, early research on self-organization, and Varela's theory of autonomous systems in order to set up conceptual tools supporting the production of non-reductionist descriptions of social systems capable of capturing the specific features of their different configurations – from the least to the most complex ones.

These three research axes converge in one aim: contributing to actualizing the innovational potential of autopoietic biology, which is still far from being fully exploited by contemporary scientific and philosophical research.

References

Bich, L. and Damiano, L. (2012a). Life, autonomy and cognition. *Origins of Life and Evolution of Biospheres*, 42(5), 389–397.

Bich, L. and Damiano, L. (2012b). Riscoprire la teoria dell'autopoiesi nella caratterizzazione dei sistemi sociali. In I. Licata, ed., *Sistemi, Emergenza, Organizzazioni. Complessità e Management*. Roma-Messina: CoRiSCo – EDAS, 83–111.

Brooks, R. A. (1997). From earwigs to humans. *Robotics and Autonomous Systems*, 20 (2–4), 291–304.

Damiano, L. (2009). *Unità in dialogo*, Bruno Mondadori, Milano.

Damiano, L. (2012). Co-emergences in life and science. *Synthese*, 185, 273–294.

Damiano, L. and Cañamero, L. (2010). Constructing emotions. Epistemological groundings and application in robotics of a synthetic approach to emotions. In Chappell *et al.*, eds., *AI-Inspired Biology*, SSAISB, 20–28.

Damiano, L., Hiolle, A., and Cañamero, L. (2011). Grounding synthetic knowledge. In T. Lenaerts *et al.*, eds., *Advances in Artificial Life, ECAL 2011*. Cambridge: MIT Press, 200–207.

Damiano, L. and Luisi, P. L. (2010). Towards an autopoietic redefinition of life. *Origins of Life and Evolution of Biospheres*, 40(2), 145–149.

Damiano, L. and Stano, P. (2015). Chemical selves. Synthetic Biology meets Artificial Intelligence on the pathway towards a paradigm shift in cognitive sciences. *Biosystems*, forthcoming.

Di Paolo, E. (2003). Organismically-inspired robotics: homeostatic adaptation and teleology beyond the closed sensorimotor loop. In K. Murase and T. Asakura, eds., *Dynamical Systems Approach to Embodiment and Sociality*. Adelaide: Advanced Knowledge International, 19–42.

Froese, T. (2007). On the role of AI in the ongoing paradigm shift within the cognitive sciences. In M. Lungarella *et al.*, eds., *50 Years of AI*. Berlin: Springer-Verlag, 63–75.

Froese, T. and Ziemke, T. (2009). Enactive AI. *Artificial Intelligence*, 173, 466–500.

Hejl, P. M. (1984). Towards a theory of social systems. In H. Ulrich and G. J. B. Probst, eds., *Self-Organization and Management of Social Systems*. Berlin: Springer-Verlag, 60–78.

Luhmann, N. (1986). The autopoiesis of social systems. In F. Geyer and J. van der Zouwen, eds., *Sociocybernetic Paradoxes*. London: SAGE, 172–192.

Pfeifer R. and Scheier C. (1999). *Understanding intelligence*. Boston: MIT.

Varela, F. (1979). *Principles of Biological Autonomy*. North-Holland.

Varela, F. (1995). The emergent self. In J. Brockman, ed., *The Third Culture: Beyond the Scientific Revolution*. New York: Simon & Schuster, 188–199.

Ziemke, T. (2007). What's life got to do with it? In A. Chella and R. Manzotti, eds., *Artificial Consciousness*. Exeter: Imprint Academic, 48–66.

Conversation with Evan Thompson

Evan Thompson, Ph.D., is the author of *Waking, Dreaming, Being: Self and Consciousness in Neuroscience, Meditation, and Philosophy; Mind in Life: Biology, Phenomenology, and the Sciences of Mind;* and *Colour Vision: A Study in Cognitive Science and the Philosophy of Perception.* Thompson is the co-author with Francisco J. Varela and Eleanor Rosch of *The Embodied Mind: Cognitive Science and Human Experience.* He is also the co-editor with Philip David Zelazo and Morris Moscovitch of *The Cambridge Handbook of Consciousness*, and with Mark Siderits and Dan Zahavi of *Self, No Self? Perspectives from Analytical, Phenomenological, and Indian Traditions.* In addition, he is the author of more than 70 articles, chapters, and reviews in the fields of philosophy and cognitive science. He received his B.A. in Asian Studies from Amherst College (1983) and his Ph.D. in Philosophy from the University of Toronto (1990). He held a Canada Research Chair at York University (2002–2005), was Professor of Philosophy at the University of Toronto (2005–2013), and is now Professor of Philosophy at the University of British Columbia. In 2014, he was the Numata Invited Visiting Professor at the Center for Buddhist Studies at the University of California, Berkeley. Thompson is also a Fellow of the Royal Society of Canada.

PLL: You collaborated with Francisco Varela during his last years. What were the issues that you mostly discussed together?

EVAN THOMPSON: During his last years we worked together on how to relate the large-scale neurodynamics of cognition, which was the experimental focus of Francisco's lab just before he died, to embodied cognition research, on the one hand, and to phenomenological analyses of consciousness, on the other hand. These interrelated themes were the focus of the last paper we wrote, "Radical Embodiment: Neural Dynamics and Consciousness," published in *Trends in Cognitive Sciences* in 2001, just after Francisco's death. That paper also contained a short theoretical analysis and definition of the concept of "emergent processes" in the context of nonlinear dynamical systems. We devised that analysis in an intensive work session in my father's apartment in Manhattan over a few days in December 2000. That was the last time we worked together, and the last time I saw him until visiting him in Paris a few days before he died. One notable memory from that Manhattan meeting is that Francisco explained to me that neural "phase synchrony," which he was a pioneer in researching, is just one limited measure of brain dynamics, and that nonlinear, cross-frequency phase-amplitude and phase-phase couplings are what is really important. He drew a picture of these kinds of couplings on the back of a restaurant napkin. At that time, there were virtually no published experiments on these kinds of neurodynamics. Now, almost 20 years after his death, this kind of experimental work has really only just begun. Francisco was always way ahead of his time.

PLL: How would you define Francisco's neurophenomenology? In what sense can we consider this really as a new discipline?

EVAN THOMPSON: Francisco presented neurophenomenology in two ways. One was methodological and experimental; the other was more philosophical.

The methodological way was to use individuals who have phenomenological training and expertise, especially as a result of Buddhist meditation practice, in order to expand and enrich the neuroscience of consciousness. His term for this way of relating phenomenology and neuroscience was "mutual constraints." The basic idea is that refined phenomenological reports can help to pick out and give significance to previously unnoticed neural patterns or neural activity that might otherwise have been treated as noise, while neuroscience findings can reveal aspects of experience that might otherwise have been overlooked. In this way, neuroscience and phenomenology refine each other.

Neurophenomenology is not the same thing as the neuroscience of meditation. The neuroscience of meditation is concerned with the neural correlates of meditative states, or with measuring the state or trait effects of meditation on the brain and behaviour. Neurophenomenology, however, uses meditation as a phenomenological method for getting at basic aspects of cognition. Neurophenomenology proceeds (1) by using experienced contemplative participants, (2) who give first-person reports on a trial by trial basis of phenomenal features of their awareness, (3) in conjunction with second-person interview methods for explicating their tacit experience, (4) in experimental protocols that address basic questions about mental function (affect, attention, memory, cognitive control, spontaneous cognition, consciousness), (5) using third-person measures that examine large-scale neural integration and coupled brain-body-world dynamics.

The second way that Francisco presented neurophenomenology was as a methodological solution to the "explanatory gap" or "hard problem" of consciousness. This is the philosophical problem of how consciousness, in the sense of subjective experience, is related to the physical world. Francisco thought that pursuing neurophenomenology would eventually dissolve this problem or render it obsolete.

In my view, neurophenomenology is not a new discipline, but a new methodological approach in cognitive science. It's compatible with the best efforts in the history of modern psychology and neuroscience to pay careful attention to first-person experience – consider Ernst Mach's *The Analysis of Sensations*, William James' *Principles of Psychology*, the writings of the Gestalt psychologists, or the work of the early psychophysicists, such as Gustav Fechner or Ewald Hering. What neurophenomenology adds to these efforts is the distinct and important type of phenomenology we find in the Buddhist meditative traditions, together with the sophistication of modern cognitive neuroscience methods.

As a philosopher, I've never found convincing the additional claim that this approach suffices to close the explanatory gap or dissolve the hard problem. Although phenomenology and neuroscience can be brought into much closer collaboration, there remains a conceptual and epistemological gap between our understanding of the phenomenal and our understanding of the physical, a gap that neurophenomenology as a methodological approach does not close.

PLL: To bring the "first person" into science was one of the interests of Francisco in his later years. Do you think that this operation has been having some success? How do you see this, in the current literature of philosophy and epistemology?

EVAN THOMPSON: A fair amount of progress has been made. Phenomenology now plays a much larger role in philosophy of mind than it did when the volume *Naturalizing Phenomenology*, edited by Francisco with Jean-Michel Roy, Jean Petitot, and Bernard Pachoud, was published in 1999. Buddhist philosophy is also increasingly visible in these discussions. There are a handful of experimental neurophenomenological studies, and recently a special issue of *Frontiers in Human Neuroscience* was devoted to neurophenomenology. Still, it must be said that there are few, if any, neurophenomenological experiments that fully meet all five of the criteria that I listed above. I suspect the reason is that this kind of experimental work is very demanding and time-consuming, and that people mistakenly think that doing the neuroscience of meditation is doing neurophenomenology.

PLL: The position of Francisco about "consciousness" was not always clear to many of his followers. This is partly because in his writing, as in those of several neuroscientists, the difference between consciousness as a "hard problem" (personal experience), and consciousness as an easy problem (neuronal process of cognition), was not always clearly stated. Then there is in Buddhism the notion of "subtle consciousness," which has no material base, and this is at an even higher level. This, in turn, is related to the Tibetan Buddhist notion of reincarnation. It is a complex, difficult kaleidoscope. Can you sketch, albeit schematically, the position of Varela in all this?

EVAN THOMPSON: Francisco held together a number of important commitments or perspectives that one rarely sees held together by a working scientist. First, he clearly recognized that there is a basic conceptual and epistemological gap between any neurofunctionalist model of consciousness and consciousness as lived experience. Second, he argued that the biological processes crucial for consciousness depend not just on the brain but also on the body or the organism as an autonomous agent. For these two reasons, he rejected the neuroreductionist statement that consciousness can be explained as nothing but a certain kind of brain process. Third, on a philosophical level, Francisco understood the fundamental epistemological point that science can never exhaustively explain consciousness from the outside, as it were, because science always takes place within the horizon of consciousness as lived experience, and all scientific observation presupposes consciousness in the form of the lived experiences of scientists. This epistemological point comes from Kant, Husserl, and Merleau-Ponty, and Francisco knew their writings well. I believe this point also conditioned his approach to his fourth commitment, which was to the value of Tibetan Buddhist meditation practice and Buddhist philosophy. According to the Tibetan Buddhist view, consciousness in its most basic and "subtle" form, which is sheer awareness, distinct from any cognitive content, is not a biological phenomenon (not even an emergent one). Indeed, according to classical Indian Buddhist philosophy, which strongly shapes Tibetan Buddhist philosophy, sheer awareness is not in any sense physical, so this view is dualist. According to the system of Tibetan Buddhist practice called "Tantra" or "Vajrayana," however, which Francisco practiced, sheer awareness or subtle consciousness is considered to have a subtle energetic base, but one that transcends the workings of the living body or biological organism. This view, in Western philosophical terms, is a kind of dual aspect view – consciousness and matter/energy are two, indissociable aspects of reality. Nevertheless, as either a dualist or a dual aspect view, Tibetan Buddhism clearly stands in tension with the view from cognitive science and biology that consciousness is contingent on brain function. As a biologist and cognitive scientist, Francisco worked from within the framework of thinking that consciousness is an embodied, biological (and social) phenomenon. From this perspective, he was critical of Buddhist philosophical arguments for mind-body dualism; he saw those arguments as outmoded conceptually and scientifically. At the same time, because he understood the epistemological, phenomenological, and methodological primacy of consciousness as lived experience (that we cannot step outside of lived experience and that lived experience is that with which we are acquainted first and foremost), and because he knew that becoming existentially and cognitively attuned to lived experience as such is a skill that takes practice and training, he took very seriously the Tibetan Buddhist statements, made in the context of meditation practice, about the experiential primacy of sheer awareness, especially the importance of orienting oneself to this primacy at the time of death. As to whether such awareness is still present after death, Francisco's attitude was to refrain from judgment, in the sense of the ancient Greek term, *epoché*, which the Skeptics used to describe the resolute withholding of assent or dissent, and which Husserl later adopted to name the suspending of all "positing," a suspension required in order to undertake phenomenology. Francisco called this attitude, "staying with the open question." As he said in an interview for the film, *Montegrande*, "Don't neglect the Buddhist observations, and don't dismiss what we know from science. Instead of trying to seek a resolution or an answer, contemplate the question and let it sit there. Have the patience and forbearance to stay with the open question." The last conversation I had with Francisco, during our meeting in Manhattan in December 2000, was about this open question. Francisco explained to me – both from his own experience and from his understanding of Buddhist ideas – why it wouldn't be unreasonable to think that there could be a basal or baseline form of awareness that is invariant across any perceptual or cognitive state. From the experiential perspective of encountering this basal "luminous" awareness, which is what it is whether you're awake or dreaming or aware under anesthesia (he had experienced being aware under anesthesia), it would be natural to think that it is not dependent on the body and its changing states. At the same time, as I argued and he agreed, although that line of thought may be compelling from a first-person perspective, it does not logically imply that there really is a form of awareness that persists in the absence of the brain or the body. I like to think that this conversation may have helped him to stay with the open question, as it has helped me since his death.

PLL: What is your present work in relation to the legacy of Francisco?

EVAN THOMPSON: Francisco was an unusually philosophical scientist, but he wasn't a philosopher, so he didn't try to put together all of his theoretical commitments in the systematic and rigorous way that a philosopher would demand. My book, *Mind in Life: Biology, Phenomenology, and the Sciences of Mind*, which was written mostly in the immediate aftermath of Francisco's death, was an attempt to put together autopoietic biology, enactive cognitive science, neurophenomenology, and philosophy of mind in a rigorous, philosophical way. This book reflects my having been imprinted by Francisco's life and work, going back to my earliest years as a teenager when we first met. My most recent book, *Waking, Dreaming, Being: Self and Consciousness in Neuroscience, Meditation, and Philosophy*, is an extended neurophenomenological investigation of consciousness in the waking state, dreaming, lucid dreaming, deep sleep, out-of-body experiences, and dying. It weaves together cognitive science, Indian philosophy, philosophy of mind, and personal stories from my own experience, including ones about Francisco going back to when I was a subject in one of his early EEG experiments from the late 1970s. A lot of the book derives from my participation in the Mind and Life Institute Dialogues with the Dalai Lama, which Francisco created, but which I started to participate in only after Francisco's death. So this book, too, bears the strong imprint of Francisco's life and work.

6.7 What autopoiesis does not include

The molecules DNA and RNA are considered in autopoiesis only as participants in the cell's self-production and not for their ability to self-reproduce and/or evolve. Varela and Maturana often emphasize that before one can talk about reproduction, one has to have a place to host such a mechanism. The container and the logic must be there first. Likewise, in describing a car, before talking about the nature of the fuel, one has to have a logical scheme describing how a car works and how motion is transmitted from the carburetor to the wheels. A classic citation by Maturana and Varela (1998) reads:

In order to reproduce something, the unit must first be constituted as a unit, with an organization that defines this unit itself. This is simple common sense logics.

Adding:

A living organism can also exist without being capable of self-reproduction.

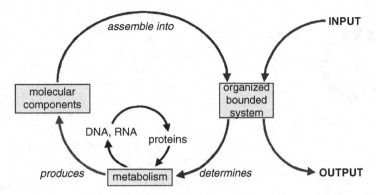

Figure 6.5 An extension of the autopietic scheme in order to specifically incorporate nucleic acids and proteins. Here, life is outlined as a system confined by a chemical boundary, which contains a network of nucleic acid- and protein-based reactions. This network (metabolism) is able to transform external energy/nutrients into self-maintainance and self-generation.

Also, Varela asserts that to take reproduction into the definition of life would be ontologically wrong (Varela, 2000), as reproduction is a

... consequence of the existence of individuals. The difficult thing is to create an organism that is capable of self-reproducing with its own boundary. To divide it up in two is easy.

As already discussed, this notion, that reproduction is a consequence of the internal logic of life, can be visualized in Figure 6.3 (see also Luisi, 1996), which is an extension of the drawing of our Green Man (Figure 6.1), discussed previously.

The lack of emphasis on DNA, self-reproduction, and evolution in the theory of autopoiesis was certainly a reason for its lukewarm reception in the community of molecular biology – a difficulty that might have been avoided had its authors been less rigid about the matter. In fact, it is not difficult to incorporate nucleic acids and enzymes into the autopoietic scheme. This was proposed more recently (Luisi, 1993, 1997) and the corresponding modification is rather simple, as Figure 6.5 shows. But this is not the main reason for the lack of acceptance of the theory of autopoiesis by most of the established literature. The next section goes into a little more detail regarding this issue.

6.8 Chemical autopoiesis: the case for self-reproduction of micelles and vesicles

The term *chemical autopoiesis* indicates the experimental implementation of autopoiesis in the chemistry laboratory. It must be said from the onset that, generally, this work has to do with surfactant aggregates, micelles, or

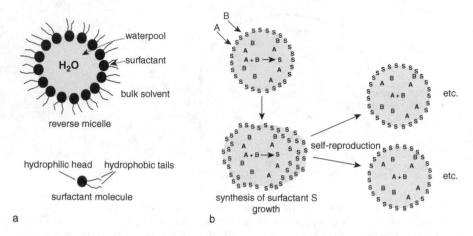

Figure 6.6 Theoretical implementation of autopoiesis in reverse micelles. More details are in the text. (**a**) Schematic view of a reverse micelle. (**b**) Scheme of the self-reproduction mechanism of reverse micelles.

vesicles, which are forms of self-organization whose detailed structures and physical properties will be discussed in Part III of this book. However, for understanding the general principles of chemical autopoiesis, the qualitative picture given in this section should be enough.

And let me go then to the original intuition that came about through a discussion with Francicso Varela, now worth mentioning also as a historical note: looking for a way to implement experimentally autopoiesis, we came up with the idea of using reverse micelles with two reagents that could react inside the boundary (but not outside) to yield as a product the very surfactant that builds the micelle itself (Luisi and Varela, 1990), and we drew a cartoon which corresponds to Figure 6.6(b).

This idea assumed that the two components – **A** and **B**, in the picture – could permeate inside the reverse micelle and give the surfactant **S** by synthesis. However, this proved to be difficult to implement from the experimental point of view.

Thus, the successful working implementation of this idea was based on the use of a precursor **A-S**, whose incorporation in the micelle and eventual hydrolysis leads to **S**, the membrane forming surfactant, see Figure 6.7.

Both in the case of reverse and aqueous micelles, fatty acids were used as micelle-forming surfactants. In the case of reverse micelles, formed in isooctane, the surfactant was octanoic acid with octyl alcohol as co-surfactant; **A-S** was octnanoyl-octyl ester, whose hydrolysis at the surface of the micelle yields more surfactant, the octanoic acid. Thus, we are following the scheme of Figure 6.7 (Bachmann *et al.*, 1990), which resulted in a multiplication of the

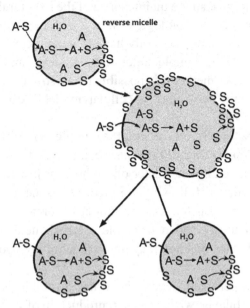

Figure 6.7 The implementation of the idea introduced in the previous scheme of Figure 6.6. It is based on the use of a precursor **A-S**, whose incorporation in the reverse micelle and eventual hydrolysis leads to **S**, the membrane-forming surfactant. More details and experimental results are in the text.

number of micelles. Octanoic acid also forms aqueous micelles, thus a similar scheme was used in water (Bachmann, 1991; Bachmann *et al.*, 1992). Later on, the same principle was used to achieve self-reproduction of fatty acid vesicles (Walde *et al.*, 1994b).

Whereas all this will be discussed later on, here it is important to emphasize that the schemes of Figures 6.6 and 6.7 are typically autopoietic, as the hydrolysis, and eventually the self-reproduction, take place from within the micellar structure.

6.9 Chemical autopoiesis: a case for homeostasis

We have illustrated above some simple chemical autopoietic systems. Here we will consider a further step in chemical complexity, dealing with the relation between autopoiesis and homeostasis.

Homeostasis is a self-regulating process by which biological systems tend to maintain stability while adjusting to conditions that are optimal for survival; and this is indeed a complex issue, particularly when applied to higher animals (see, for example, Recordati and Bellini, 2004; Kitano, 2007; Cooper, 2008). In the simpler case of unicellular organisms that are not subjected to environmental changes, it may simply mean the maintenance of the structural and organization system's feature for a certain biological time – with the caveat that for homeostasis, all properties do not necessarily have to remain constant.

This is the case for a simple autopoietic system, and looking back at Figure 6.3, one can ask whether it is possible to construct a system experimentally – one that would be a kind of cellular model – capable of observing homeostasis.

In fact, there has been one attempt to implement this experimentally. The experiment (Zepik *et al.*, 2001) has been realized using oleate vesicles (which will be discussed in detail later on in the book), with oleate anhydride, which binds to the bilayer (following in practice the scheme of Figure 6.7).

This system was constructed as a continuous bioreactor, in the sense that material was continuously added from the external medium by a microstep syringe motor, so that the reaction may go on indefinitely upon external supply of fresh reagents.

In this kind of experiment, where the water insoluble oleate anhydride is binding to the oleate bilayer, we have self-reproduction of vesicles (Walde *et al.*, 1994b) due to the hydrolysis of the anhydride, which affords fresh oleate monomers. However, in the experiment described by Zepik *et al.* (2001), there is a competitive reaction, an oxidation reaction that destroys oleate.

Clearly, this system corresponds to that of Figure 6.3. In fact, by regulating the relative concentration of the reactants and consequently the two basic velocities, the three conditions of Figure 6.3 can be obtained, as shown in Figure 6.8.

There are several questions that this kind of experiment opens up. The first observation is that it is possible to prepare in the lab simple autopoietic systems responding to the general scheme of Figure 6.3, and in particular displaying continuous homeostasis. This, in my opinion, is very important for the general discourse of cell models and life in the lab.

On the other hand, it is also interesting to observe that this paper has seldom been cited in the literature, and actually one referee, after a lukewarm acceptance, asked as to why I had undertaken such a lengthy and boring job. This referee's observation is indicative of the little impact of autopoiesis in general, as well as model homeostatic systems, represent for today's general establishment.

There is instead an important question stemming from this kind of experiment: whether one can have cells in homeostasis, namely displaying a sort of simple metabolism, prior to or independently from DNA existence. Do we really have to start, in the origin of life, from a self-replicating DNA-centered system?

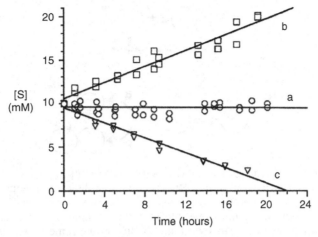

Figure 6.8 The time course of oleic acid generation in the Zepik *et al.* system. (**a**) Hydrolysis and oxidation. The velocity of hydrolysis and the velocity of oxidation (see Figure 6.3) are numerically equal. (**b**) Only hydrolysis. The velocity of oleate production (and therefore the formation of vesicles) is predominant. (**c**) Only oxidation (!). The decay velocity (the rate of oleate oxidation and breakdown) is predominant, which leads to the destruction (death) of the whole system. (The oxidation reaction is the OsO_4-catalyzed cis hydroxylation of olefins with the hexacyanoiron(III) ion as a co-oxidant. Osmium tetroxide reacts first with the olefin to give a cyclic diester, which is then hydrolyzed to the diol and osmate (OsO_4). The latter is continuously regenerated to OsO_4 by an excess of K_3 $[Fe(CN)_6]$. The oxidation affords 9,10-dihydroxystearic acid, which does not form vesicles).

6.10 Second order autopoietic structures

We have focused until now on a single autopoietic unity, which corresponds biologically to a unicellular organism. Many organisms, however – even simple ones – are constituted by many different types of cells and can be seen as integrated aggregates of different autopoietic units. Maturana and Varela (1998) consider the case of meta-cellular organisms as *second-order autopoietic systems*. The different cell aggregates are in close coupling with each other, and actually the different cells interact with each other (a case of mutual cognition) in the ontogeny; this occurs so as to eventually form an integrated organism, which has a distinct biological identity, although constituted by distinct classes of cells (autopoietic units).

Maturana and Varela utilize, to describe these concepts, the illustration given in Figure 6.9. This depiction shows two autopoietic units that recursively interact with each other and simultaneously with the environment, whereby the structural coupling can go in two directions: on the one hand towards inclusion of the boundaries of both units, leading to symbiosis and

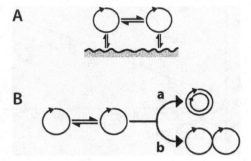

Figure 6.9 **A:** Schematization of two autopoietic units interacting with the environment and with each other. **B:** Structural coupling throughout recurrent interactions may drift into two directions, both indicated in this panel: either (**a**) towards inclusion of the units one into the other (symbiosis); or a case (**b**) in which each unit maintains its own individuality, at the same time establishing a metacellular unit, a higher hierarchic system with novel emergent properties. (Adapted from Maturana and Varela, 1998.)

symbiogenesis, a notion dear to Lynn Margulis (Margulis, 1993). This corresponds to a particular type of emergence, as discussed in the previous chapter – the emergence coming from the melting of two different complex things into a comprehensive complexity with new features.

The other alternative leads to the establishment of a novel structure in which each component maintains its boundary and its individuality, but the two (or more) units also constitute a novel, higher level of organization (also a case of emergent behavior). In fact, in a way, the fly can be seen as an emergence from a multitude of different classes of cells, which positively and recursively interact with each other and recognize each other, building an integrated system.

In their 1998 book, Maturana and Varela are wary about the question of whether a meta-cellular structure, constituted by several autopoietic units, can also be considered as an autopoietic unit. They recognize, however, that such meta-cellular structure is characterized by an operational closure, as its identity is specified by a network of dynamic processes, which are due to the internal rules of the system. In most of the following literature, this caution is considerably attenuated, and it is then licit to consider second order autopoietic systems as autopoietic systems. As already mentioned, the human being has also been in due time considered by Varela as an autopoietic structure (Luisi *et al.*, 1996).

Consider a complex organism, like a mammal (and the same is true for any multicellular organism, be it a mushroom, a bee, or a vascular plant), in terms of autopoiesis. It is probably useful to depict this organism as a stratified ensemble of various self-organized autopoietic subsystems. There are the cells, the tissues, the organs, the entire macroscopic structure, and each of these hierarchic levels is *per sé* an autopoietic system characterized by a network of processes that involves smaller components. Each of these subsystems is linked to each other: the life of a horse is determined by a network of processes involving all organs – heart, lungs, brain, kidneys, etc. – and it is the organization and the mutual relation of all these organs that determines the life of the animal (or mushroom, or plant). It is again life as a systems view.

6.11 Social autopoiesis

We return now to basic autopoiesis, in order to consider what emerged as an interesting, and unexpected, development of this theory. By "unexpected," I mean that the authors of the Santiago School had not foreseen it. This is *social autopoiesis*.

In fact, when the social sciences picked up on this idea they stirred up a great deal of intellectual excitement. The German sociologist N. Luhmann constructed an entire field based on social autopoiesis (Luhmann, 1984), and autopoiesis was also applied to the judicial system, to literature, and in systemic family therapy (see the work by John Mingers, 1992, 1995, and 1997).

Varela remained somewhat skeptical about these extensions of autopoiesis. He says in this regard (Varela, 2000):

These ideas are based, in my opinion, on an abuse of language. In autopoiesis, the notion of boundary has a more or less precise meaning. When, however, the net of processes is transformed into one 'interaction among people', and the cellular membrane is transformed into the limit of a human group, one falls into an abuse. . .

Recently, Maturana has expressed similar thoughts (see Poerksen, 2004).

Luhmann's central point in developing social autopoiesis is to identify the social processes of the autopoietic network as processes of communication (for more detail see Capra and Luisi, 2014); while Mingers, as mentioned above, emphasizes family and judicial aspects, among others. All this is very interesting; I would like to emphasize, however, that the main point of social autopoiesis in general is quite simple.

To understand this, consider that the main feature of autopoiesis is self-maintenance due to a process of self-generation from within. And the point is the following: that although this concept came from the analysis of a living cell, it can be applied to social systems.

Consider, for example, a political party, or a family, or a hospital, or a church, but also an enterprise, like Monsanto or Volkswagen. In each of these systems, there are rules that define and characterize the system itself, which can be seen as a kind of boundary generated by the social structure itself. These are all open systems, in the sense that they have to interact with the social medium around

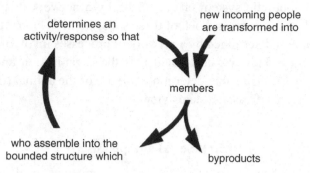

Figure 6.10 Schematic diagram of social autopoiesis. The cyclic logic of autop-oiesis applied to social systems; notice the analogy with Figures 6.4 and 6.5. The transfer from biology to social science assumes that "human relationships" substitute for the chemical interactions among the cell constituents and that the definition rules of the social community substitute for the membrane boundary.

them, and should do so, without losing their own identity. Thus, Monsanto or Volkswagen has to permit the entrance of novel market products in their chain, but at the same time, they must maintain the characteristics of their trademark organization. And in a political party, or in the hospital, there are new entrances and exits, a dynamic equilibrium that does not change the intrinsic inner organization of the political party or of the hospital. The new people who enter are transformed into steady members by the binding rules of the party or of the hospital. There is regeneration from within, and there is the defense of the self-identity; the metaphor of the living cell applies. (See Figure 6.10 and notice the analogy with previous Figures 6.4 and 6.5.)

In all that, certain characteristic features of biology can be recognized, such as the notion of emergence – the family being an emergent property arising from the organization of single individuals, and so it is for the hospital or for the political party.

In the case of the political party, or the hospital, where we have a continuous flux of people entering and exiting, we have a case for homeostasis. When an automobile company would have to substitute fuel-driven cars with electric cars, then we are rather in the presence of an adaptive kind of mechanism.

It is also apparent how the notion of autonomy can be applied to such social systems as we do for biological systems. In fact, they are characterized by their own internal laws, which are valid within their boundaries – and in this sense they are characterized by operational closure.

As I mentioned, Maturana and Varela have reacted lukewarmly to the spread-ing of the concept of social autopoiesis. Despite the authors' doubts, the fact that something has grown out of their ideas and is producing new viewpoints and fields of inquiry is certainly a testimony of the vitality of their original concept.

6.12 Autopoiesis and the chemoton: comparison with the views of Tibor Ganti

One advantage of the idea of autopoiesis is its extreme simplicity. This comes at the cost of a lack of structural and mechanistic details. Ganti's chemoton provides a more detailed and complex view of the internal features of cellular life (Ganti, 1975, 2003). However, as we will see, this more detailed, mechanistic picture does not allow for an expansion into epistemic or social concepts.

Let's now sketch the basic theory of the chemoton in order to draw a comparison with autopoiesis.

Interestingly enough, the papers on autopoiesis and the chemoton both appeared in 1974–1975, and they share a similar history, that both had difficulty in gaining acknowledgment in the mainstream of international biochemistry. In Tibor Ganti's case, one reason was his isolation in Hungary, which at that time did not readily permit contact with the scientific world at large. In addition, diffusion of the chemoton idea was delayed for the same reasons mentioned for autopoiesis, most notably the fact that they are both systemic views where nucleic acids are not presented as the main heroes.

The chemoton scheme is illustrated in Figure 6.11. It consists of three sub-systems: a metabolic autocatalytic network, a bilayer membrane, and a replicable information carrier molecule, or template. Here, as in autopoiesis, the membrane is an important player (compare it with Figure 6.3). However, what happens within the membrane is depicted in more detail. In particular, looking at Figure 6.10, note that the entire system is fed by nutrient X, which first feeds the metabolic network with the various A_i molecules. This is an auto-catalytic chemical cycle, autocatalytic because two A_1 molecules are produced from the original single A_1 molecule.

From this first metabolic cycle, two products are formed, which feed the other two cycles: T^1 is the precursor of the membranogenic molecule T, and the association of several T molecules gives rise to the membrane self-assembly depicted as T_m. This membrane self-assembly can grow and divide sponta-neously. From the A cycle stems the product V^1 as well, which makes a polymer pV_n of n molecules of V^1. The polymer pV_n undergoes template replication; R is a condensation by-product of this replication and it is needed to transform T^1 into T.

At first sight, the system may appear rather cumbersome and complicated; Ganti's claim (Ganti, 1975, 2003) is that this scheme represents the simplest rendering of a minimal living chemical system. In fact, the model is arranged such that the various components interlock with each other: the growth of the membrane occurs only if the template is replicated, and the membranogenic T is formed from T^1 only in the presence of R, which is a by-product of replication.

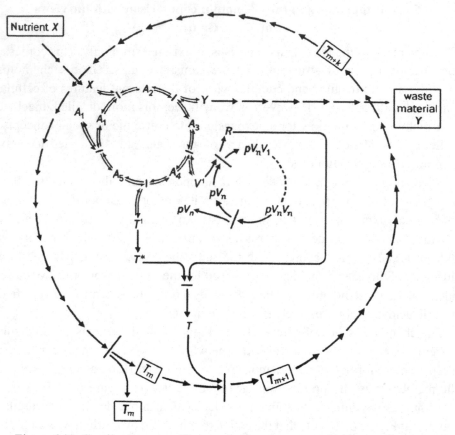

Figure 6.11 Ganti's chemoton. The metabolic subsystem, with intermediates A_1 $\to A_2 \to \ldots \to A_5$, is an autocatalytic chemical cycle, consuming X as nutrient and producing Y as waste material; pV_n is a polymer of n molecules of V^1, which undergoes template replication; R is a condensation by-product of this replication, needed to turn T^1 into T, the membranogenic molecule; the symbol T_m represents a bilayer membrane composed of **m** units made of T molecules (redrawn from Maynard-Smith and Szathmáry, 1995, based on the original by Ganti, 1984).

In other words, there is a coupling between replication and membranogenesis. (See also Maynard-Smith and Szathmáry, 1995).

As already mentioned, this is a more detailed view of the cellular features. On the other hand, the great simplicity of autopoiesis has allowed the implementation in the chemical lab of autopoietic systems; its generality has permitted the rise of social autopoiesis; and the philosophical tissue in which it is embedded has permitted the link with the concepts of emergence, biological autonomy, and referentiality, up to the notion of cognition, and even a bridge to cognitive science. Autopoiesis and the chemoton are two different and brilliant ideas, with different potentialities; and they were born out of the same "Zeitgeist," namely to look at cellular life from the point of view of systems biology.

Concluding remarks

The theory of autopoiesis is based on taking a picture of the actual behavior of a living cell. As such, it is not an abstract theoretical model for life – there are many of these – but a phenomenological analysis of life as it is on Earth. It is, in a way, a picture of the blueprint of cellular life, and it is fascinating to see how many concepts related to the process of life – emergence, homeostasis, biological autonomy, operational closure, open systems, interaction with the environment, cognition, evolutionary drift, etc. – pour forth from this analysis in a coherent way. We will see some of these concepts in the next chapter. In addition, autopoiesis permits the construction of chemical models, as seen in chemical autopoiesis; and it pertains also to the social sciences, with the notion of social autopoiesis. A bridge between biology to the cognitive domain is also made possible. This richness is not present in the chemoton or any other autocatalytic networking.

The main ingredient of this unity is the fact that all is seen "from within," that is, from the logic of the internal organization of the living system. As soon as the autopoietic unit reaches the complexity of biological autonomy, everything that happens within the boundary, as well as the perturbing events from the outside, are interpreted and elaborated in order to maintain the identity of the living.

We have also touched on some of the philosophical implications of these views, and added that the developments of autopoietic thinking have in some cases diverged from the original statements of Maturana and Varela. We will see that particularly in the case of the important notion of cognition, discussed in the next chapter. And we will see then that the notion of cognition permits a bridge between the biology of cellular life and the cognitive sciences. I mention this here just to make the point that autopoiesis is the only available simple theory that is capable of providing a unified view of life from the molecular level up to the level of human perception.

Despite the richness of this envelope, autopoiesis does not have a large impact on mainstream biological science. Why is this so? This question has been partly answered already: autopoiesis originated in a time-window (the early 1970s) when the world of biology was completely dominated by a vision of DNA and RNA as the "holy grail" of life. Alternative views about the mechanism of life didn't have much chance of appearing in mainstream journals. This argument also holds for Ganti's chemoton theory. In the case of autopoiesis, this situation also reflects the intellectual choice of the authors of autopoiesis, and partly perhaps their rigidity, as it would have taken relatively little to make autopoiesis more harmonious with the DNA/RNA world.

There are some signs in the modern life sciences literature to indicate a return to a systems view, which emphasizes collective, integrated properties. In this new – perhaps more philosophically – mature Zeitgeist, autopoiesis could re-emerge as a very useful conceptual framework.

Chapter 7
Cognition

Introduction

As already mentioned, along with the question "What is life," there was another question on Maturana's agenda, namely "What is cognition?" In general, autopoiesis is concerned with organization, and cognition with the "doing" of the organism in its environment. In investigating the relationship between these two questions, Maturana and Varela arrived at the conclusion that the two notions, life and cognition, are indissolubly linked to each other in the sense that one cannot exist without the other. The strong point about the notion of cognition in Maturana's and Varela's work, is that each living organism is considered to be cognitive – including bacteria.

I must say at this point that my view and treatment of cognition in autopoiesis may be seen as departing somehow from the strict original view of the Santiago School. The strict view is clearly given by Maturana in the conversation reported below, where he states:

Cognition is something that an observer says about a system, not a feature defining the system. Therefore, cognition is not a defining condition of a living system, not a defining condition of life.

My departure is due to the consideration that the theory of autopoiesis needs to handle explicitly the relation between the living and the environment; and for that one needs a "third person" phenomenology, without repeating all the time that we are in fact dealing with the view of the observer. We come back to this point more in detail in the next section. For the sake of this introduction, let me now return to a general overview of the subject.

The importance of the interaction between the living and the environment has been known since the early time in biology. One important reference in this regard is Claude Bernard, who worked in the middle of the nineteenth century. He introduced the notion of "milieu interieur" (Bernard, 1865), that is, internal milieu.

The French physiologist, who is accredited with the discovery of glycogen hydrolysis in the liver, is also accredited with the introduction of the notion of homeostasis, meant as resistance to change (although the term was coined later by the American physiologist Walter B. Cannon in the twentieth century). Noting the constancy of chemical composition and physical properties of blood and other fluids, Bernard proposed that living organisms were capable of auto-regulation, namely that they contained substances responsible for the maintenance of the internal chemical equilibrium and the function through which the organism interacts with the environment. This was the discovery of a mechanism of auto-regulation of the living organism, manifested throughout the network of reactions and exchanges with the environment (see also Lazzara, 2001).

From Claude Bernard, we can jump to a modern biology author, the well-known scholar of evolution and genetics Richard Lewontin, who has been working well outside the realm of autopoiesis. Mentioning that the atmosphere we all breathe was not on Earth before living organisms, he wrote (Lewontin, 1993):

... There is no "environment" in some independent and abstract sense. Just as there is no organism without an environment, there is no environment without an organism. Organisms do not experience environments. They create them. They construct their own environments out of the bits and pieces of the physical and biological world, and they do so by their own activities.

And it is interesting to observe that this way of thinking is not only proper among biologists, but it is also present in certain European philosophers such as Maurice Merleau-Ponty. Consider the following statement by the French author (Merleau-Ponty, 1967):

It is the organism itself – according to the proper nature of its receptors, the thresholds of its nerve centres and the movements of the organs – which chooses the stimuli in the physical world to which it will be sensitive. The environment emerges from the world through the actualization or the being of the organism.

And further (p. 63):

A living organism at any moment in its life is the unique consequence of a development history that results from the interaction of and determination by internal and external forces. The external forces, what we usually think as "environment," are themselves partly a consequence of the activities of the organism itself as it produces and consumes the conditions of its own existence.

It is now time to discuss the notion of cognition within autopoiesis. As I said, my view is somewhat different from the original thinking of Maturana, and for that, it is a good exercise to read first the following conversation with Humberto Maturana.

Conversation with Humberto Maturana

Humberto Romesín Maturana studied at the University of Chile, medicine then biology. In 1954, he obtained a scholarship from the Rockefeller Foundation to study anatomy and neurophysiology at University College London. He obtained a Ph.D. in biology from Harvard University in 1958. Later he worked at the University of Chile in Santiago, where he had Francisco Varela as his student, who later became his colleague at the same university. Humberto Maturana collaborated with Heinz von Foester at the Biological Computer Center at the University of Illinois (1968), and later on with Fernando Flores and Ivan Ilich. The first important book, published in Spanish, was *De máquinas y seres vivos: Autopoiesis, La Organización de lo vivo* (1973). The other two important books, coauthored by Francisco Varela, are *Autopoesis and Cognition* (1980) and *The Tree of Knowledge* (1984), translated in very many languages. As of the year 2000, Maturana established his own reflection and research center: the Instituto de Formación Matríztica, together with Ximena Dávila (http://www.matriztica.cl).

PLL: Dear Humberto, let us begin with some historical notes. You told me once that an important starting point for your search on autopoiesis was a question of one of your students at the university.

HUMBERTO MATURANA: Yes, let me tell a bit of my story. In 1960 I was teaching to medical students about the organization of living systems, and one of them asked the following question: "You say that living systems begun on the Earth some 3800 million years ago. What begun then, so that you can say now that living systems begun then?" At least, this was the question that I heard. I could not answer it then. However, pondering about it, the following points came to my mind. "**1.** Whatever begun when living systems begun, must have been conserved since then, and must be observable now. **2.** Whatever begun in the origin of living systems must be some process, such that when it occurred it constituted some discrete entity, such that as I looked at it I would see now a living system; and **3.** I would also see that that entity would be able to give rise directly or indirectly through it operation to all biological phenomena." And I found molecular autopoiesis observing the description of the metabolic cycles. I made my reflections, of course, thinking in terms of molecules, molecular interactions and molecular processes. Ximena Dávila led my attention to realize that, to avoid confusion, we should be saying that living systems are *molecular autopoietic systems*. In fact, some confusion has happened using the notion of autopoiesis as a synonym of self-organization, which it is not.

The question about cognition appeared, when I realized that it was not possible to understand the operation of the nervous system in the process of perception, if one did not accept the fact that we do not distinguish, in the experience that we may be living, between what we call perception and illusion. In our daily living, we call perception any sensory experience that we live as valid, and that we confirm with another experience whose validity we do not doubt. Conversely, we call illusion a sensory experience that we live as valid, and that we invalidate as we compare it with another experience of which we chose not to doubt. As I accepted that, I realized that what we distinguish when we talk of knowing was an operational coherence between interacting systems, which was the result of a history of structural coupling.

All that lead me to realize that, as an observer, I could only speak about what happened to me in the realization of my living, alone or with others, in a medium about which I could only talk by talking about what I did. So, yes, to say something about anything, I need myself, but I must be careful to not confuse domains: I must not confuse the domain in which I speak as I perform an operation of distinction, with what I say that is happening with my operation of distinction.

For example: we find ourselves living or doing whatever we do when we begin to reflect about our doings. Under these circumstances, if we find an animal that we had not met before, and we want to learn about its living, we let it be, we let it show us what it does, and we use what we see it does, to describe its living. We may even say that it knows the world in which it lives. Other example: if we are interested in chemistry and we find some substance unknown to us, and we want to know about it, we let it tell us of its "properties" by making it interact with other substances whose "properties" we know. After doing that, we describe the "properties" of the new substance by referring to what happens to it in its interactions with the other substances that we knew. However, we do not speak of the cognitive domain of the new substance. . . as we might do so when talking about the new animal that we discover.

This is the matter: we are observers of the worlds that we bring about with our doings, and we are at the same time observers of ourselves as we distinguish what we do. . . and we do not want to confuse domains of our understanding with the domains of our doings. I do not like to use metaphors in writing because they are usually not understood, and become misleading. Nonetheless, in our conversation I could say metaphorically that a living system "knows how to live in some particular circumstances as it realizes its molecular autopoiesis in it." However, when I say that a molecular autopoietic system is a living system, I am not speaking metaphorically; I am saying that a molecular autopoietic system exhibits directly, or as a consequence of their interactions in their natural evolutionary drift, all those phenomena that we may observe in living systems now. A computer program that models a molecular autopoietic system in the display of its screen by imitating molecular processes does not produce a molecular system, and the display in the screen is not a molecular autopoietic system, so it is not a living system.

I admit that as I was talking of molecular systems when I first proposed the notion of autopoiesis, I should have talked of molecular autopoietic systems explicitly, as I emphasize since many years now under the suggestion of Ximena Dávila. I repeat, discrete molecular autopoietic systems are living systems, and living systems are discrete molecular autopoietic systems.

PLL: When you presented autopoiesis, originally, you and Francisco wrote that every system displaying autopoiesis should be living. In other words, autopoiesis is the necessary and sufficient condition for life. However, you also wrote a little later that there is no life without cognition. The question then is: shouldn't then cognition be present in the definition of autopoiesis as a condition for life? Why do we see cognition as a different chapter, almost detached from autopoiesis?

HUMBERTO MATURANA: Talking about autopoiesis, we were speaking of molecular systems, and I left implicit the molecularity of the autopoietic system. I should have said, as Ximena and I say now, that: *"a molecular autopoietic system is a living system, and every molecular autopoietic system is a living system."* In the *Tree of Knowledge*, I say that when an observer sees that a system operates in a manner that is adequate to the circumstances of its existence, he "knows," has knowledge of how to operate in the circumstances of its existence. And this is why I say that when a living system lives, it lives because it knows how to operate in the circumstances in which it realizes its living. Cognition is something that an observer says about a system, not a feature defining the system. Therefore, cognition is not a defining condition of a living system, not a defining condition of life. An observer that hears you speaking adequate Italian would say that you know Italian.

PLL: What you say about the observer makes sense at the human level. However, let us take the case of a bacterium, which is cognitive, because with its cognitive tools, matured in evolution, brings forth its own world by interacting with its specific environment. Can we say that the bacterium "knows" about the world and acts accordingly, acting and doing being the same? Even if the bacterium does not "see" anything?

And if it is so, the question arises regarding the link between autopoiesis and cognition. I propose the following argumentation, and I would be glad if you tell me where the mistake is.

A molecular autopoietic system, being an open system, necessarily implies some exchange with the environment, namely some form of metabolism (otherwise it would be a perpetuum mobile). Since metabolism is specific for each organism, it seems to be an expression of cognition. Seen in these terms, a molecular autopoietic system would automatically imply cognition. Isn't this so?

HUMBERTO MATURANA: I could say metaphorically that a living system "knows how to live in some particular circumstances as it realizes its molecular autopoiesis in it. . ." but I would not say "cognition is the capability of a living system to realize its own molecular autopoiesis." To say that would mean: "cognition **is a property** of living systems." Would it be right of me to say that hydrogen and oxygen know how to interact with each other to form water? Metaphorically, I could say so. But could I say that cognition is the capability of hydrogen and oxygen to unite to form water?

I would not choose to say the latter.

PLL: Of course, I would never talk about cognition for inanimate matter, like hydrogen or oxygen. However, for me it is different for the notion of property. I could say to my students that hemoglobin has the property of binding and carrying oxygen. I think this is not a metaphor, but a bit of scientific sound information. Now, you say that all living organisms are cognitive. Why can't I say that cognition is a property of living organisms?

The reason why I try to find a link between autopoiesis and cognition lies mostly in the necessity to find a basis to teach autopoiesis to college students. I would like to link autopoiesis to the biochemical notions they know – like metabolism, homeostasis, and so on. And let me repeat the following piece of my way of teaching all that. As I say to my students: living cells are molecular autopoietic systems; they are thermodynamically open systems; this means that they interact with the medium; they do so thanks to their specific metabolism; metabolism permits then the recognition of the environment, and therefore is the basic element of cognition (it is not cognition, but its "pragmatic arm"). And in this way, I link autopoiesis, cognition, metabolism, and homeostasis in the greater design of biology.

. . . But you would say that all that is not right . . . that cognition is just in the mind of the observer.

Of course, you need an observer to recognize cognition – but you need an observer anyway also to recognize molecular autopoiesis. Therefore, we can talk in third person and try to have a consistent third person view. . .

HUMBERTO MATURANA: We cannot say that cognition refers to something existing independently of what we do when we distinguish it. The same is the case with metabolism and homeostasis, which are names for processes that we distinguish in the realization of the living of living systems. Since I cannot say anything about something that exists independently of what I do as I distinguish it, I prefer to talk by describing what an observer does or should do, when he or she claims that something is the case. Examples. If I say, I know Spanish I show or describe the doings that constitute the doings of knowing Spanish that is talking Spanish. If I say that a given organism is alive, I wish to show the doings that validate that assertion. Therefore, when I say, "a bacterium, or a bird knows how to live," I am saying that it operates in a way in which it realizes it's living as a molecular autopoietic system in the circumstances in which I distinguish it. Indeed this is why Ximena Dávila and I say: "the living of an organism occurs in the ecological organism-niche unity in which it realizes its molecular autopoiesis." And we also say that if the niche in which it realizes its molecular autopoiesis does not arise as it realizes it's living, it would die. In *The Tree of Knowledge* we say: "to live is to know," not that living is the same as knowing, or that knowing is a feature of living. And we do so meaning that if I see it living, I, as an observer, would be inclined to say that it operates in a manner that results in the conservation of its molecular autopoiesis.

PLL: When you write that "a molecular autopoietic system is a living system" – you write something that is particularly relevant for my experimental work: I was left thinking again whether or not my chemical autopoietic systems should be called living. As you probably remember, I have made long ago in my lab micellar systems (a micelle, like a soap micelle, is constituted by a surfactant S); and then these micelles bind and engulf a molecule, which is then hydrolyzed inside, producing from within the boundary the very surfactant S that constitutes the micelle. This process can lead to autocatalytic self-reproduction, or to a homeostatic system, depending upon conditions. I believe this is a molecular autopoietic system, but I hesitate to call a micellar system "living." This brings us again to the question of the origin of life on Earth.

HUMBERTO MATURANA: If you were to say that a micellar system, in general, is a molecular autopoietic system, I would say that it would be inadequate. But if you were to say **that kind of micellar systems** that produce themselves as closed molecular systems that produce themselves as discrete autonomous entities, are *molecular autopoietic* systems, I would say yes, agreeing with you. Furthermore, if you were to show that that kind of micelles undergo systemic reproduction through a fracture that results in two or more micellar systems of that kind, I would say that those molecular autopoietic systems would give rise to a lineage and to the evolution of that lineage of living systems through natural drift. When living systems arose in the Earth so many million years ago, what arose were molecular discrete entities, and if we had been there, we would have seen that they were closed networks of molecular processes, that formed molecular autopoietic systems together with the medium in which they could realize them in an ecological organism-niche unity. If that happens with the micelles that you produce in your laboratory, then what you have done is the creation of the conditions for the spontaneous arising of molecular autopoietic systems, that is, of primordial, or primary living systems. If you were to go around claiming that you have created life, this would be something that I would consider a not-recommendable mistake; but if you say that you have created in your laboratory the conditions for the spontaneous arising of molecular autopoietic systems, ergo of living systems, I would consider it adequate. The next question would be what to do now, if anything. First, I congratulate you, and then I say that I would like to "know" what you did or do. Next, I would like to ask: since natural drift will be occurring from what you have done, what would I like to be conserved in that process?

PLL: Let me go back to the question of cognition, just to make a general point. The point, for me, is that the notion of cognition generally accepted in the literature, even in the secondary wave of autopoiesis literature, views things in a way, which is somewhat different from your original proposition. The metaphor for me, looking at the literature on autopoiesis at large, is that you, with your original thinking, gave a kick to a big rock in the top of a mountain, and the rock began to roll down, but acquiring its own dynamics and features, also independently from the original kick. So it is, for example, for social autopoiesis, which you may not like much, but it is now something existing as an autonomous reality, independent from your original kick.

And so it is partly, in my opinion, for the notion of cognition. The way this notion has crystallized coming down the hill, is the following: cognition has to do with the interaction between the living and the environment, this is a cognitive interaction in the sense that each organism creates its own world, thanks to its specific sensorial tools. It became then a kind of "third person" epistemology. In addition, trying to operate a synthesis about the notion of cognition between the common saying – expressed above – and what you are saying, can I say the following (again in poor words): that cognition is the capability of a living system to realize its own molecular autopoiesis?

HUMBERTO MATURANA: We human beings frequently confuse operational domains in our reflections. What I am saying is that the domain in which a living system exists as a molecular entity is the domain of the realization of its *molecular autopoiesis*. The realization of the molecular autopoiesis results in the constitution of the living system (the molecular autopoietic system as a discrete autonomous entity) in a different domain in which it operates interacting as an organism (as a totality) with other entities. The processes that take place as the organism interacts as a totality occur in the relational domain, not in the anatomy or physiology of the organism. The relational domain, or what happens in it, cannot be reduced to the molecular processes of the operation of the organism (the living system) as a molecular system, although those processes make its operation as a totality possible. Language in us human beings occurs as a relational dynamics in our interactions as organisms, and does not occur as an aspect of the molecular processes that take place in us and make possible our interactions as language beings. Cognition is a reference that we make as observers about what we think happens, as an answer to a question that we make about his or her knowledge in some domain. He or she answers with a behavior that we observe, and if we think that behavior is adequate (adequate in the domain in which we presented our question according to what we think is adequate behavior in that domain) we say: thanks, you know.

PLL: I understand well what you are saying. The role of the observer is central. But this is also the point: we always need an observer, also for the laws of thermodynamics or of gravity. Perhaps then, we could say at the beginning and at the end of a given text, or of a given lecture, that he is always there. Then forget him and present things as our third person observation. . .

HUMBERTO MATURANA: Yes, we forget what we are doing and confuse domains, in a reductionist way. Now let me add something more general. Since we are the present actors of an evolutionary history, whatever appears in our distinctions together with its ecological niche (or domain of existence) has coherences that are not arbitrary. In fact, they correspond to the evolutionary present that we are living at that moment, and that may not be fully apparent for us, but which we may find if we are interested. The beauty of our human living is that, although we cannot claim that we distinguish things as they are in themselves, we know that they belong to the domain of coherences of the realization of our living. If we accept that, we can say, through the coherences of the realization of our living, that we exist as *molecular autopoietic systems;* and that whatever we distinguish, arises as some aspect of what we can observe through what we do in our daily living in the cosmos.

PLL: Let me go to something else. The next question is more on epistemology, and goes back to the discussion you had with Michel Bitbol a couple of years ago in Sardinia. It was about the role of the observer, and you presented a new vision of your classic position – or so we understood – and Bitbol was actually defending your older view. Maybe here we missed something, but people remained a little confused, and it would be nice to have your clarifying answer.

HUMBERTO MATURANA: My difference with Bitbol with respect to the observer was, I think, both epistemological and biological. I was talking about a human observer that was aware that he or she could not sensibly talk about anything that existed independently of what he did. We human beings cannot pretend to talk in a manner that makes sense in our living about anything that does not pertain to the sensory, operational, and relational domain of the realization of our living. In this sense, I felt that Bitbol thought that he could. The observer is fundamental, but as a biological being can only talk about what he or she does as a biological being. This is why I said that any reference to a transcendental reality was an explanatory invention. All this in the understanding that we human beings always follow without trepidation a path that we accept as valid, otherwise we do not do so, or act with doubt.

PLL: Something more personal: Francisco went in his life a good deal towards Buddhism and related eastern philosophical and spiritual views. You also had a lot of rich experience beyond the realm of classic science, but you never had particular interest in these eastern traditions?

HUMBERTO MATURANA: Francisco and I were very different. He always said that it was good to belong to a community defined by a manner of thinking, and in his youth he moved from Catholicism to Marxism and finally to Buddhism. I never did that because I wanted to understand living systems in their living, and to follow that path open to reflect without fundamentalism of any kind. So I have been interested in looking everywhere, and although Buddhism I like very much because of its grounding in human experience, I have remained a biologist without attaching myself to any mystical, metaphysical, or political doctrine. I respect western traditions, but they have not influenced as such my thinking.

PLL: What is your present trend of research – and which direction would you suggest to younger researchers interested in autopoiesis?

HUMBERTO MATURANA: My present research in Matríztica is with Ximena Dávila, and is oriented to understand and answer the question "how is it that we human beings want to avoid the negative consequences of what we do on other human beings, other living beings and the biosphere, and sometimes... many times... we do not do so unconsciously and consciously?" Indeed, Ximena and I created Matríztica in relation to that question as it arose in the work of Ximena in relation to human organizations. In Matríztica we are also concerned with following the consequences of understanding that we are molecular autopoietic systems, and of the biological fact that we do not distinguish in the experience between what we call in our daily living perception and illusion, in relation to the matter of cognition.

We do not know whether what we live now as valid, it will be treated later as a perception or as an illusion. This is not a failure or insufficiency in the operation of our nervous system, but this is how we are. So, our concern is oriented to the question, how to show that we create the worlds that we live by living them, in a way that we become inspired through our understanding of our total ecological responsibility to choose consciously to act accordingly, or to consciously choosing not to do so without being able to claim ignorance? Knowledge is not sufficient to act, what is needed is understanding that we understand, to know that we know, so that we cannot pretend that we do not know without feeling guilty. To act with social ethical responsibility we need *knowledge, understanding, and adequate action at hand*: we call these "the three pillars of spontaneous ethical social responsibility."

What I would suggest to young researchers interested in autopoiesis? Open yourselves to understand molecular autopoiesis as the basic biological happening, because this is the origin of all biological phenomena and is not a theory, not a doctrine. It is an observation of the molecular processes that realize and constitute the living. A molecular autopoietic system is a living system. . . a living system is a molecular autopoietic system. . . living is the happening of molecular autopoiesis. Different kinds of living systems are different manners of realization and of conservation of the realization of different forms of molecular autopoietic systems.

7.1 The notion of cognition

Biologists and philosophers agree in general on the fact, that each living system is a complex of circular interactions with its own environment, and this ensemble can be viewed as a continuous flow of mutual and coherent changes that have the aim of maintaining the equilibrium of self-identity. From this very general background, the notion of cognition as presented in Maturana and Varela's work stands as its own particular, quite original view. Let us cite first the words of Maturana and Varela (1998, p. 29):

. . .our starting point to get an explanation that can be scientifically validated is to characterize cognition as an *effective action*, an action that will enable a living being to continue its existence in a definitive environment as it brings forth its world. Nothing more, nothing less.

In other words, again a quote from the same page of the cited book:

. . . cognition as the phenomenon to be explained, is the effective action of a living being in its environment.

During our conversation, Maturana stated something from which, as I observed before, I may need to depart. This is when he says:

Cognition is something that an observer says about a system, not a feature defining the system. Therefore, cognition is not a defining condition of a living system, not a defining condition of life.

I agree with Maturana that, from an ontic point of view, we should accept the idea that each living organism is characterized by its own operational closure, and that therefore its dominion of "doing" – cognition – is *per sé* something different from the dominion of the observation, which is simply an observation. However, for heuristic reasons mostly, I prefer here to assume instead a "third person" view, namely observing and describing what cognition is for the observer. Thus, for me, cognition becomes a kind of property of the living. In fact, if the "doing" of the organism brings forth a world, which is specific for that organism, this is due to a particular and specific kind of interaction between the organism and its environment. As it is clear from the conversation, Maturana does not like the notion that cognition is a property of the system – for him, it is a kind of abstraction in the mind of the observer. I believe that the two points of view may converge with each other, when there is the necessity to explain things to students or to a broader public. In fact, the two authors also state (Maturana and Varela, 1998, p. 28) that among the four essential conditions to propose a scientific explanation, the first is:

... Describing the phenomenon (or phenomena) to be explained in a way acceptable to a body of observers.

And asking then the question "... how can we tell when we have reached a satisfactory explanation of the phenomenon of knowing?" the answer is (Maturana and Varela, 1998, p. 30):

... when we have set forth a conceptual system that can generate the cognitive phenomenon as a result of the action of a living being.

And this is actually what I do in this field – since my first writing on autopoiesis in the 1990s – recognizing namely that the bringing forth of a world is due to the cognitive interaction of the living with the environment, and that this interaction can be linked to the biochemistry of the living organism (see also my points in the conversation with Humberto Maturana).

Thus, my position is to explain cognition in terms of the interaction between the living and the environment, and then to link cognition to the basic biochemistry of cellular life. As I mentioned already, this "third person" view is not to the complete liking of Humberto Maturana, as is apparent from the conversation.

Therefore, let me go a little further in this relation between cognition and biochemistry.

It seems clear that in the biological cell, the link between the internal and the outside world is metabolism itself. Metabolism is the result of the internal organization of the cellular system, and is implemented by the interaction with the environment that feeds the cell and accepts expelled by-products. The living system and the environment change together in a congruent way, and induce changes and adaptation in each other. In this dependency, "biological autonomy" can again be seen, in the sense that in an autopoietic system all changes and adaptations serve the maintenance of the structural identity.

One important point in the theory of autopoiesis, as we have seen, is this: the interaction with the environment – actually the "perturbation" – in order to be accepted, must be consistent with the internal logic of the living. In other words, the interaction between an autopoietic unit and a given molecule X is not primarily dictated by the properties of the molecule X, but by the way in which this molecule is "seen" by the living organism.

As Varela puts it (Varela, 2000):

There is no particular nutrient value in sugar, except when the bacterium is crossing the sugar gradient and its metabolism utilizes the molecule so as to permit the continuity of its identity.

Actually, the compounds that the living organism extracts from the environment can be seen as something that the organism itself lacks for implementing its life. The interaction with the environment is characterized by what Maturana and Varela call *structural determinism*, namely it is determined by the internal organization of the living. In turn, as already mentioned, such a structural determinism for each particular organism is due to the tools developed in the biological evolution, and in fact we can see the environment and the living as co-evolving.

The appropriation of these missing parts is what gives "meaning" and links the autopoietic unit with its world. I believe that it is at this point that the notion of cognition is logically inserted into the world of autopoiesis. (To be noted: Varela, in his book *El Fenómeno de la Vida* of 2000, recognizes that the choice of the term *cognition* perhaps was not an ideal one, as it has a strong anthropomorphic connotation).

Thus, let me repeat that the statement "each living organism is a cognitive system" may be taken to indicate that the interaction between the living organism and its environment is highly specific, as it depends on the sensorial tools evolution has developed in order to permit the recognition and proper interaction with its environment. In fact, another passage of Maturana and Varela's book (1998, p. 29) indicates, when explaining cognition, also:

... the phylogenetic and ontogenic drift with conservation of adaptation (structural coupling)...

This gives the specificity of each living being in creating its own world, as clearly, the sensorial tools of the fish are different from those of the earthworm, and these are different from those of the butterfly.

7.2 The co-emergence between the living and the environment

The link between autopoiesis and cognition must be completed by the observation that the cognitive interaction brings about a new reality: the environment is created by the living organism – and the living organism is also created by this interaction with the environment – and this through a series of recursive interactions, which in turn have been produced during the mutual co-evolution. In this process of *enaction*, a term proposed by Varela (2000), or co-emergence, as we can say more generally, the organic living structure and the mechanism of cognition are two faces of the same phenomenon of life (Varela, 2000).

The term *create*, as used in the previous paragraph, may sound forced, but it is not. The emphasis and overall concern here is not to define cognition in terms of an input from the external world acting on the perceiver, but rather to explain cognition and perception in terms of the internal structure of the organism. In this view, autopoiesis and cognition are closely linked. The important feature of both is that they represent a general pattern applicable to all levels of life.

Cognition, then, operates at various levels, and as the sophistication of the organisms grow in evolution, so does its "sensorium" (sensorial tools) for the environment, and so does the extent of co-emergence between organisms and environment. Thus we go from unicellular to multi-cellular organisms, where we can have flagella and light – or sugar – sensitive receptors, to the development of sensitive tentacles in the first aquatic organisms, and up to the higher cognitive functions in fish. In all these cases, the organisms contribute to the "creation" of their environments. For example, the onset of photosynthetic organisms may have indeed created a novel oxygen-rich environment. Similarly, the spider web, the woody constructions of the beaver, and the cities constructed by mankind modify the structure of their environments. In all these cases, the environment is created by the organism, and this creation permits the being of the living organism. The illustrations in Figure 7.1 give clear examples

Figure 7.1 The spider web, the termite nest, and the city of humans as mutual co-emergence of living beings with their own specific environments.

of this. In particular, the spider has created its own web, but then it is this web that permits the life of the spider to exist. And this relation of reciprocity is present in the termites nest, as shown in the figure, but also in a bee hive, in an ant nest, in a bird nest, and, quite similarly, in the cities of men. Here too, the city, with its houses, avenues, parks, and with all its structures, as hospitals, schools, stations, and churches, has been created by humans, but then it is this very construction which permits the life of people.

With humans, the sophistication of the sensorial tools acquires a particular level of functional complexity; there are first of all our senses, but there is also the development of the central nervous system and with it, the emergence of mind and eventually of consciousness. However, from the flagella up to the brain, the same basic mechanism is operative: acts of cognition and mutual co-emergence with their own specific environments.

At this point, having recognized that life has no meaning without cognition, we need a way to represent the whole of this complex situation. In a first approximation, this can be done with the trilogy shown in Figure 7.2; this is, again, something which is not present in the primary literature of autopoiesis and cognition.

Here we see life represented not only in terms of the autopoietic unit, but in terms of a trilogy, where the living organic structure (autopoietic unit) interacts cognitively with the environment. As said above, each specific organism has its

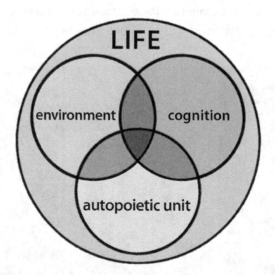

Figure 7.2 The trilogy of life (from Capra and Luisi, 2014). The organic, living structure interacts with the environment via a cognition sensorium, which is a specific product of its development and evolution. It does not make sense to consider each of these three domains as independent of one another. Life is the synergy of the three domains, as the notion of the "embodied mind" implies.

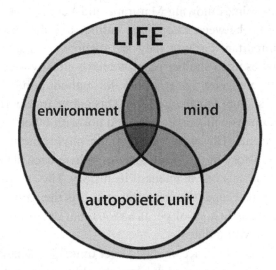

Figure 7.3 The trilogy of the organic structure, environment, and cognition (at the level of mind) interacting with each other to form the process of life.

own cognitive set of sensory apparatus. When one substitutes the term *cognition* with *mind*, one arrives at the notion of the "embodied mind" (see Figure 7.3).

This notion, proposed by Varela already in the 1990s (Varela *et al.*, 1991) is now widely accepted in cognitive science (see also Capra and Luisi, 2014, Chapter 12). It means that it makes no sense to talk about mind in an abstract way. Mind is always present in a bodily structure; and vice versa, in order to have truly living organisms, it must be capable of cognition (the process of knowing). The same holds for human consciousness. Consciousness (at least at the level of sensorial consciousness) is not a transcendent entity but is always manifesting within an organic living structure; the same notion applies as for mind/embodied mind, and embodied consciousness.

In this way, we have described a design that goes from the cell to the realm of consciousness, while remaining in the realm of biology and without using any transcendent or metaphysical aspects. This entire wide spectrum appears as a product of immanence, namely a construction from within. In closing the present section, we should mention that, at the human level, the interactions between organisms and their environments include both interactions among humans (the domain of the social sciences) and interactions with nature (the domain of ecology).

Still, cognition is a notion that applies only to living entities, and not to the inanimate world – the chemical reaction of carbonate with hydrochloric acid is not cognition. We have mentioned above the structural determinism as one basic element for the specificity of the interaction between environment and the living.

We can add at this point, following Maturana and Varela, that among all these cognitive interactions between autopoietic entities and the environment, some are particularly important because they are recursive – that is, they happen repeatedly. This takes us to another point we have already mentioned, the link with evolution and ontogeny. For example, throughout a membrane there is a continuous flux of sodium or calcium ions. This active transport is selective, in the sense that it happens with certain ions and not with others. Where does this specificity come from? The answer lies in the phylogenesis (Maturana and Varela, 1998), the history of the living species, where each state of the system at any point in time is only one moment of its history. The environment does not prescribe or determine changes to the organism; it is the structure of the living system and its previous history of perturbations that determine what reactions the new perturbation will induce.

These coupled interactions, accumulated over time, give a particular historical perspective to the autopoietic system. It becomes a historical product, the result of a long series of coupled interactions. As Varela says (Varela, 1989a, p. 64):

If one may consider the environment of a system as a structurally plastic system, the system and its environment must then be located in the intricate history of their structural transformations, where each one selects the trajectory of the other one.

There are several other aspects related to this notion of cognition. For example, Andreas Weber discusses the relation between cognition and biological constructivism at large, and considers the semiotic aspects of this philosophy of cognition, particularly in relation to meaning (Weber, 2002, and references therein). Regarding the relation between cognition and semiotics (broadly defined as the science of signs and symbols), and from semiotic to "meaning," consider that any perturbation can be seen by an autopoietic system as a "sign" interpreted according to its inner structure.

We will come back to some of these implications later on, talking about epistemology, but for now I would like to quote Weber, as he summarizes very well some of the most important concepts discussed so far (Weber, 2002):

... the Varela's school is emphasizing that the external world acts as a mere "kick", which motivates the system to establish a new equilibrium characterized only by the necessities of self-support. For a bio-semiotic approach this means that it is no longer concerned with the constraints of the mind–body-problem. Dualism becomes obsolete by the material circularity of autopoiesis. In a self-referential system, meaning is the "inner side" of the material aspect of the system's closure.

7.3 The link with classic biochemistry

In this investigation and dicussion about cognition, I have somewhat departed from Maturana's original view, where he says that cognition is not a defining condition for living, but simply what the observer says about the organism's behavior. And I have tried to see cognition as a property of the organism.

Let us now try to connect the notion of cognition, seen in this way, to the more classic concepts of biochemistry. Let us begin by imagining an autopoietic system that satisfies all criteria of autopoiesis, in the sense of being a thermodynamically open system that is self-maintaining due to an internal regeneration of components from within. Once that is said, in particular that autopoiesis implies an open system, it is also said that the autopoietic system must be alimented by the external medium, and there is therefore a coupling between the autopoietic organization and the external medium. In biochemical terms, such a coupling corresponds, or is part of, the cell metabolism. In turn, metabolism is directly linked to the process of homeostasis discussed previously.

This is equivalent to saying that metabolism of some sort is always implied in the very definition of autopoiesis.

It can be further said, by stretching the above statement slightly, that metabolism is the biochemical rendering of cognition at the cellular level. A living entity "knows" how to act because of its internal organization and the corresponding specific interaction with the milieu. In fact, each different cell has a different metabolism, as this is due to the specific series of sensorial tools (coming from evolution), which are different from cell to cell – for example, from the amoeba to the liver cells.

The consequence of this series of statements is that, from a classic biochemistry point of view, autopoiesis and cognition are tightly correlated with each other. There is no autopoiesis without cognition. We should, however, specify which kind of metabolism/cognition we are talking about.

The simplest, and more general, form of metabolism is the one which leads to homeostasis. We have talked about homeostasis in the preceding chapter, illustrating one example of chemical autopoiesis. Homeostasis, as the maintenance of the same properties during the life cycle, is not a simple concept, as one should specify if all properties of the system, or which ones, are kept constant.

We know that several types of cells in our own body – like the mature neural cells, or some of the heart muscle cells – do not self-replicate during our own life. However, in addition to these, it can be said also that most of our body cells, and most of the life we see around us, is characterized by homeostasis, and possibly growth.

In the case of an autopoietic system, homeostasis means maintenance of both structure and internal organization. In the case of the perturbation by a foreign molecule, there will be a different structure, but the same overall autopoietic organization.

We can say at this point that autopoiesis corresponds to metabolism and therefore to cognition, in the form of life as homeostasis.

Here, another question arises. In particular: whether autopoiesis is the necessary and sufficient condition, or only the necessary one, is a question which has been asked by Bourgine and Stewart (2004); in the same year by Bitbol and Luisi (2004); and earlier, in another context, by Weber (2002). We have seen that, according to Maturana and the original thinking of the Santiago School, autopoiesis is equivalent to life, and therefore we would not need further speculations about this point. However, as mentioned repeatedly, current literature is not always in agreement with that.

For example, the equivalence between metabolism and cognition has been suggested (Bitbol and Luisi, 2004), with the addition that autopoiesis is the necessary condition, but not the sufficient condition for life (one should also have cognition). This specification is necessary if we want to hold that cognition also implies adaptation – and in fact, that paper was based on the concepts and terminology originally presented by Piaget (1967). Of course, after the adaptive changes, the living system remains autopoietic – the autopoietic organization, as the necessary and most general condition for life, does not change. The discrimination between homeostatic metabolism and adaptive metabolism was implicit in this paper, but not explicitly illustrated. If we restrict ourselves to homeostatic life, then autopoiesis corresponds automatically to life.

Finally, it should be mentioned that the interaction between organisms and their environment may be part of the more general scenario of ecology. If we accept that living organisms make and continuously change the environment in which they live, and vice versa, we must also accept the idea that the world is constantly changing and cannot exist without changing. To this aim, let us cite Lewontin (1993) again:

> . . . living organisms make and continuously change the environment in which they live, and vice versa, so that every act of consumption is also an act of production; also, that we must forget the idea that there is a constant and fixed world – as we are constantly changing it and cannot live without changing it. From that, the difficulty of finding a healthy equilibrium that preserves as much as possible the identity of the living.

This implies that ecology is also tightly connected to the defense of biodiversity. All this cannot be dealt with in this book; the reader is referred to the recent book by Capra and Luisi (2014), where questions about ecology in our (dangerously) changing world are considered in detail.

7.4 About epistemology in autopoiesis

In order to better qualify the epistemology of cognition, let us repeat the main message of the Santiago School, according to which cognition "cannot be taken as though there were 'facts' or objects out there that we grasp and store in our head" (Maturana and Varela, 1998). The experience of anything out there is validated in a special way by the human structure, which makes possible the thing that arises in the description. This general statement holds true also for cognition seen in the "third person."

In other words, in keeping with the original version of the notion of cognition, knowing is doing, which means that the "bringing forth" of the world, as seen in Section 7.2., takes place by each action we perform. It is the doing of the observer that determines what is going to be seen, and therefore, citing again Maturana and Varela (1998, p. 27): "All doing is knowledge and all knowing is doing." And further (p. 29): "Knowing is effective action that is operating effectively in the domain of existence of living beings."

This goes back to the insistence of Maturana, apparent also in the conversation earlier in this chapter, to discriminate the two levels: the level of operation of the living, and the level of description, which is done by the observer.

In that conversation we can find again a statement that Maturana keeps repeating, "all what is said is said by an observer." Let us see what he means exactly, and for that let us cite him directly from *Autopoiesis and Cognition* (Maturana and Varela, 1980, p. xxii):

... the fundamental cognitive operation that an observer performs is the operation of distinction – by which the observer specifies a unity as an entity distinct from a background [...] Thus, although a distinction performed by an observer is a cognitive distinction and strictly, the unity [...] exists in his cognitive domain as a description [...] the observer specifies a reference that allows him to speak as if the unity [...] existed as a separate entity.

We have mentioned that Maturana and Varela describe the interaction between the living organism and the environment as one which "brings forth" a particular world. Because of the specificity of cognition, the world of the bat must be different from the world of the fish or of the earthworm. This means that there are many different worlds, and while this is generally acceptable and in a way almost trivial when referred to different organisms, it becomes more subtle at the level of the human being. In fact, the human beings differ from each other in terms of cognition, so that, in principle, each of us "sees" a different world.

It must be added here that Maturana's position, by which we cannot have access to any pre-constituted reality, cannot be simply defined as constructivism – at least Maturana would not accept this terminology for his work, preferring the phrase "bringing forth" to the term "construction" (Glaserfeld, 2005, p. 11, cited by Mascolo, 2011). See for all this, the *Biology of Cognition* by Humberto Maturana (1970), and then the well-known *Autopoiesis and Cognition*, written with Francisco Varela, initially published in Spanish as *De Máquinas y Seres Vivos. Una caracterización de la organización biológica*, in April 1973.

Concerning the role of the observer, and his description: this is, in fact, what I have done in his treatment of autopoiesis and cogntion, namely to put myself on the side of the observer who makes a description. Clearly, this is not at the level of the operation of the living autopoietic structure; it is a "metadomain," as Maturana calls it, but it permits me to develop certain concepts, which, although being in a way descriptions, help to clarify and develop certain important points.

It is worthwhile to bring in a citation of Maturana and Varela about the consequences, also at the ethical level, of this type of cognition (1998, p. 245):

The *knowledge of knowledge compels*. It compels us to adopt an attitude of permanent vigilance against the temptation of certainty. It compels us to recognize that certainty is not a proof of truth. It compels us to realize that the world everyone sees is not *the* world but *a* world which we bring forth with others. It compels us to see that the world will be different only if we live differently. [. . .] That is why everything we said in this book, through our knowledge of our knowledge, implies an ethics that we cannot evade, an ethics that has its reference point in the awareness of the biological and social structure of human beings. . .

Then cognition brings also to tolerance, as cognition is also a mind process, with the recognition that the mind of different individuals may be different from each other, due to different background, religion, education, and tradition. There is no objective world outside, one which is the same solid external world for all of us; instead, each of us creates its own world.

For a lucid account of the epistemology of Maturana, see the book by Rossella Mascolo (2011), from which I actually take some of the relevant citations, as well as the book of Letizia Nucara (2014).

At this level, the importance of language emerges again. About "saying," Maturana and Varela (Maturana, in particular) ascribe a great importance to language as our distinctive way to be human, and also as our cognitive instrument. And this goes, more in general, with the observation that all reality, in Maturana and Varela's world, is contextual, since cognition is seen not as the knowing of an objective external world, but as a continuous structural coupling between the living and his medium – operated in general through language.

Considering this obligatory circularity brings to mind the thinking and doing of the famous painter and engraver, Maurits Cornelis Escher, and in fact, in Maturana and Varela's book (1998), this is amply recognized. There are a couple of drawings from Escher, in particular *The Picture Gallery* (p. 243), which I would like to include here, as well (Figure 7.4.), with the comments of Maturana and Varela:

Knowing how we know is rather like the boy in Escher's Picture Gallery. The picture he looks at is gradually and imperceptibly transformed into. . . the city where the gallery and the boy are! We are unable to locate the starting point: Outside? Inside? The city? The boy's mind?

Figure 7.4 Maurits Cornelis Escher, *The Picture Gallery*, 1956. © 2015 The M.C. Escher Company-The Netherlands. All rights reserved.

We cannot dwell here on this point; I would rather go back to the beginning of the history of cognition, as this permits me to introduce a further epistemic concept, related to the nervous system, which is very important for today's philosophy of science and cognitive science in particular.

The beginning of this manner of thinking about cognition can be found in the old work with the frog's retina by Maturana, who based some of his earlier, important observations on the previous researches by R. W. Sperry (1945). Sperry took a tadpole (frog larva), cut the edge of the eye, and rotated it by 180 degrees. With this adult animal, he made his famous experiments: he covered the rotated eye, and showed a worm to the frog – which made a perfect hit with its tongue. When he repeated the experiment covering the good eye, the frog shot out its tongue with a deviation by exactly 180 degrees. Citing the authors (Maturana and Varela, 1998, p. 125):

... the frog shoots out its tongue as if the retinal zone, where the image of the prey is formed, were in its normal position. This experiment reveals in a very dramatic way that, for the animal, there is no such a thing as up and down, front and back, in reference to an outside world, as it exists for the observer doing the experiment. There is only an internal correlation between the place where the retina receives a given perturbation and the muscular contractions that move the tongue...

From this, the important – and rather dramatic – conclusion, that behavior arises because of the nervous system's internal relations of activity. The nervous system can therefore be considered as an operationally closed system – it works with operational closure.

We have introduced the notion of operational closure, quite in general, in the previous chapter, talking about autopoiesis. And this is true, according to Maturana, also at the level of the nervous system. But the nervous system as part of an organism operates with structural determination, and therefore, the structure of the environment cannot specify its changes, but can only trigger them – as we have already discussed with the above citation by Weber. It is a trap, continue the authors (Maturana and Varela, 1998, p. 133), to assume that the nervous system operates with representations of the world – and they add that there is a trap in assuming that the nervous system functions as a case of solipsism – the view that only the interior exists. The solution proposed by Maturana and Varela is based on the role of the observer to cognitively correlate these two extreme positions.

I would also like to cite Rossella Mascolo (2011, translation from Italian is mine) who wrote a lucid account of the concept of Maturana's cognition:

... to recognize the observer as constitutive for the observed, yields an ontological unity which permits to explain all phenomena associated with life, including the very human experience, posing them in a relational space. In fact, for Maturana, we human beings exist in a space which is constituted by the relation of those elements, structurally determined, which we are. In other words, human condition is expressed in the relation to the others and to the world in which we live. Also culture is a closed network of conversations, which is actualized on an emotional basis, so that each particular way of life is defined in each case by a particular configuration of individual emotions; and the course human history follows is the course of emotions, which may cause cultural changes, like networks of conversations that a given community nourishes by its own living.

The refusal of an objective reality out there brings, as a consequence, to emphasize and offer subjectivity – the personal experience – into the realm of science. And, of course, this step is not an easy one. There is here the whole question of the "first person" experience, and whether and to what extent this may be part of science. Particularly Varela, in his last years, has dedicated intense efforts to this question and he introduced the concepts of neuropheno-menology, based on the "first person science," in which observers examine their own conscious experience using scientifically verifiable methods. (See also, in this regard, the conversations with Evan Thompson and Amy Cohen Varela, both in the previous chapter). In turn, this is linked to the even greater problem, that of consciousness, a question to which Varela also worked experimentally with the tools of neurobiology.

In this interest on consciousness one can see also the particular link that Francisco had with Buddhism. I mention this because the life and thoughts of Francisco Varela were indeed significantly influenced by Buddhism. *The Embodied Mind* (Varela *et al.*, 1991) is in fact a book, which, among other things, proposes an ambitious synthesis between Buddhism and cognitive science.

The last years of Francisco's activity were almost completely dedicated to neurobiology, and in this field, actually, his name began to be accepted also by mainstream science, as he was asked to write a review for *Nature* (Varela *et al.*, 2001).

Francisco Varela (Figure 7.5) died in May 2001. To better understand the following figure, you may refer – again – to my conversation with his wife, Amy Cohen Varela, in the previous chapter.

Figure 7.5 Francisco Varela speaking in the 1985 Cortona-week (the first one).

7.5 Ontogeny, evolution, information: the view from within

We have seen at the start of this chapter that the theory of autopoiesis, via the correlated notion of cognition, links the biological domain with the cognitive one – and we have, in fact, seen that life can be considered as a trilogy which, at the level of human being, encompasses mind and eventually consciousness (restricting us, for the time being, to sensorial consciousness). Mind is a process of the cognitive domain (see Capra and Luisi, 2014), and sensorial consciousness is related to cognition, in the sense that cognition is the general background on which the particular process of consciousness may develop. The relationship between brain-mind-cognition-consciousness is a complex one, and not completely clarified in the literature, also because of the multiplicity of the meanings of consciousness. This is discussed at some length in the recent book by Capra and Luisi (2014), and it would be out of place to review this here. Rather, I dwell here on a couple of "minor" aspects, which are also related to cognition.

In terms of autopoiesis, how does a cellular structure deal with ontogeny, its development in morphogenesis? The changes in the form during development after birth are due to the internal program, and are therefore consistent with the organization from within. Likewise, the ontogeny of a multicellular system is determined by the domains of interactions specified by the total unity, and not by the individual cell components.

And... what about evolution? Although evolution is not emphasized in the basic definition of autopoiesis, for the same reason DNA mechanism is not, it is indeed an important part of the thinking of Maturana and Varela, who in fact devote significant space to this concept (Maturana and Varela, 1998).

Evolution, almost paradoxically, is seen as a consequence of self-maintenance: the unit tends to maintain its own identity and has no urge to change. When there is the need for adaptation, the cell adjusts with minimal changes, so as to disturb its own identity as little as possible. Adaptation is a consequence of the interaction with the environment, and it happens with the same rules as cognition, namely responding from within. If this is not possible, if the perturbation is too large, there may be the disintegration of the autopoietic structure – meaning death (see the next section). By way of this mechanism of adaptation, the autopoietic unit is in constant dialogue with the environment during its life pattern.

Since the dynamic of the environment may be erratic, the result in terms of evolution is a natural drift, determined primarily by the inner coherence and autonomy of the living organism. In this sense, Maturana and Varela's view is close to Kimura's theory of natural drift (1983) and to Jacob's notion of "bricolage" (1982). Evolution does not pursue any particular aim – it simply drifts. The path it chooses is not, however, completely random, but is one of many that are in harmony with the inner structure of the autopoietic unit: here comes the great importance of the notion of structural determinism; and, of course, in keeping with thermodynamics and the natural laws.

Adaptation, being generally due to a stimulus from the environment, corresponds to a kind of information. This consideration warrants devoting a couple of words, more in general, on the notion of information as seen from the edge of autopoiesis. There is much written about biological information, and most of us perhaps agree that this is often too much, and confusing and/or un-useful to say the least. Particularly strange and confusing, in my opinion, is the notion of information as an entity having an intrinsic objectivity, a thing in itself. I like to take the stand, which appears obvious to me, that information is always relational – relative to the capability of cognition from the receiver. A drug *per sé* contains no information; it does only for a specific cell that is capable to recognize it. A male pheromone is information for a specific butterfly receptor, yet is no information at all for a canary. And a software program is information for a computer capable of recognizing it, and it is not recognizable for a fax machine. Oxygen is an information signal for hemoglobin, and not for keratin. Likewise, DNA is information when recognized by a living cell capable of transcribing it – it is not information when it is added to a synthetic polymer.

In terms of autopoiesis, a chemical or an adaptation signal is information only if it resonates with the inner structure of the system, namely if the autopoietic system recognizes it as belonging to its own vocabulary. Then it can trigger an interaction, and may elicit a response.

Thus, although the interaction with the environment may be seen in different modes and by way of different semantics – information, cognition, adaptation, and so on – the unifying concept, as deriving from autopoiesis, is the coherence with the internal organization of the autopoietic system, or the view from within.

In this section, we have touched on evolution and adaptation, concepts which connect with a very important pillar of the life science, Darwinism. We cannot dwell on this complex subject in the narrow frame of this book. Yet, it is important to note that Darwinism today finds considerable objections and update requests. An idea of this is in the conversation with Denis Noble, the author of *The Music of Life* (2006), which you can read below.

Conversation with Denis Noble

Denis Noble is Emeritus Professor of Cardiovascular Physiology and Co-Director of Computational Physiology at Balliol College at Oxford University. Noble is the author of *The Music of Life* and 10 other books as well as 500 scientific papers. Noble has received several honorary PhDs and numerous other awards, including the British Cardiovasculal Society's Mackenzie Prize and the Russian Academy of Sciences' Pavlov Medal. He is a fellow of the Royal Society and an honorary member of both the American and Japanese Physiological Society. He is particularly famous for his studies on proteins and cells that generate the rhythm of the heart.

PLL: There seems to be an anti-Darwinian wind in the life science of today. Can you briefly summarize what is going on?

DENIS NOBLE: I would call it anti-neo-Darwinian rather than anti-Darwinian. Darwin was rather more inclusive than the neo-Darwinists. The breaks with neo-Darwinism are:

  i Mutations are not always random.
 ii Proteins did not always evolve by gradual accumulation of point mutations.
iii There have been successive episodes of what Shapiro calls Natural Genetic Engineering.
 iv There are mechanisms by which acquired characteristics can be inherited, including transmission of RNAs, DNA marking, by-passing of the germ line through behaviour, Baldwin Effect. . .
  v Symbiogenesis has played a major role in some stages of evolution, particularly the transition from prokaryotes to eukaryotes.

Darwin actually praised Lamarck (in preface to the 4th edition of *Origin of Species*) as a great naturalist who championed the transformation of species. He also acknowledged the existence of inheritance of acquired characteristics, including a theory of gemmules as a possible mechanism.

PLL: It seems to me that the objections are not so much against the basic pillars of Darwinism, but rather on the notion of gene and genome as used and interpreted by most of the standard literature of Modern Synthesis.

DENIS NOBLE: Yes, that is correct. It is the MS (neo-Darwinism) that is in question. I think the concept of a gene has changed. It was originally defined in terms of the phenotopye (Johanssen's definition) and was therefore necessarily the cause of that phenotype. Defined as a DNA sequence, the status is very different. Most gene knockouts are buffered by the organism, so KOs do not reveal regulators.

PLL: I appreciate very much, of course, the dimension of systemic view given to this modern conception of evolution, according to which the latter is shaped by many mutually interacting genetic streams, including the environment at various levels. Here I see clearly the importance of physiology, because these interactions are mostly due to the functions which are at the end of the chain in each case. But: there is more about physiology than this?

Denis Noble: Yes, physiology moves back onto centre stage as a contributor to the causes of variation. I hope this gets the dialogue going.

7.6 What is death?

We have begun this book with the question: "What is life?" As we near the end of this chapter, it is proper to consider the question, "What is death?" – still from the edge of autopoiesis and cognition.

The starting edge can be seen in the general statement, made repeatedly by Maturana (see also the conversation with him in this chapter) that the domain in which a living system exists as a molecular entity is the domain of the realization of its molecular autopoiesis. The bacterium just does what it does in its molecular dynamic, and either conserves its molecular autopoiesis and lives, or it does not and dies.

Thus, death is the loss of the molecular autopoiesis – and on this we all agree. If there is no molecular autopoiesis, if and when autopoiesis disappears, life also disappears.

But let's take a more general view, including cognition, and for that, let's go back to Figure 7.2, in which life was depicted in terms of a trilogy: autopoietic unit, cognition, and environment. This view, as I mentioned already, is not present in the original literature on autopoiesis, and was presented as a way to connect autopoiesis and cognition to the external medium.

We said there that the specific interaction of the living with its specific medium is a prerequisite for life. In fact, if we eliminate the interactions/ perturbations coming from the outside, the living is no longer acting as a thermodynamically open system, the metabolism does not exist anymore, and the system then does not receive anymore nutrients and energy. And then, based on the Figure 7.2, a system without cognition cannot be considered living.

This is a kind of trivial point if applied to lower animals and to plants – no acting for acquiring and metabolizing nutrients, no life. It becomes less trivial when applied to humans who have lost their sensorial tools to interact with the

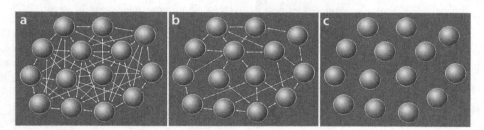

Figure 7.6 The life seen as an integrated system (a), whereas progressive frag-
mentation corresponds to decay (b), and death (c). The mutually interacting parts
give rise to self-maintenance by regeneration of the components, which is the
basis of the autopoietic view of life.

environment: for example, people in a coma, when their physical senses, and
their mind, are no longer active. The mechanism of molecular autopoiesis of
regeneration from within these people may be kept alive by specific drugs,
oxygen, and forced feeding, but the "doing" in terms of relations with the
environment and with other people is no longer there.

Here one opens up a complex discourse, about bioethics, about whether and to
what extent one should maintain the plug-in; and it is still an ongoing discussion
due to different religious, ethical, and social points of view. About the various
aspects of "death," I refer to a book by Father Alfonso Aguilar (2009), where
several of these aspects are taken into consideration.

Here, let us instead consider the question of death from the perspective of a
systems view, using for this purpose simple-minded illustrations (see
Figure 7.6).

Figure 7.6, *panel (a)*, represents a living system, where all parts are interacting with each other. This can be the metabolism of a cell, where all chemical
components are linked to each other by the complex biochemistry of metabolism; but the circles in the figure can also represent the various organs of our
body – heart, lungs, brain, kidneys, stomach, and so on – which in a living
system are all connected to each other, in the sense that none of these components can "act" independently from the others. In fact, in this example as well as
in the case of the cell metabolism, each component depends on and determines
the state of the neighbouring components, so that life is indeed an integrated
network of linked interactions, as we have already expressed. So, the question of
where is life localized in a bacterium, or where is life localized in the human
body, have the same answer: there is no localization; life is the entire network of
interactions. Life is the integrated system.

Moreover, the progressive elimination of these links and interactions brings
about a fragmented situation, as that depicted in *panel (b)* of Figure 7.6: here the

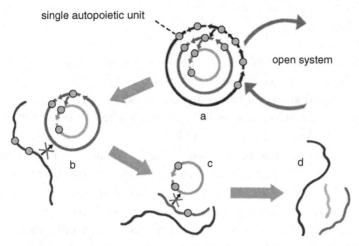

single autopoietic unit

open system

a

b

c

d

Figure 7.7 A schematic illustration of death of a higher order autopoietic structure. It is a process of disintegration of the self-organization, with disappearance
of the emergent properties of the ensemble. In the figure, is indicated a complex
structure formed initially (**a**) by three levels of organization, each containing
more autopoietic units, each one feeding the units at lower level. In this simplified
description, the first level corresponds to the entire animal. The second one
corresponds to the organs, and third one to the single cells. Each step of this
process is indicated by the large grey arrows: firstly, the outer level is disassembled (**b**), and cannot feed anymore the lower level – this lack is indicated by
the crossed arrow. Then, progressively, the second level (**c**) and the third one
(**d**) are in turn disassembled. This process is the opposite of the progression of
emergence – a kind of "negative emergence" (see Damiano and Luisi, in Aguilar,
2009).

network is less organized and the organs of the body do not "talk" anymore with
each other. When the process goes on and there are no more interactions between elements, then there is death, as represented in *panel (c)*.

The same consideration can be applied to a social system (see also
Section 6.11 on social autopoiesis): for example, a society, or a political party,
can be described as *alive* when all parts and components interact positively with
each other, when there are conversation relations, and the structure dies when the
parts do not communicate anymore with each other.

However, we have to take into account that, for example, in our body, after the
severing of the network links, a single organ can remain alive for a certain time –
alive in the autopoietic sense, namely a system which regenerates its components from within; and this is so from a strictly biochemical viewpoint, as long as
there are nutrients and chemical energy inside the system. This means that the
cells inside the organs can remain alive beyond the death of the whole organism.
Then, the single cells also begin to decay and die. Death of the organs may come

after the death of the horse; and the single cells of the tissues and organs may be the last to die.

Thus, the process of death of a complex living organism can be seen as a succession of breaking down of levels of organization, as naively schematised in Figure 7.7. Here again, death is seen, just as life, as a process of self-organization and mutual cognition among parts – in an inverse, negative sense.

Each one of these subsystems is linked to each other (following the examples of hierarchic levels as already traced in Figure 6.9).

There is, of course, also the question: "why do we have to die?" In fact, if we consider a colony of bacteria, each cell divides at a certain point in two, and there is no "corpse" attending the formation of the two daughter cells; and the same can be said for stem cells and gametes as well as for hydra (see also Part III). Can these systems be said to be biologically immortal? The fact that there is no dead corpse left behind can indeed be taken as a criterion for lack of death and hence immortality. On the other hand, one might say that the individuality of the parent organism disappeared – and that is indeed a criterion for death.

In terms of autopoiesis, one can refer again to Figure 6.3: an autopoietic structure would be one in which the cell can indefinitely display homeostatic self-maintenance, as $v_p = v_d$, and it continues to do so – as long as nutrients are provided.

In the case of an active, proliferating autopoietic structure $v_p > v_d$, then the cell would grow, accumulating materials, and eventually divide. Also in this case, it would continue to do so – as long as nutrients are provided. Moreover, if extremized, this would be the case of an immortal autopoietic structure.

Cancer cells are considered biologically immortal, in the sense that they can proliferate forever. The case of HeLa cells is well known, the cells of a patient who died by cancer in 1951: her tumor cells are still proliferating[1] more than 60 years afterwards (Sharrer, 2006; Masters, 2002).

Concluding remarks

We have seen in this chapter the close connection between autopoiesis and cognition and we have made the point that, quite in general, autopoiesis and cognition are strictly related to each other, and when there is autopoiesis there is also cognition, at least in the form of homeostasis metabolism.

We have mentioned the notion of embodied mind – that mind is not a transcendent property coming from some higher place, but is instead an immanent process that comes from within.

---

[1] Paul Cantalupo (University of Pittsburgh) and his coworkers described how HeLa sequences have muscled their way into the Cancer Genome Atlas (*J Virol*, 89: 4051–57, 2015).

The picture is consistent throughout from the very beginning: once we accept that life comes from inanimate matter, and we accept the Darwinian view, according to which the higher forms of life come from the evolution of simpler ones, then all complexity comes from immanence – also, from within. We have mentioned the notion of embodied mind, and, in one step further, we can say the same about sensorial consciousness – that this kind of consciousness is not a transcendent property coming from some higher place, but is instead an immanent property that comes from within. Note, however, that I am talking here about sensorial consciousness, the so-called "easy problem" of consciousness (Chalmers, 1995); the notion of consciousness as subjective experience (the sense of being, the knowing that I know, my feeling of red, and so on, which is the so-called "hard problem" of consciousness), may be altogether another, more complex, business (Chalmers, 1995; Bitbol and Luisi, 2011; Thompson, 2014).

This emergence of mind and sensorial consciousness are not, however, a mechanical outlet of the brain complexity (the more traditional view), but the result of the co-emergence (enacting) discussed above.

If mind comes *from within*, then, of course, the same can be said about all "by-products," such as ethics, spirituality, and the very notion of God. They all would be, in this perspective, self-generated values. Hardly a novel idea, but what is new in this case is that such a derivation of the very spiritual values comes straight from the biology of autopoiesis. We arrive at a form of "lay spirituality" or "natural spirituality." Thus, spirituality is not based on a mystical delivery from transcendent levels, but comes from the immanence of physical life: the embodiment of mind and consciousness with organic structure in the very process of life – a scenario which also eliminates the Cartesian conflict between mind and body.

It is now necessary to ascend the ramp of complexity, and see the various levels of biological structure, and how the increase in size and hierarchic order are attended by the arising of functionality: the important keywords are then self-organization and emergence. To this, Part III of the book is dedicated.

Questions for the reader

1. Can you think of any living system that does not comply with the criteria of autopoiesis?
2. Let us consider the following two situations: First, a synthetic vesicle that absorbs a particular molecule from the medium, and by so doing, reproduces itself via an autocatalytic process – all this is clearly an autopoietic system. Then, second case, a bacterium that "recognizes" and absorbs sugar from the environment – a case of elementary cognition. Would you consider, using the

criteria of autopoiesis, both systems as "cognitive"? Or would you put a qualitative difference between the two? And which one?

3. A question, and also a research proposal. Consider Figure 6.3, the fundamental discourse in chemical autopoiesis – and the corresponding examples of chemical autopoiesis given in Chapter 6. All these are based on reactions taking place at the boundary of the compartment, micellar or vesicular. Can you conceive a chemical autopoietic unit in which we start from a pre-existing internal metabolism?

Part III

Order and organization in biological systems

Chapter 8

Self-organization

Introduction

At the beginning of Part I, we mentioned Oparin's bold idea that the transition to life would be based on a gradual and spontaneous increase of molecular complexity – and we added that indeed, there are quite a few processes of self-organization in nature that bring about an increase of molecular complexity in perfect agreement with thermodynamics. This is what we will see in this third part of the book, devoted to order and organization in biological systems.

Self-organization has to do with the ordered increase of structural complexity – but obviously increasing the size alone would not be enough to proceed towards life. Another property must be considered when smaller parts assemble to constitute a larger entity: emergence – namely, the arising of novel properties. Novel, in the sense that they are not present in the constituting parts. Although self-organization and emergence go hand in hand, for heuristic reasons they will be discussed in this book as separate chapters. We will see also that the notions of self-organization and emergence do not refer only to static, equilibrium systems, but are also present in dynamic systems far from equilibrium – a quite particular and very important aspect of life. A particular combination of self-organization and emergence gives rise to self-reproduction, which will also be discussed in Part III.

After the general comments, let us start with self-organization processes, considering first the terminology. The terms *self-assembly* and *self-organization* are often used synonymously, depending on the author. For example, Whitesides uses the term *self-assembly* (Whitesides and Boncheva, 2002; Whitesides and Grzybowski, 2002). I consider this term too general, and I will use instead the term *self-organization*.

Some of the self-organization processes take place with a negative free energy change without any hindrance from kinetic activation energy barriers – and as such they can be considered "spontaneous." The attribution of spontaneity to processes that occur because of the concomitance of a negative free energy change and low activation energy barriers may be controversial for some: in fact, in some classic physicochemical textbooks, the term *spontaneous* characterizes a reaction that has a negative free energy change regardless of the kinetic constraints – i.e., regardless of whether it happens – a definition that I will ignore in this book. In particular, I will refer to these spontaneous processes as being "under thermodynamic control" – knowing also that not everybody may agree with this terminology[1].

All biology is characterized by self-organization, from protein folding, protein-protein interactions, the duplex of nucleic acids, protein-nucleic acid complexes, the assembly of actin and the formation of microtubules or the assembly of tobacco mosaic virus, as well as the assembly of ribosomes, of a cell and organs. Among the many books devoted to this field, see, for example, those by Riste and Sherrington (1996); Birdi (1999); Westhof and Hardy (2004) and also an editorial by Glotzer (2004).

However, there is from the onset another important clarification to be made, and this concerns the difference between static and dynamic aspects of self-organization. Actually, for many physicists, the "out-of equilibrium" situation, which we will see in the following pages, is a most important aspect of self-organization.

There is a vast amount of literature on self-assembly and/or self-organization, as this notion is used in practically all fields of science – including social sciences – and at all scales. [See at this regard the clear illustrations presented in the reviews by the already mentioned Whitesides and collaborators (Whitesides and Grzybowski, 2002, Whitesides and Boncheva, 2002), or by J.-M. Lehn (2002); and then the basic book by Kauffman (1993).] Other examples range from classic organic chemistry to polymer chemistry (Lindsey, 1991; Pope and Muller, 1991; Lawrence *et al.*, 1995; Zeng and Zimmermann, 1997), to the new frontiers of nano-technology, nano-robotics (Whitesides *et al.*, 1991; Bissel *et al.*, 1994; Whitesides and Boncheva, 2002), surface chemistry, and electrochemistry (Miller *et al.*, 1991; Dubois and Nuzzo, 1992; Ulman, 1996; Decher, 1997). There is much literature on self-organization in the social and economic domains (self-organizing communities, self-organizing financial markets), in the study of semiotics and language, and in brain research. It is not the aim of this chapter to discuss all of this literature, but to give a few specific examples and references relative to classic biological systems.

---

[1] For example, Albert Eschenmoser, as a fine tuning to the above, states that "The criterion for stating that a chemical reaction proceeds under *thermodynamic control* is **not** whether under given conditions the more (or most) stable of the possible products is formed, but rather whether the chemical equilibrium between product(s) and starting material(s) remains *established* under the conditions of product formation, such that alternative products are in (established) equilibrium with each other. [. . .] Thermodynamic control **must** lead to the more stable prooduct; kinetic control **may** lead to it" (Eschenmoser, 2003).

8.1 About the term *self*

Self, in the connotation of this chapter and in the field of life science in general, defines a process that is endogenous, namely dictated by the "internal rules" of the system. The term can be applied to the case of kinetic control as well, provided that the agents responsible for the kinetic control are considered as part of the system's "internal rules." Conversely, when the structure is organized by external, imposing forces, we do not have self-organization. All this sounds rather trivial, however; in the literature there are several cases where the term *self-assembly* is misused (for example, where the assembly is actually man-made, as in the example of layered nano-structures organized by a series of external manual operations).

The assembly of a TV set, or the page numbering of a manuscript, are certainly not self-organization processes, as they are imposed from the outside. Protein folding, subunit interactions in oligomeric proteins, DNA duplex formation, and micellization are all examples of self-organization under thermodynamic control. Conversely, the examples of more complex biological systems (e.g., the assembly of a virus, of a bee hive, and all other social-insect constructions), being the result of genomic and enzymatic activity, should be considered self-organization processes that are mostly under kinetic control.

The application of thermodynamic and kinetic concepts becomes difficult in the macroscopic and social domains; however, the criterion of whether the organization is the fruit of the internal rules of the system is generally always applicable. For example, the organization of a political party, or of a company, or the patterns of the swarm intelligence, can be considered as the outcome of the system's internal rules. We have seen the same concept, in a different terminology, when talking about social autopoiesis in the previous part of this book.

8.2 Self-organization of simpler molecular systems

Micelle formation is a nice example of self-organization under thermodynamic control. Following the addition of some soap to water at a concentration higher than the *cmc* (critical micelle concentration), spherical micellar aggregates spontaneously form. This process takes place with a negative free-energy change – and the process is attended by an overall increase of entropy.

This is due to the fact that amphiphilic molecules tend to aggregate in order to decrease unfavorable contacts with water molecules – thus, upon aggregation, water molecules are set "free" and this brings about an increase of entropy. This is qualitatively illustrated in Figure 8.1.

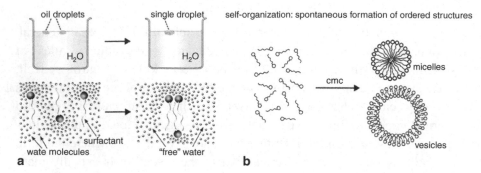

Figure 8.1 (**a**) Two droplets of oil in water tend to fuse in a single droplet, and the decrease of the total surface makes more "free" water molecules. In the same way, the aggregation of surfactant molecules brings about an increase of entropy due to the "liberation" of water molecules. (**b**) Hydrophobic forces are the main factors leading to the association of surfactant molecules and their self-organization in ordered structures, like the assembly of micelles and vesicles (*cmc* is the critical micellar concentration).

The example illustrates beautifully, in its simplicity, that formation of *local order* and *increase* of the overall entropy can take place simultaneously. The organized, ordered structures are thus a kind of by-product of the overall increase of entropy ("disorder").

Of course, self-assembly of this kind occurs in water, and not, say, in ethanol. Any self-organization process must be defined in a given set of initial conditions. Initial conditions, as always in thermodynamics, determine the outcome of the process, and in particular whether the process is also under thermodynamic control or not. Aside from that, it is well known that amphiphilic molecules tend to aggregate in a large series of structures, as shown in Figure 8.2.

Part IV of the book will consider in detail the physical and chemical properties of vesicles and liposomes; here it is important to anticipate that the formation of such aggregates can be attended by an additional type of ordering process: compartmentation and segregation of guest molecules. This is qualitatively illustrated in Figure 8.3: during formation of vesicles, depending upon their chemical nature, solute molecules can be sequestered in the bilayer, or in the aqueous interior, or on the polar bilayer surface.

Crystallization is another example of self-organization. The form of the crystal lattice, say of NaCl, is not imposed by external forces, but is the result of the internal structural parameters of the NaCl system under the given conditions (temperature, etc.). Protein folding in a given solvent is also a self-organization process, determined by the internal rules of the system (the primary structure). As is well known, Chris Anfinsen and coworkers demonstrated in the

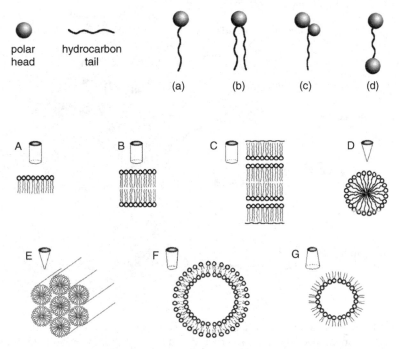

Figure 8.2 *(Upper panel)* Diagrammatic representation of detergent molecules:
(**a**) Single tailed; (**b**) double tailed; (**c**) zwitterionic; (**d**) bolamphiphilic. *(Lower
panel)* Different types of surfactant aggregates in solution: (**A**) monolayer; (**B**)
bilayer; (**C**) liquid-crystallin phase lamellar; (**D**) normal micelles; (**E**) cylindrical
micelles (hexagonal); (**F**) vesicles (liposomes); (**G**) reverse micelles.

1960s that protein folding is mostly a process under thermodynamic control
(Sela *et al.*, 1957; Anfinsen and Haber, 1961; Anfinsen *et al.*, 1961).

The criterion resides in the reversibility of folding in vitro: the protein is
put in a glass vial, far removed form the magic of the cell, denatured with
urea or another mild reagent, and the denaturing agent is then removed by
dialysis – to see whether the original native conformation is regained. If it
is, the process of folding is under thermodynamic control, since only
thermodynamic factors can be present in the *in vitro* system (see
Figure 8.4).

A lot of proteins also associate in complex self-organized structures – such as
the assembly of hemoglobins. (Adult mammalian hemoglobin, coming from
self-organization of two α and two β-dimers, will be shown later on, in
Figure 9.1, B.) Other complex protein structures, again the result of astounding
self-assembly processes, are illustrated in Figure 8.5.

Not only proteins and their oligomers, but also nucleic acids give beautiful
examples of self-organization – think of the formation of the DNA duplex,

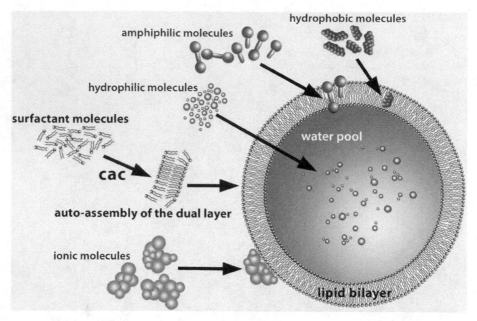

Figure 8.3 Surfactant molecules, once the *critical aggregate concentration* (**cac**) is reached, spontaneously form an ordered structure, in this case a vesicle. Self-assembly of a vesicle (look also at Figure 4.3) can be attended by an additional ordering process: compartmentation and segregation of molecules. Water-soluble molecules can be entrapped inside the water pool of the vesicle during its assembly. Other molecules, depending upon their chemical nature, can be sequestered: ionic molecules on the polar head groups of the bilayer surface; hydrophobic molecules inside the lipid bilayer; amphiphilic molecules in the hydrophobic bilayer (polar heads towards the surface and non-polar groups in the hydrophobic bilayer).

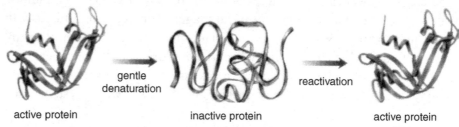

Figure 8.4 Protein denaturation and reactivation. When a protein is denatured, it loses its native shape and activity. If denaturation is gentle and if the denaturing conditions are then removed, proteins may regain their native conformation. This "reversibility" shows that the folding into the native conformation is a process under thermodynamic control, dictated by the primary sequence in that particular environment. (Adapted, with some modifications, from Mader, 1996.)

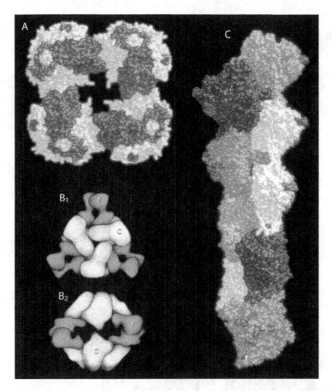

Figure 8.5 Some examples of self-organization in oligomeric proteins. (**A**) The transacetylase core of the pyruvate dehydrogenase complex. The core consists of 24 identical chains (12 can be seen in this view). (**B**) The aspartate transcarbamoylase, formed by six catalytic (lighter subunits) and six regulatory chains (darker subunits): (**B₁**) view showing the threefold symmetry; (**B₂**) a perpendicular view. (**C**) The helical assembly of several identical globular subunits in F-actin polymer. The helix repeats after 13 subunits. (All adapted from Stryer, 1975.)

where the primary structures of the two strands determine the rules for self-assembly. Similarly, in the folding of t-RNA, the primary structure of the RNA molecule (single strand, in this case) determines the self-organization of functional 3D conformation.

Some classic examples of self-organization are DNA self-assembly (see Figure 8.6), as well as the t-RNA folding (see Figure 8.7). Quite interesting is the folding dynamics in proteins: once the first intermolecular interactions have been established, the next ones are more easily realized within the structure, as several chain segments are now closer to each other. These examples may well introduce the next argument, the relation between self-organization and autocatalysis.

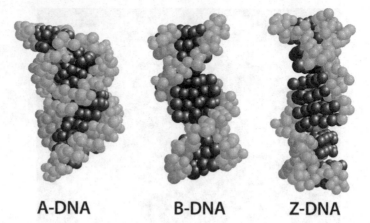

A-DNA   **B-DNA**   **Z-DNA**

Figure 8.6 The well-known self-assembly of the DNA duplex (main double helix conformations are shown).

Figure 8.7 Folding of t-RNA, another classic case of self-organization.

8.3 Self-organization and autocatalysis

Let's consider now a concept that will be useful in later stages of this book – the relation between self-organization and catalysis.

When surfactant molecules solubilize in water, generally the process is slow at the very beginning, and gets faster with time: the more surface bilayer is formed, the more the process speeds up, because there is more and more active surface where the next steps of aggregation can take place. The same behavior is observed in crystallization; in both cases we are looking at a particular kind of an *autocatalytic process*: the product of the reaction (organized surface bilayer or crystals) speeds up further self-organization. Actually, the point can be made, that self-organization in chemical systems is generally attended by some kind of autocatalytic behavior.

The relation between self-organization and autocatalysis is discussed in some detail by Burmeister (Burmeister, 1998), and the relation between chirality and self-organization/self-replication in biopolymers was considered from the theoretical point of view by Avetisov and Goldanskii (1991).

8.4 Polymerization

Polymerization appears to be a simple method to increase molecular order: starting from a gaseous or liquid monomer mixture, long covalent macromolecules with a vast series of emergent properties can be obtained. At first sight, there is a strong similarity between polymerization and two other processes mentioned earlier, crystallization and surfactant aggregate formation. In all these cases, in fact, starting from a disordered mixture, the low-molecular-weight components are stringed together in a compact ensemble – a crystal, a micelle, and a polymer chain. As shown in Figure 8.8, there is, however, an interesting difference between a polymerization process and a self-assembly process of surfactant aggregates: whereas the surfactant self-assembly is attended by an overall increase of entropy (because of water being made "free"), polymerization is generally attended by a decrease of entropy – as free monomers are being attached to each other and immobilized in a covalent string.

Is polymerization a self-organization process? Namely, is the polymer the result of the internal rules of the system? The answer is not as clear as in the case of the formation of a micelle or of a crystal. I believe that a positive answer can be given in the case of spontaneous polymerization, as in the formation of nylon starting from a mixture of dicarboxylic acid chloride and alkyl diamines; or the polymerization of styrene induced by heat or by light; or the formation of oligopeptides starting from N-carboxyanhydrides. However, in the case of stepwise polymerization – for example, in the Merrifield synthesis – this is not the case, and I would not include this polymerization process in the category of self-organization.

The simplest case of polymerization is the formation of a linear chain. There are additional stages of complexity in polymer chemistry, which include

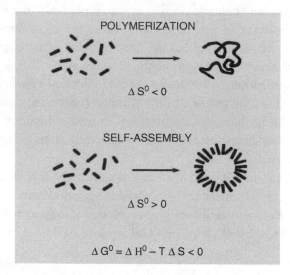

Figure 8.8 Two main ordering processes: comparison between the process of polymerization and the self-assembly process of surfactant aggregation (water molecules in the self-assembly are not shown for simplicity).

branching, cross-linking, polymer networks, and dendrimers (for reviews, see Hawker and Frechet, 1990; Föster and Plantenberg, 2002; Bucknall and Anderson, 2003; Pyun *et al.*, 2003). Not all these stages of complexity qualify as self-organization processes, as some are imposed by external forces and are not caused by the internal rules of the system. Perhaps, the most interesting cases of self-organization are those obtained with aggregations of di-block polymers, as indicated by Bucknall and Anderson (2003). A di-block polymer has a chain consisting of two chemically different long sequences, of the type:

...**AAAAAAAAAAA–BBBBBBBBBBBBB**...

As in the case of surfactants, consisting of a polar and an apolar moiety, such block-polymers tend to phase-separate due to the immiscibility of the two blocks. A variety of structures can then be obtained by regulating the chemical nature and the length of the two ...**A**... and ...**B**... blocks. (See specific literature in Bucknall and Anderson, 2003; see also Ma and Remsen, 2002).

8.5 Self-organization and kinetic control

Thus far, various cases of self-organization under thermodynamic control have been considered. There are also several cases in which self-organization is the result of kinetic control – in which the reaction product is not necessarily the most stable, but forms because of kinetic constraints (see Figure 3.2 and related text in the previous Section 3.3 for the definition of kinetic control). The most general case of kinetic control is given by enzymatic reactions, and in fact, once the intermediate (complex enzyme-substrate) has been formed, there is only one way to go, the one with the lowest activation energy. We have discussed this point in Part I, and the case of the biogenesis of macromolecules is not conceptually different.

A very particular and important case is given by those reactions under kinetic control that lead to stereoregular biopolymers. Polysaccharides, polypeptides, and nucleic acids are all stereoregular macromolecules. Clearly the linear, stereoregular chains are not the most stable products thermodynamically; an equilibrium random mixture of all possible enchainments would be entropically more favored, as it would correspond to billions of possibilities. Thus, the constitutional order and the stereoregularity of biopolymers results from a clear case of kinetic control. The same holds for stereoregular synthetic polymers, such as isotactic or syndiotactic polypropylene, where stereoregularity is due to a particular class of Ziegler-Natta catalysts (see Figure 8.9).

Also on the basis of this example, we can go back to the question of whether stereoregularity is a self-organization process. Again, the answer is positive, if the catalyst is considered as one of the determinants of the "internal rules" of the system. For example, if the growing polymer is considered as a complex with the catalyst, then it can be defined as a self-organizing system. However, as in the general case of polymerization mentioned above, this is clearly a gray zone, where the notion of self-organization becomes less clear.

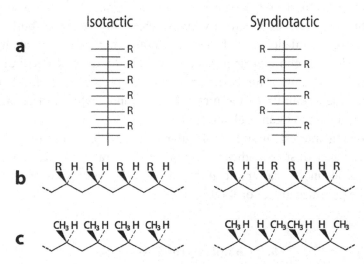

Figure 8.9 Stereoregular vinyl polymers in the Fischer projection (**a**) and the same structures represented in tridimensional conformation (**b**). With **R** = methyl, we have isotactic and syndiotactic polypropylene, as shown in (**c**).

8.6 Self-organization and breaking of symmetry

Self-assembly of chiral molecules may result in organized aggregates displaying a remarkable enhancement of optical activity. The best known examples are amino-acid residues that assume a periodic conformation – α-helix or a β-sheet chain. In this case, the enhancement of optical activity is due to the onset of a particular "rigid" conformation.

There is a less trivial, actually quite fascinating area of inquiry in this field of chirality of high molecular weight compounds: this is symmetry-breaking brought about by self assembly. One case in point is the aggregation of porphyrin-like compounds, as investigated independently by the research group of Ribo (Ribo *et al.*, 2001; Crusats *et al.*, 2003) and by Purrello and coworkers (Purrello, 2003; de Napoli *et al.*, 2004).

Some of these phenomena may have a link with the origin of life. As Ribo says (personal communication, 2004):

The effect probably can only occur when the process of growth of the mesophases is a cluster to cluster process and not a molecule to aggregate process. This work suggests that in supramolecular systems chiral long range forces can transfer information to the bond-length scale of size. This suggests that the role of vortices should be taken into account as one of the chiral polarization forces directing the chirality of spontaneous symmetry breaking.

Ribo also suggests that this might be relevant in a prebiotic scenario at the early stages of the origin of life, since there may well have been vortices determining permanent sign directions in primordial times.

A beautiful example of this memory effect due to kinetic factors has been reported by Raymond and coworkers (Ziegler *et al.*, 2003). These authors describe the synthesis of kinetically stable chiral architectures built entirely from achiral structural elements. To induce chirality in this non-covalent assembly, they exploit the chiral arrangement of specifically designed achiral ligands surrounding metal ions. This complex is able to preserve its chiral memory despite the kinetic lability of the metal–ligand bonds, while at the same time allowing for dynamic ligand substitution.

The growing interest in supramolecular chirality stems not only from the intrinsic relevance of such studies for the origin of chirality in life, but also from the potential technological applications, such as the separation of optical isomers for the pharmaceutical or food industries.

8.7 Complex proteic systems

Usually, a complex interplay between thermodynamic and kinetic control is at work to guarantee the complexity of larger biological structures. In addition,

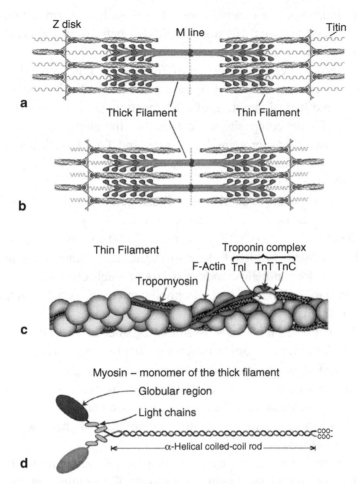

Figure 8.10 Schematic diagram showing interaction of thick and thin filaments in skeletal muscle activity: (**a**) relaxed muscle fiber; (**b**) muscle fiber during contraction; (**c**) thin filament elements (actin, tropomyosin, and the troponin complex); (**d**) myosin structure (not in scale) showing the 2 heavy chains (each one characterized by a globular region and a long helical rod) and the 4 light chains. See text for more details. (Adapted, with modifications, from Stryer, 1975.)

many such syntheses *in vivo* take place on a matrix – pre-existing fibers or membrane structures or organelles – so that steric factors also play a role in the assemblage. These steric factors can also be seen as determinants for the kinetic control. All this can be evidenced by the example of muscle-fiber organization. In Figure 8.10 are shown a relaxed fiber (**a**) and a contracted fiber (**b**). Muscle fibers consist of thick and thin filaments, and each type of filament is a conglomerate of different proteins (Stryer, 1975; Alberts *et al.*, 1989).

Let us consider first the thin filaments. The globular F-actin polymerizes to give a long double helix [it was displayed before, in Figure 8.5 (c)], a polymerization that can take place in the test tube and is therefore *per sé* under thermodynamic control. Panel (**c**) of Figure 8.10 shows more details about the complex structure of the thin filament: around the double helix of actin there is another double helical structure of a filamentous protein, the tropomyosin. In the grooves of this composite helix, there are the elements of the troponin complex (formed by three proteins, **TnC**, **TnT**, and **TnI**, as indicated in figure).

Thus, the thin filament is really a quite complex interplay of self-organized structures locked one into the other.

Take a look, now, to the thick filaments, illustrated in panel (**d**) of Figure 8.10. The main element is myosin, which comprises two heavy chains (each one has a globular head and a long α-helical rod, about 2,000 amino acid residues long) and four light chains, localized close to the myosin heads (two per head). The two myosin long α-helical rods coil around each other to form a coiled coil, which is stabilized mostly by hydrophobic interactions between the two α-helical heavy chains.

Hundreds of myosin molecules assemble together to build thick filaments, mostly due to ionic interactions between the tails of the individual molecules. It is possible to induce spontaneous aggregation and disaggregation of myosin molecules by varying the salt concentration. The overall assembly process of thick and thin filaments to build the muscle fibers is certainly very complex, and it is difficult to discriminate the relative importance of thermodynamic and kinetic effects (Alberts *et al.*, 1989, 2007).

Thermodynamics has, in this case, to do with the specificity of the protein structure that determines the binding selectivity, for example, actin to tropomyosin. However, the assemblage is also determined by the sequential steps of the enzymatic syntheses, as well as by steric factors due to fiber matrices.

The complexity of this structure results in the important emergent property of mechanical movement.

The same consideration – mechanical movement – holds true for another beautiful object, depicted in Figure 8.11: it may look like an oriental mandala – but it is an axoneme, from an eukaryote flagellum (Stryer, 1975). This is really a wonderful construction, a molecular motor that implies the coordination of dozens of different specialized parts, each part being the organized structure of a bundle of specific proteins. The perfect juxtaposition of all these components is the result of a calibrated, sequential ordering and organization process, where again thermodynamic and kinetic factors exert a concerted mechanism of utmost complexity. Here, too, as in the muscle proteins, the final result is mechanical movement, in this case the propeller action of cilia (human example: the ciliated cells of the trachea) and flagella (human example: spermatozoon).

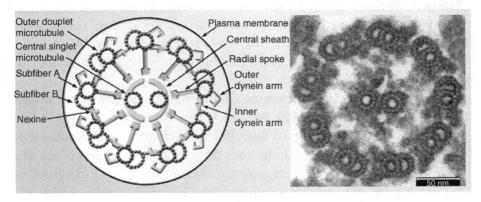

Figure 8.11 *On the left*, schematic diagram of the structure of an axoneme – or an oriental mandala? (Adapted from Stryer, 1975.) *On the right*, the same structure in a TEM micrography.

8.8 Self-organization of ribosomes

Another complex macromolecular aggregate that can reassemble from its components is the ribosome (Alberts *et al.*, 1989). It appears at any rate that ribosomes are indeed able to spontaneously self-organize from RNA and proteins. The information for the correct organization is contained in the structure of the components.

Ribosomes from bacteria, archaea, and eukaryotes (the three domains of life on Earth) differ in their size, sequence, structure, and the ratio of protein to RNA. Figure 8.12 illustrates the difference between bacterian (70S) and eukaryotic (80S) ribosomes. (The parameter S, from Svedberg, measures the sedimentation velocity of the particles).

The ribosome is a cellular machine that is highly complex. It is made up of dozens of distinct proteins (the exact number varies a little bit between species) as well as a few specialized RNA molecules known as ribosomal RNA (r-RNA). Ribosomes consist of two major components, the small ribosomal subunit that reads the RNA, and the large subunit that joins amino acids to form a polypeptide chain. In bacterian cells, the larger subunit is called 50S, the smaller 30S. In turn, subunit 30S is composed by a molecule of r-RNA called 16S (1542 nucleotides, in *E. coli*) and contains 21 small different proteins, called S1, S2, S3, ... S21.

The 50S subunit of *E. coli* contains (see Figure 8.12) a 5S and a 23S r-RNA structures (120 and 2904 nucleotides, respectively) and contains also 31 different small proteins (Alberts *et al.*, 2002), indicated as L1, L2, L3, ... L34. Their location in the ribosomes is only partially clarified (Green and Noller, 1997).

The two subunits of eukaryotic ribosome are called 40S and 60S, respectively, whereby the smaller one is similar to the bacterian 30S subunit, but contains 31–33 proteins and one molecule of r-RNA 18S, as shown in Figure 8.12. The larger subunit contains about 50 proteins and 3 different molecules of r-RNA, namely 5S, 5.8S, and 28S.

The 30S subunit of *E. coli* was disassembled and successfully reconstituted starting from RNA and proteins by Traub and Nomura (1968). In 1975, Cohlberg and Nomura, for the first time, did the same with the 50S subunit of *Bacillus stearothermophilus*, starting from the purified components (Cohlberg and Nomura, 1976). Later on, Mangiarotti and Chiaberge (1997) succeeded in re-assembling the 40S and 60S subunits of the eukariotic *Dictyostelium discoideum*. The important work by Alberts and associates (Alberts *et al.*, 1989) should be mentioned again in this regard. More information on the self-organization of ribosomes will be given later on.

In eukariots there are two kinds of ribosomes: the cytoplasmatic ones, as described before, and those in mithocondria/chloroplasts. Since most of the protein synthesis takes place in the cytoplasm, cytoplasmatic ribosomes are those which are commonly considered proper of eukariots. By the way, the fact that mithocondria ribosomes are similar to bacterian ribosomes, and even more so for chloroplast ribosomes, is taken as an indication that mithocondria and choloroplasts are derived from symbiotic bacteria (Rawn, 1989).

Questions of reconstitution of ribosomes from their components is a fascinating chapter, pertaining to the field of self-organization. This question is elicited by the great complexity of the ribosomal particles, and by a series of experimental observations, which have shown that several of the ribosomal proteins are not essential for the ribosomal function (Zhang and Cech, 1998; Nissen *et al.*, 2000; Calderone and Liu, 2004).

A theoretical analysis of the minimal ribosome has been undertaken by Mushegian (2005), emphasizing the ribosomal proteins, with the idea of organizing the latter in seven groups. In this way the author could define which proteins interact more significantly with certain regions of r-RNA, and argued that by this procedure one can arrive at describing the situation for LUCA, the last universal common ancestor (Mushegian, 2005).

Experimental studies carried out with ribosomes lacking specific proteins complement these theoretical studies.

The peptidyl transferase activity has been studied often by using a model reaction, described by Gilbert *et al.* (2004). According to this assay, when ribosomes charged with poly-Phe-tRNA (bound to the 50S unit) are reacted with puromycine, poly(Phe) is released from t-RNA.

Mostly using this reaction, Maden, Traut, and Monro (Monro and Marcker, 1967; Maden and Monro, 1968) and later on the school of Nierhaus (Nierhaus

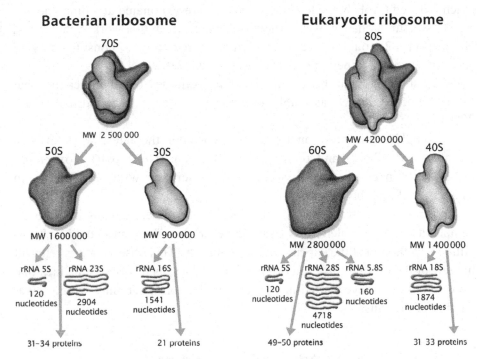

Figure 8.12 Schematic structure of bacterian (70S) and eukaryotic (80S) ribosomes. See text for more details.

and Montejo, 1973; Nierhaus and Dohme, 1974; Hampl *et al.*, 1981) have begun a systematic investigation of the relative importance of the L proteins in the ribosomal function. They have shown that the omission of L7/L12, L26, L31, L33 had no effect on the peptidyl transferase activity; instead, the elimination of L16 reduced the activity by 90 percent. It could also be shown that a partial reconstitution of the 1.5c core (obtained treating 50S particles with 1.5M LiCl) with the four proteins L6, L10, L11 and L16 produced particles with a substantial activity. A large excess of L16 induced an increment of activity – confirming the importance of this protein for the peptidyl transferase activity of the ribosomes.

In general, the experiments by Nierhaus' group confirmed that several proteins could be omitted, in particular L1, L5, L7/L12, L9, L19, L23, L29, L31–33, but not L2, L3, L4, L15, L16; and the 23S r-RNA was also essential (Hampl *et al.*, 1981). Later on, L15 and L16 were also shown to be non-essential (Franceschi and Nierhaus, 1990). It is, however, important at this point to mention that the remaining components, 23S r-RNA, L2, L3 and L4, have never been assembled and tested as a minimal ribosomal particle.

Those experiments were carried out initially by taking out one protein at a time; when, however, all proteins were eliminated at a time, activity was very much reduced (probably, when many proteins are eliminated at a time, there is a large structural change of the whole ribosome...). In this work, Schulze and Nierhaus (1982) could obtain a "minimal ribosome particle" constituted by 16 components only, namely by 15 proteins plus the 23S r-RNA.

Other studies could be mentioned, for example, those indicating that L2 is involved in the assembly of ribosomal subunits (Diedrich *et al.*, 2000).

To complete this view on ribosomes, consider the latest work of Ada Yonath (2010, 2012) and of her group (Bashan *et al.*, 2010; Fox *et al.*, 2012; Krupkin *et al.*, 2014). See also the interview with Ada Yonath in Chapter 4 of this book.

As mentioned above, these few studies mentioned here already indicate the complexity of the problem. It is clear, however, that ribosomes could be "simplified" – although at the expense of some activity. This observation is relevant in view of the origin of life and origin of ribosomes, as it conveys the idea that ribosomes might have started in a much simpler form, certainly with a lower number of proteins.

8.9 Self-organization in viruses

Fraenkel-Conrat and Williams (1955) demonstrated for the first time that Tobacco Mosaic Virus (TMV) could be reversibly reconstituted in vitro starting from the two main components, the purified genome and the capside proteins [see Figure 8.13 (a)]. There is a spontaneous assembly at pH 6, which forms a virus with a genomic length that is sufficient for infection. Later on, Okada and Ohno (1972) studied in detail the mechanism of this self-assembly. Butler (1999) proposed a mechanism to explain the elongation of TMV.

However, this cannot be generalized. For example, the formation of some other viruses is determined by key enzymatic steps; see the complex sequence of reactions given in Figure 8.13 (b), which illustrates the fascinating assembling process of the T4 phage, a bacterian virus.

Here again there is a combination of kinetic and thermodynamic control: in fact, the single components (the head, the tail, the tail fibers, etc.) are being made via kinetically controlled syntheses, and in this case, the mechanical mixing in a test tube of the single components would not form the original infectious virus – although most of the following assembly steps are determined by thermodynamic control.

This short section has presented only a bit of information about the structural self-organization of some viruses, and was not intended as a discussion at large about viruses. This is a large and complex argument, which has witnessed

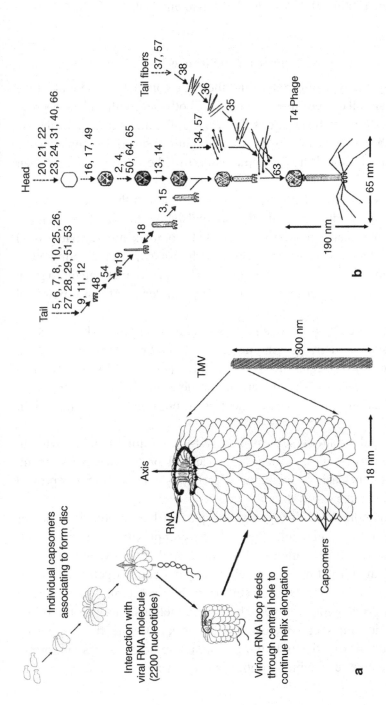

Figure 8.13 Examples of assembly among viruses; (**a**) the assembly of Tobacco mosaic virus (TMV) and (**b**) the assembly of a T4 phage from its constituents. The numbers next to the arrows refer to gene products that are required for a particular step in assembly. (Adapted, with some modifications, from Wood, 1973.)

a virulent revival of interest in the last few years. I only mention here the initial work of Villarreal (2009, 2011; see also Villarreal and Witzany 2010).

8.10 Swarm intelligence

Let's now change the type of example and the scale completely and look at the formation of an anthill or a beehive: the design of both the anthill and the beehive is not imposed externally. Instead, it is the result of the internal work of the ant and the honey bee social systems. It is self-organization, if we consider the individuals themselves and their genome as part of the internal rules of their systems.

There is also a very complex genomic determination in the case of "swarm intelligence." Some birds, fish, and other animals show this kind of phenomenon. For example, there are some coordinated, complex motion patterns of groups constituted by several hundreds (or even thousands) of small individuals. Synchronized movements of so many bodies can "create" something like a great figure that continuously changes, frightening and/or confusing potential predators.

These patterns are typical among birds, such as starlings (Figure 8.14, left), and fish, such as herrings. Another example is the very regular pattern formed by some bird species during migration (Figure 8.14, right). It is known that the flying-formation pattern is not determined by a leading bird, because all birds, in turn, change their position regularly in the flight formation.

Thus, these regular patterns are cases of self-organization realized without any imposing external rules, similar to the anthill or the beehive – again, understandable on the basis of evolution. Clearly this configuration has proved the best for the animal's survival.

Moreover, this complex self-organization pattern has no localized center that determines the organization: the ordered complexity is an emergent, collective property, and one might even draw an analogy between the swarm intelligence and the formation of a micelle: the parts come together and form an ordered ensemble. In one case the self-assembly is determined by simple thermodynamic driving forces (hydrophobic interactions, etc.); in the other, complex genomic and social factors are at work. Despite the staggering differences in complexity, the underlying principle in both cases is the formation of a collective ensemble without an ordering center or localized intelligence[2].

---

[2] For a more detailed view of swarm intelligence, see http://www.swarm.org.

Figure 8.14 Swarm intelligence is frequent among birds. On the left, a flock of common starlings, flying over Rome. Flocks of starlings are formed by several thousand and even more birds that fly in a synchronized way, so as to form large images – continuously changing shape – in the sky. They use this behavior to protect themselves against predators. On the right, the regular flight pattern of migratory birds (cranes), an effective way to have some benefits during long flights (better aerodynamics and good visuals), sharing duties among all the birds. Both the examples shown are consistent with a collective self-organization without a localized center of organization.

We will come back to this interesting point in the next chapter, which considers emergent properties.

8.11 Can living cells be reconstituted *in vitro*?

An important series of experiments, which links self-organization to the very core of living biological structures, is reconstitution. If you were to disassemble a complex biological structure into its components, and then ask whether by mixing them all together, would you obtain the original structure/function? The argument is analogous to the re-folding of the proteins *in vitro* after their denaturation. As we have seen before, the fact that a protein refolds *in vitro*, far removed from the magic of the living cell, suggests that the folding is under thermodynamic control, namely that it is the most stable of the folding possibilities under that set of conditions. Likewise, if it would be possible to reconstitute a living cell from its disassembled components, this might be taken as an indication that the cellular structure is also, to some extent, under thermodynamic control.

(Note that the term *reconstitution* is used in biology also for experiments of biological formation of cells, e.g., from stomatocells or T-cells regeneration. Here, I will use the term only in the meaning of self-organization).

The interest in cell reconstitution began in the 1970s. The reassembling of *Amoeba proteus* was attempted starting from the major constituents, namely nucleus, cytoplasm, and cellular membrane (Jeon *et al.*, 1970). When the reconstitution experiment is carried out with cells of the same kind, the self-assembly is rather successful and *c.* 80 percent of the reconstituted cells behave like normal amoebae.

Experiments were carried out also with *Acetabularia mediterranea*, a green alga that was reassembled from the cell wall, the nucleus and the cytoplasm (Pressman *et al.*, 1973). And only one year later (Veomett *et al.*, 1974) tried with mammal cells (mice) starting from citoplasmatic component without nucleus (cytoplasts) and a nucleated subcellular fraction (karioplasts) in the presence of cytochalasin B and of the Sendai virus as fusion agent. This experiment was also successful.

At this point, we could say that cellular life can be reconstituted from non-living components. However, this statement should be taken with some caution, as that the re-assembly is generally "guided" by microsurgery, i.e., it is not a totally spontaneous process.

The same method (using cytochalasin B and Sendai virus) was utilized (Ringertz *et al.*, 1978) to reconstruct cells via the fusion of myoblast nuclei and cytoplasmatic fibroblasts. Since the diversification of the muscle fiber could be observed in the reconstituted cells, the authors argued that the cells maintained their functionality after reconstitution.

Kobayashi and Kanaizuka (1977) re-assembled *Bryopsis maxima* from its two dissociated components, chloroplasts and the protoplasmatic fraction. This, in turn, consisted of nuclei, mitochondria, endoplasmatic reticulum, and other components. The reassembly produced protoplasts, cells without walls, from which mature plants eventually originated.

Particularly interesting has been the work by Newport (1987) on the reconstitution of the nucleus, a work revisited later by Zhao *et al.* (2000), utilizing eggs of Xenopus, a kind of primitive frog, bound to the aquatic environment.

Life from non-living components can also be obtained by cloning. This term, as is well known, indicates those molecular biology techniques to reproduce clones (organisms with identical genomes) in the laboratory. Cloned multicellular organisms have been obtained starting from an egg cell deprived of the nucleus, on which the nucleus of a somatic cell (non reproductive) from another individual of the same species has been inserted. The first experiments were carried out with frogs in the 1950s, followed by experiments with mammals, and in 1997, Dolly was born – the famous experiment by the group of Ian Wilmut. A step forward (?) was the announcement from the University of Kyunghee in South Korea of a human clone, which was apparently destroyed after the fourth cellular division.

See also the review by Liu and Fletcher (2009), which begins with the statement:

We are much better at taking cells apart than putting them together.

In the case of cloning experiments, two complementary observations can be made: that they demonstrate the emergence of life from non-living components; that the process is not a spontaneous assemblage, but rather is guided by microsurgery techniques. Note also that in all these cell reconstitution experiments, the starting reagents were not the molecular components; they were already complex systems, like the cytoplasm, the nucleus, or the cell membrane.

Concerning the importance of structural integrity, an article by Morowitz dealing with organisms brought to temperatures close to zero Kelvin (Morowitz, 1974) comes to mind. Under those conditions, all molecular movements are stopped, and the organism loses "the memory" of all processes prior to the cooling, except for the structural and organization information. The organism can be considered dead – and immortal at the same time, in the sense that once it is again brought to normal temperature, it is functional and living again. This shows that the activity of a cell is due to the structural information within the system, which is kept intact when the temperature becomes close to absolute zero.

This consideration is relevant for understanding why the spontaneous assemblage from all single molecular components does not work for complex cells, contrary to the TMV: the problem is to regain the original structural organization. And the cell organization is not based on thermodynamic control; it is due to a series of enzymatic reactions (kinetic control) as well as reactions that take place on a kind of rigid matrix – controlled then by the geometry of the matrix itself. A simple example of this argument is given by the difference in assemblage between TMV and the T4 virus, as already illustrated above.

In conclusion, then, while all these experiments of cell reassembling and cloning demonstrate that life is an emergent property, they also show that the structural organization of cells and complex viruses is not a spontaneous process; it is not dictated by thermodynamics alone. Not a big surprise, but a good clarification point.

A particularly interesting and fascinating case of self-organization at a multicellular level is offered by the hydra, one cnidarian hydrozoan. Side Box 8.1, prepared by Giorgio Venturini, describes the most important features of this case.

Side Box 8.1

Phenomena of self-organization in Hydra

Giorgio Venturini

Giorgio Venturini was born in Ascoli Piceno (Italy) in 1944. He received his medical degree *Summa cum Laude*, at the University of Roma (1969). Now retired, he was a full professor of Cytology and Histology at the Department of Biology, Faculty of Sciences of the University Roma Tre (Rome, Italy).

Previously, he taught general biology at the Medical School of the University of Roma and at the Somalia National University of Mogadishu, and biology, anatomy and physiology in the Graduate School in Sanitary Physics, University of Rome.

Giorgio Venturini was awarded the *Camillo Golgi* Prize in 1988, by the National Academy Lincei, for nervous system physiology.

His research activity mainly concerns the study of the nervous systems of lower invertebrates such as platyhelminthes and coelenterates.

The morphogenetic gradients play an essential role in determining the body pattern during the earliest stages of the embryonic development of many organisms and play a key role in the processes of self-organization. However, there are few usable animal models for the experimental analysis of self-organization during development.

The Hydra is particularly useful in this context, in as much as its body arrangement and the positional information can be completely destroyed and reformed with experiments of cell dissociation and re-association.

The Hydra, a member of the Cnidaria, is a primitive metazoan characterized with a simple body plan, a radial symmetry, in which along the axis oral-aboral are present a head equipped with tentacles, a body column, and a foot. These are shown in Figure 8.15.

Some details of the head morphology of Hydra, whose role in development, self-organization, self-regulation, and regeneration has been studied in detail, as described later on, are shown in Figure 8.16.

The extraordinary regenerative capacity of the Hydra (as also suggested by its name, as noted in Figure 8.17) have been known since the mid-eighteenth century, when the Swiss naturalist Abraham Trembley demonstrated that this small freshwater hydrozoan, characterized by a reproduction through budding, was able to quickly reform

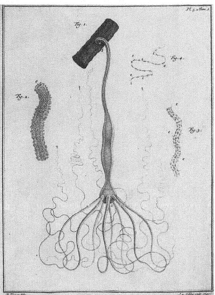

Figure 8.15 The body arrangement of the Hydra in a micrograph (left), taken in Venturini's lab, and in an engraving (right), taken from the original work by Abraham Trembley (*Memoires pour servir à l'histoire d'un genre de polypes d'eau douce, à bras en forme de cornes*, 1744).

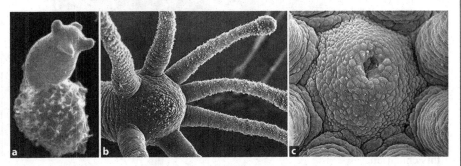

Figure 8.16 SEM micrographs of *Hydra vulgaris*. In (**a**), the production of a hydra from an egg (in addition to asexual reproduction by budding, this small cnidarian also shows sexual reproduction). On the oral pole (top) is evident the hypostome surrounded by bumps of the tentacles. In (**b**), the head of a Hydra adult with the hypostome and the tentacles. In (**c**), at a higher magnification, a detail of the oral pole showing the open mouth. ((**a**), from Martin *et al.*, 1997; (**b**) and (**c**), original material prepared and photographed by G. Venturini, A. Merante and A. Severati.)

amputated parts, and, even from small fragments of the organism, could reform complete individuals (Trembley, 1744).

In Hydra, the processes which determine the body pattern are constantly active even in adults, due to the dynamic characteristics of its tissues. These processes are activated by an organizer region in the head (*HO*), which continuously produces and

Figure 8.17 *Hercules and the Hydra* (a painting by Antonio del Pollaiolo, *c.* 1475, Galleria degli Uffizi, Florence). It was Linnaeus, inspired by the mythical monster that could regenerate its amputated heads, who gave the name Hydra to this cnidarian.

transmits two signals that are distributed in gradients along the axis of the body. A first signal defines the gradient of activation of the head (*HA*), that is, the ability to produce a head with tentacles from the cells of the body column, during either budding or regeneration. The second signal defines a gradient of inhibition of the head (*HI*), which inhibits the production of new heads and therefore limits the production of the buds to the lower portion of the body column in adult Hydra (see Figure 8.18). The molecular bases of the two gradients are not well known, yet the path of canonical Wnt[3] (HyWnt, Hydra ortholog of Wnt) is known to play a central role in the activity of the organizer of the head.

---

[3] The *canonical Wnt pathway* is a signal transduction pathway that leads to regulation of gene transcription. More details are in Note 5.

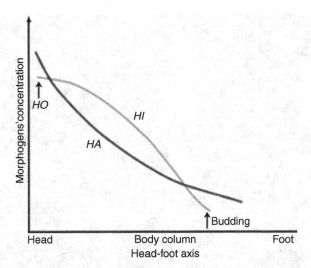

Figure 8.18 The gradients of concentration of the head activator (**HA**) and the head inhibitor (**HI**) along the body axis of Hydra, generated by the head organizer (**HO**). The budding is inhibited by **HI** and, therefore, can only occur on a limited portion of the body column where the relative concentration of **HA** is greater than **HI**.

In a normally fed specimen, the cells of the column body are constantly proliferating: the animal maintains a constant body size as the expanding tissue is continuously displaced towards the apical or basal ends, where it is continuously sloughed. Therefore, the cells are in a steady state of production and loss. The differentiation of cells to produce head or foot structures and the position of the region in which the budding occurs depend on the gradient of competence (relative concentrations of HA and HI) that is established along the body axis (as shown in Figure 8.18). Since Hydra doesn't have a defined lifespan, the activities of proliferation and differentiation continue indefinitely.

Following amputation of the head, with the elimination of the head organizer (HO), HA and HI gradients would be destined to disappear. But since HA is more stable than HI (HA half-life = 36 h, HI half-life = 3.2 h), the formation of a new organizer, able to restore the morphogenetic gradients, will be induced. This will ensure the differentiation of a new head and its tentacles will reform.

The most striking demonstration of the regenerative capacity of Hydra is probably its ability to produce complete organisms from cell aggregates (Gierer *et al.*, 1972). Suspensions of Hydra cells, obtained by mechanical disruption of Hydra tissues, can be centrifuged to produce pellets, or cell aggregates, in which the gradient of activation of the head is completely destroyed. Within two to three days in these aggregates new heads appear that organize the surrounding tissues to reconstitute entire normal animals.

In more detail, as shown in the four panels of Figure 8.19, each aggregate produces within 24 hours a hollow sphere [panel (a)], whose cells are organized in two layers similar to the ectodermal and endodermal layers that are typical of

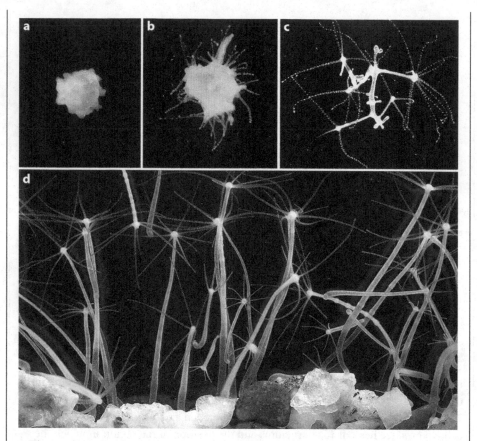

Figure 8.19 Production of several *Hydra vulgaris* specimens starting from a single cell aggregate, obtained by centrifugation of a cell suspension. The various steps of the process are described in the text. (Photographs by G. Venturini.)

the Hydra body. Later, some heads appear, each of which will organize the surrounding tissues, by first generating the tentacles [panel (b)], then the body columns [panel (c)]; in the meantime some foot structures will also be formed. Finally complete and functional Hydra organisms will detach [panel (d)]. For further information, see Gierer *et al.*, 1972.

The same phenomena of self-organization and self-regulation that are active in Hydra regeneration, in budding and in maintaining body size and shape, also play a central role during the differentiation of new organisms from cellular aggregates.

Since cells with different levels of competence for the head will be randomly distributed in the aggregate, it is assumed that the formation of the heads will occur at the regions where, by chance, a sufficient number of cells with a high competence for the head are close together. These regions, defined as centers of activation, will form an organizer (HO), and will be able to create new gradients of head activator (*HA*) and inhibitor (*HI*). These gradients now define the organization of the body axes. The gradient of *HI* inhibits the production of additional heads in the proximity of the first one.

Experimental data suggest that the centers of activation require a *cluster* of 5–15 cells with high competence for the head. Molecular analysis shows that, in the areas corresponding to these *clusters*, the genes HyBra1 and HyWnt (orthologs of Bra1[4] and Wnt[5]) are expressed.

Some observations suggest moreover that the signaling pathway Nitrogen Oxide – cGMP – protein kinase G, whose presence in the Hydra had been described previously (Venturini *et al.*, 1995; Colasanti *et al.*, 1997), is also involved in the control of the proliferation/differentiation of regenerating structures of cnidarians (Colasanti *et al.*, 2009).

References

Colasanti, M., Mazzone, V., Mancinelli, L., Leone, S., and Venturini, G. (2009). Involvement of nitric oxide in the head regeneration of *Hydra vulgaris*. *Nitric Oxide*, 21, 3–4, 164–70.

Colasanti, M., Venturini, G., Merante, A., Musci, G., and Lauro, G.M. (1997). Nitric oxide involvement in *Hydra vulgaris* very primitive olfactory-like system. *J. Neurosci.*, 17, 1, 493–99.

Gierer, A., Berking, S., Bode, H.R., David, C.N., Flick, K.M., Hansmann, G., Schaller, H., and Trenkner, E. (1972). Regeneration of hydra from reaggregated cells, *Nature New Biol*, 239, 98–101.

Martin, V., Littlefield, L., Archer, W. E., and Bode, H. R. (1997). Embryogenesis in *Hydra*. *Biol. Bull.* 192, 345–363.

Trembley, A. (1744). *Memoires pour servir à l'histoire d'un genre de polypes d'eau douce, à bras en forme de cornes*. Leida.

Venturini, G., Colasanti, M., Lauro, G.M., and Merante, A. (1995). Nitric Oxide is involved in the control of the feeding response of *Hydra vulgaris*. *Rend. Fis. Acc. Lincei*, s. 9, v. 6, 179–88.

---

[4] The gene *brachyury* (Bra1) appears to have a conserved role in defining the midline of a bilaterian organism, and thus the establishment of the anterior-posterior axis; this function is apparent in Chordates and Mollusks. Its ancestral role, or at least the role it plays in the Cnidaria, appears to be in defining the blastopore.

[5] The Wnt signaling pathways are a group of signal transduction pathways made of proteins that pass signals from outside of a cell through cell surface receptors to the inside of the cell. Three Wnt signaling pathways have been characterized: the *canonical Wnt pathway*, the *noncanonical planar cell polarity pathway*, and the *noncanonical Wnt/calcium pathway*. All three Wnt signaling pathways are activated by the binding of a Wnt-protein ligand to specific receptors (*Frizzled* family receptor, a family of G protein-coupled receptor proteins), which pass the biological signal to other specific receptor proteins (*Dishevelled* proteins) inside the cell. The *canonical Wnt pathway* leads to regulation of gene transcription, the *noncanonical planar cell polarity pathway* regulates the cytoskeleton that is responsible for the shape of the cell, and the *noncanonical Wnt/calcium pathway* regulates calcium inside the cell. Wnt signaling pathways use either nearby cell-cell communication (paracrine) or same-cell communication (autocrine). They are highly evolutionarily conserved in animals; that means they are similar across many species of animal, from cnidaria to humans.

8.12 Touching on the "divine proportion": Φ and the golden mean

We have considered until now the self-organization in biological systems, seen under the perspective of natural laws – thermodynamic and kinetic factors. In this section, we will deal with some form of regularity in nature that appears at times to challenge or be above such natural laws. One of these is the so-called "golden mean," also named "golden section" or "golden ratio."

The geometrical proportions in nature have attracted the attention of thinkers since antiquity. It started with the study of the pentagon and the pentagram at the time of Pythagoras (Livio, 2002), down to the discovery, by his disciple Hippasus of Metapontum, of the irrational numbers. Irrationality, by infinite reciprocal subtraction, can be seen easily in the golden ratio of the regular pentagon. The golden section was better defined, later, by Euclid (and it is the first written description we have at present) as a proportion derived from the division of a line into two unequal segments (see Livio, 2002). Euclid called it "extreme and mean ratio."

The numerical value of the golden mean can be expressed in the simplest way as follows (see its graphic representation in Figure 8.20):

given a line with three points, A, B, C, where AC is the total length and BC the longer segment, the golden section is obtained when AC/BC = BC/AB.

This number, called Φ, is the irrational number 1.61803 39887, etc. (which has been calculated till the 4,599th decimal figure). Its reciprocal (therefore, equivalent to *BC/AC* and to *AB/BC* in the previous relation) is also considered a "golden" irrational number, called φ, equal to 0.61803 39887, etc. (note that the zero is followed by the same decimal series calculated in Φ).

One of the classical constructions of the golden section is by inscribing a square into a semicircle (see Figure 8.21). The radius of the semicircle cuts the extended baseline of the square in the proportion of the golden ratio on both sides.

The golden ratio plays a crucial role in the symmetry properties of two regular solids, the dodecahedron (with 12 pentagonal faces) and the icosahedron (with 20 triangular faces). In both cases, these properties are based on a remarkable symmetry of the pentagon: each of its five diagonals cuts two diagonals with a golden section. In other words, the golden section is displayed in the well-known regular pentagram, the figure in which each intersection divides both

Figure 8.20 A graphic representation of the golden mean: a line *AC* is divided by a point *B*, so that the ratio between the total length and the longer segment (*AC/BC*) is equal to the ratio between the longer and the shorter segments (*BC/AB*).

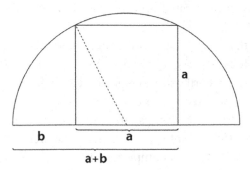

Figure 8.21 Classical construction of the golden section: $\Phi = (b+a)/a = a/b$ and $\varphi = a/(a+b) = b/a$.

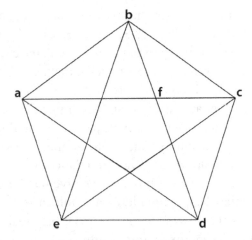

Figure 8.22 The construction of the golden section from a regular pentagle's geometry: $\Phi = ac/af = af/fc$ and $\varphi = af/ac = fc/af$. Moreover, it can be demonstrated that $af = ab$ (so that the side and the diagonal of a regular pentagon are in golden ratio).

lines in the golden ratio. In addition, the proportion between the pentagon's diagonal and its edge is again equal to Φ (see Figure 8.22).

The golden section has fascinated mathematicians, philosophers, and artists throughout the ages. In the Renaissance, it was defined as "the Divine Proportion"[6] and was exalted, together with the square and the circle, as one of the three classical symbols of perfection. Renaissance artists and architects considered it as the proportion most pleasing to the eye, and many of them,

---

[6] *De Divina Proportione* (*On the Divine Proportion*) is a book on mathematics written by Luca Pacioli and illustrated by Leonardo da Vinci, composed around 1498 in Milan and first printed in Venice, in 1509. Its subject was mathematical proportions (the title refers to the golden ratio) and their applications to geometry, nature, visual art, and architecture. The clarity of the written material and Leonardo's excellent diagrams made this book well known worldwide, as a result of Renaissance ideality.

including Leonardo, incorporated approximate golden ratios into their works (see Livio, 2002).

The point is also that proportions corresponding to this value have been found in several geographical places, especially in monuments (such as the great Egyptian Pyramids, the Parthenon and several Greek temples, as well as in many other cultures' monuments); it also appears in music, in violin construction, and in nature – in the shells and snails, and even in the human figure. (See, for example, Runion, 1990; Huntley, 1970; Herz-Fischler, 1998; and the already cited book by Livio, 2002. For a comprehensive review, visit http://www.math .uiuc.edu, or refer to the book by Livio.)

Is Φ (or φ) then the result of a way by which certain natural things self-organize? And the question for us, here, is why nature should care for such geometrical constraints.

Moreover, things become even more intriguing if one considers the Fibonacci series, which is: 0, 1, 1, 2, 3, 5, 8, 13, 21, 34, etc. (each number being given by the sum of the two previous ones) – with the consideration that so many plants, in their proportions, respect such a mathematical series (Runion, 1990; Huntley, 1970). This is so in the configuration of the scales in the pineapple, or in the pine cone, and in many spirals in plants and their fruits. Well known, in this sense, is the striking growth pattern of sunflower seeds, which features two sets of interpenetrating spirals, one running clockwise and the other counterclockwise (see Figure 8.23). And the relation with the golden ratio is indeed remarkable: the ratio between two successive numbers in the Fibonacci series is the closest rational approximation to the golden section – examples are: 2/1, 3/2, 5/3, 8/5, ... 610/377, 987/610, ... so that the greater the value of the chosen pair, the better the approximation[7] of Φ.

Typically, the number of spirals in each set turns out to be two consecutive Fibonacci numbers. This means that the golden angle is the generative principle of this pattern, just as it is in the helical phyllotaxis.

Why should this be so? This has been a wonder always, and, as mentioned above, Huntley in a classical book (1970) aptly reminds one about the "divine proportions."

Things begin to get somewhat less mysterious, however, when one notices that the configuration of the leaves that gives rise to Fibonacci's series in plants brings about phyllotaxis, namely the disposition that optimizes the exposition to sunlight and permits self-similarity. Thus, at least in those cases, Fibonacci's series is a consequence of a genomic order, which in turn has been shaped by evolution, so as to permit a better life and reproduction in plants.

---

[7] 8/5 = 1.6 is better than 5/3 = 1.6667 in approximating Φ (= 1.6180339887...). Even better is the result given by greater Fibonacci's pairs, as 610/377 = 1.618037... and 987/610 = 1.618032...

Figure 8.23 Seed heads in a sunflower, packed tightly in two sets of interlocking logarithmic spirals. iStockphoto.com/ © Nicholas Belton.

Seen in this way, Fibonacci's rule can be seen as a genetically determined self-organization pattern. And the golden section between the successive numbers is thus a consequence of this primary self-organization principle, shaped by evolution.

This explanation may work for the things of nature, but not for, say, the painting and the proportions of those temples of antiquity which have been built by architects, who apparently were not aware of the existence of the golden rule. Then, according to some, the golden rule may even be inbuilt in human beings as an archetypical structure. Also, the proportions of the golden section have been recognized in the spirals of snails and several shellfish, as well as in the spirals of galaxies and other macroscopic or celestial structures. Self-organization works at different scales, possibly suggesting an underlying order or finality (see also our discussion on "finality" in the next chapter).

And in fact, regarding larger scales, there are very challenging ideas in literature about a possible "hidden order" in nature, or equivalent terms and concepts. For example, the ideas of the physicist Geoffrey West and two ecologists, Jim Brown and Brian Enquist, according to whom a series of biological patterns vary as quarter-powers of body mass – like the heart rate of mammals, life span, the height of trees, and so on (as reviewed in Whitfield, 2006).

Talking about patterns and order in nature, one cannot refrain from mentioning the work of Mandelbrot (1982) and his fractals. This is, however, out of the space/time possibility of this book; I refer the reader to the recent Capra and Luisi book (2014), where we consider this issue at some length.

8.13 Out-of-equilibrium self-organization

The chemical and biological examples of self-organization given above are familiar to chemists and biologists. As already mentioned, for a relatively large community of scientists working in the area of physics of complex systems, as well as artificial life, self-organization (and emergence) is also something else: a characteristic of dynamical systems, namely systems that change over time. Terms such as chaos and non-linear dynamics, self-organized criticality, self-organization in non-equilibrium systems, fluctuations, bifurcations, are typical in this field (Nicolis and Prigogine, 1977; Bak *et al.*, 1987, 1988; Langton, 1990; Baas, 1994; Hilborn, 1994; Strogatz, 1994). Such dynamical systems are generally out of equilibrium, and at first sight it is counterintuitive that a dynamic system out of equilibrium may form self-organized structures. This is, in fact, the challenge and beauty of this particular field.

One classic example is the formation of "beehive" structures in an oil silicon layer that is heated between two glass surfaces. Resulting convection (the so-called Bernal convections, or Bernal cells) brings about long-range hexagonal structures, instead of an increasingly disordered molecular mixing, as one would have expected. This regular, beautiful structure remains visible by the naked eye as long as the temperature difference is maintained (see Figure 8.24). The temperature at which the hexagonal cells form is a critical point, or bifurcation point, from which the system may adopt one pattern or another one (see Coveney and Highfield, 1990).

This notion of bifurcation point, connected with those of instability and fluctuations, is the basis of this branch of science of self-organization in out-of-

Figure 8.24 The artificial beehive structure obtained by heating a silicon oil between two glass plates (Bernal convection). (Adapted, with a few modifications, from Coveney and Highfield, 1990.)

equilibrium systems. The story begins with Alan Turing, who, in search of the chemical basis of morphogenesis, predicted that, while homogeneity is the normal situation near equilibrium, systems may become unstable if far from equilibrium because of fluctuations (Turing, 1952).

The next two important steps in this narrative are: (1) the description of the "bruxellator" by Prigogine and Lefever, who, following Turing's work, analyzed theoretically the ingredients that should be present in a model of chemical reactions in order to produce spatial self-organization (Prigogine and Lefever, 1968); (2) the description of the Belousov-Zhabotinsky (B–Z) reaction.

The work of Prigogine and Lefever, in total observance with the second principle of thermodynamics, showed that regular oscillations of concentration could be interpreted in terms of thermodynamics. In particular, the "bruxellator" shows how it is possible that order originates from the disorder throughout self-organization, which in turn is due to oscillations in a system out of equilibrium. Also, if the system is maintained far from equilibrium via a continuous addition of colored chemicals, the system can oscillate between two or more colored states. This is a particular case of dissipative structures giving rise to self-organization, the important concept developed by Prigogine and his school. The notion of bifurcation also comes from this school, and the relevance to what is being discussed is illustrated in Figure 8.25. Here, the property λ (in the ordinate) at equilibrium, or close to it, obeys a linearity regime and the principle of minimal entropy production (another theorem of the Prigogine school) and the system is in a stable, stationary state.

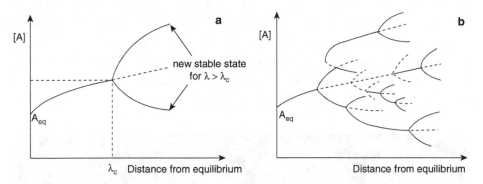

Figure 8.25 Bifurcation far from equilibrium. (**a**) Primary bifurcation: λ_c is the distance from equilibrium, at which the thermodynamic branching of minimal entropy production becomes unstable. The bifurcation point or critical point corresponds to the concentration λ_c. (**b**) Complete diagram of bifurcations. As the non-linear reaction moves away from equilibrium, the number of possible states increases enormously. (Adapted, with permission, from Coveney and Highfield, 1990.)

However, at a critical distance from equilibrium, the system must "choose" between two possible pathways, represented by the bifurcation point λ_c. The continuation of the initial pathway, indicated by a broken line, indicates the region of instability. The concentration of the species A and the value of λ assume quite different values, and the more so, the further from equilibrium. An important point is that the choice between the two branching directions is casual, with 50:50 probability of either. The critical point λ_c has particular importance because beyond it, the system can assume an organized structure. Here the term *self-organization* is introduced as a consequence of the *dissipative structures*, dissipative in the sense that they result from an exchange of matter and energy between system and environment (we are considering open systems).

It is also worthwhile mentioning that the origin of homochirality has been viewed in terms of a bifurcation scenario (Kondepudi and Prigogine, 1981; Kondepudi *et al.*, 1985). In this case, the homochirality present on Earth would be a product of contingency.

These organized structures can take the form of *oscillations*, and this is indeed the case of the above mentioned Zabotinski-Belousov (Z-B) reaction, observed in the 1950s by the Russian chemist Boris Belousov (see Winfree, 1984). It is interesting to note that Belousov – just like Turing – was investigating a biological model, the Krebs cycle in his case. The complex reaction mixture contained potassium bromate (to simulate citric acid), sulfuric acid, cerium ions (to simulate the action of certain enzymes), and, as it was shown by later studies, may involve about 30 chemical intermediates and subreactions, some of them of autocatalytic nature (Winfree, 1984). Aside from the complexity of the reaction, which I will not consider here, it is important to go back to the original observation of Belousov (later studied in detail by Anatoly Zhabotinsky): the solution began to oscillate, with great precision, between a colorless state and an orange one; and the formation of particular spatial structures was also observed (see Figure 8.26). This was in keeping with the prediction by Alan Turing (who in the meantime had committed suicide).

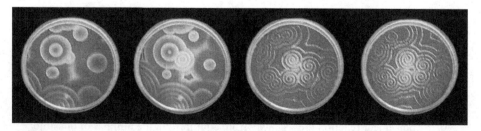

Figure 8.26 Some aspects of the Z-B reaction; details are in the text. (Adapted from Coveney and Highfield, 1990.)

As remarked by Coveney and Highfield (1990), it is too bad that the two scientists never met.

The field of *oscillating reactions*, or periodic reactions, or chemical clocks, came out of this background; indeed quite a number of chemical systems have been described, which show this oscillating, periodic, and regular behavior (Field, 1972; Briggs and Rauscher, 1973; Shakhashiri, 1985; Noyes, 1989; Pojman *et al.*, 1994; Jimenez-Prieto *et al.*, 1998).

How relevant is this phenomenology of out-of-equilibrium self-organization for life? For the scientists working in this field, the answer cannot be anything but positive, since all living systems are open, far-from-equilibrium, dynamic, non-linear, and dissipative structures. The complex adjectivation used in this last sentence is, in fact, typical for Prigogine's theory applied to living systems. To this list, irreversibly evolving systems could be added, which are characterized by an "arrow of time," as Coveney and Highfield (1990) remark.

A landmark for this field is the work of Per Bak, Chao Tang, and Kurt Wiesenfeld ("BTW") published in 1987 in *Physical Review Letters*, one of the most cited papers in the field. This paper was received with much enthusiasm by the community of scientists interested in complexity, and its ideas were applied in a large number of quite different fields, like cosmology, evolution, economics and computing, sociology, neurobiology, and plasma physics, all subjects we do not have the time to review. The key point of BTW's paper was the discovery of a mechanism by which the emergence of complexity from simple local interactions could be *spontaneous* – and therefore plausible as a source of natural complexity. Another important point is the scale-invariant behavior, which is the property that has permitted the application of the theory to so many different systems of varying complexity. Important, for example, is the applied sciences of earthquakes, forest fires, landslides, and epidemics.

We will come back to the questions of non-linearity and complexity when considering the notion of emergence, in the next chapter.

Concluding remarks

In this chapter, we have dealt mostly with the following aspects of self-organization:

1. Self-organization equilibrium (static) systems under thermodynamic control (originated from spontaneous processes with a negative free-energy change), such as supramolecular complexes, crystallization, surfactant aggregation, certain nano-structures, protein folding, protein assembly, and DNA duplex.
2. Self-organization systems arriving at equilibrium, but being under kinetic control (biological systems with genomic, enzymatic, and/or evolutionary

control), such as protein biosynthesis, virus assembly, formation of beehive and anthill, and swarm intelligence.

3. Out-of-equilibrium systems (non-linear, dynamic processes), such as the Zabotinski-Belousov reaction, and other oscillating reactions; bifurcation, and order out of chaos; convection phenomena, tornadoes, vortexes – and the self-organized criticality, as well as fractals (although not treated expressly here).

All that was restricted to biological systems, neglecting the many other aspects where self-organization has been treated in the literature, such as economy, evolution, cosmology, and geology. Needless to say, self-organization patterns play an important role in social systems, as well: human enterprises that form out of self-imposed rules, such as business companies, political parties, families, tribes, armies, churches, etc. (a part that has not been considered in this book). This entails the interesting question of whether the social organizational systems are genetically determined – sociobiology, as in the case of social insects – or induced by social and educational constraints. This theme is quite interesting, but again it is out of the limits of this chapter.

Going back to the question of the origin of life: self-organization processes give a kind of "free ticket" to move upwards in the ladder of complexity. However, is this really enough to reach the point of building macromolecular sequences and the first self-reproducing protocells?

We have already mentioned in the introduction to this chapter that self-organization is not enough. Emergence is the subject of Chapter 9 – we will see what emergence can do to move further towards the complexity of life.

Chapter 9

The notion of emergence

Introduction

The term *emergence* describes the onset of novel properties that arise when a higher level of complexity is formed from components of lower complexity, where "novel" means that these properties are not present at the lower level of the components. This is often summarized in the popular assertion that "the whole is more than the sum of the parts," and/or with the vague term *holism*.

For a long time, emergence has been an active field of inquiry in the philosophy of science. As noted by McLaughlin (1992), the work of "British emergentism" can be dated back to Mill (1872) and Bain (1870) and flourished in the 1920s with the work of Alexander (1920), Morgan (1923), and Broad (1925); and the inquiry continues up to the present (see Wimsatt, 1972, 1976a, 1976b; Kim, 1984; Klee, 1984; Sperry, 1986; O'Connor, 1994; Bedau, 1997; Farre and Oksala, 1998; Holland, 1998; Primas, 1998; Schröder, 1998; Barabási and Réka, 1999; Juarrero and Rubino, 2010; Addiscott, 2011).

The possible relevance of chemistry in the notion of emergence was realized as early as the mid-nineteenth century (Mill, 1872); and in 1923 (as quoted by McLaughlin, 1992), Broad stated that:

... the situation with which we are faced in chemistry [...] seems to offer the most plausible example of emergent behavior.

In our times, in addition to chemistry and biology, emergence is being considered in quite a variety of research fields, such as cybernetics, artificial intelligence, non-linear dynamics, information theory, social systems, music (the harmony arising from a musical phrase – something obviously not present in the single notes). In the preface to *The ECHO III Conference*, edited by Farre and Oksala (1998), applications to language, painting, memory, biological evolution, and the nervous system, are also mentioned and discussed. An insight into physics, including superconductivity and other collective phenomena, is provided by Coleman (2007). An insight into the emergence of values at the human level, including ethics and religious naturalism, is provided by Goodenough and Deacon (2006). Due to such a large variety of origins, it is not surprising that the term *emergence* often has a confusing connotation.

9.1 Ontic and epistemic

Before proceeding, let me clarify one point, which is useful to reintroduce a basic element of the philosophy of science. As many other concepts in this field, the notion of emergence can be considered on two different levels. On the one hand, there is the "ontic" interpretation, which refers to a theory about things as they really are, independent of any observational or descriptive context. On the other hand, there is the "epistemic" interpretation, which refers to our knowledge of observable patterns or modes of reactions of systems (Primas, 1998). Whereas the first level makes a direct application to practical problems rather difficult, an epistemic description allows an easier contact with empirical reality. This chapter deals with the second epistemic approach, with the expectation that this is more appealing and comprehensible to graduate students of science. This epistemic approach is based on the assumption that objects and levels of structures can be considered as *separated* – assumption that a more rigorous, ontic approach would be considered doubtful.

To this relation between ontic and epistemic approach, refer to the following citation by Primas (1998):

... the epistemic description contains a non-removable reference to the observing tools, the referent of such derived theories is the empirical reality.

This brings about an interesting relation between the ontic and the epistemic approach (Primas, 1998):

... an epistemic state of a derived theory refers to our partial knowledge of the ontic state of the fundamental theory.

In other words, a context-dependent theory still reflects some aspects of the independent reality.

Hans Primas uses this argument to introduce the notion of contextual ontology, which refers to emergent properties arising from hidden features of the independent reality. This permits a clear view of the relation between contextual and fundamental theories, and also a generalization that is relevant for philosophy of science at large (Primas, 1998):

Only if we maintain multiple sets of contextual ontologies, we can tolerate the coexistence of complementary views in our experience of reality. While an independent reality itself is directly inaccessible, the numerous inequivalent contextual descriptions allow us to get deeper insight into the structure of independent reality.

The simple examples presented in this chapter are relative to an epistemic approach, namely to a reality in terms of descriptive terms, rather than in terms of "things as they really are" (the ontological view).

9.2 A few simple examples of emergence

The trivial chemical example that is usually given in the literature on emergence since the early times by the British emergentists is water being formed from its atomic components. The collective properties of water are not present in hydrogen and oxygen; so the properties of water can be viewed as emergent ones.

This kind of simple-minded argument can be generalized to encompass all cases in which a molecule is formed from its atomic components – e.g., CH_4, CO_2, HCl, NH_3, and so on. In each case, obviously, the properties of the resulting molecule are not present in the initial components and can be seen as emergent properties. Francis Crick, in his book *The Astonishing Hypothesis* (1980), stresses the concept that there is nothing particularly new or exotic in the notion of emergence, as chemistry is full of it. He gives, as one of many, the example of benzene. The aromatic character of the benzene molecule is obviously not present in the atomic components, but is a property arising in the ensemble of the particular atomic configuration – an emergent property.

Turning to a more complex chemical structure, such as mammal myoglobin [Figure 9.1 (A)], the specific binding properties towards the heme group are

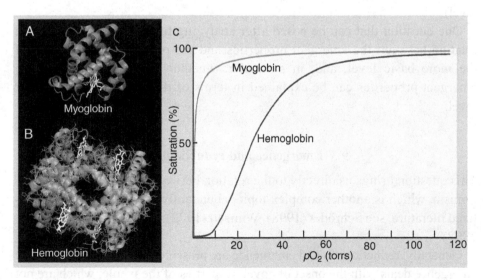

Figure 9.1 The single chain of myoglobin (**A**) and the four chains (two α and two β) of hemoglobin (**B**). The cooperative behaviour of hemoglobin (sigmoid curve), as indicated in the right panel (**C**), can be seen as an emergent property, coming from the interaction of the four chains.

obviously not present in the single amino acids – the binding specificity is an emergent property of the particular molecular ensemble.

Going from mammal myoglobin to hemoglobin [Figure 9.1 (B)], a higher level of hierarchic structure is found, as mammal hemoglobin is formed by four chains, each of them very similar to that of myoglobin. In this case, a new quality arises from this assembly, the cooperativity: whereas the binding of oxygen to myoglo-bin (single chain) gives a normal hyperbolic saturation curve [Figure 9.1 (C)], in the case of mammal hemoglobin, the binding isotherm is sigmoid. The difference is the very basis of respiration in mammals and is due to the cooperativity of the four chains in hemoglobin, which can be viewed as an emergent property arising from the interaction of the four chains.

The emergence of novel properties due to self-assembly is also present in much simpler systems. Consider, for example, the formation of micelles and vesicles from surfactants, as already seen in previous chapters (Figures 4.3 and 8.1). The compartmentation, and the constitution of an internal water pool which is distinct from the outside, are such properties. There are also collective motions of aggregates, as well as changes in the physical properties – such as the pK of fatty acids constituting an assembly – that are clearly emergent properties of the aggregate. Collective properties, which are characteristic of whole aggregate systems, have also been noted by Menger (1991).

There is also emergence at the level of geometry, for example, from lines to angles to surfaces to cubes, where at each level of complexity, novel, emergent properties arise (surface, volume, or angle, which are not present at the lower level).

One question that can be posed after analyzing these few examples is the relation between the emergent properties and the properties of the parts of the more basic level, and, in particular, whether and to what extent the emergent properties can be explained in terms of the elements of the lower level.

9.3 Emergence and reductionism

This question brings us directly to the relation between emergence and reduc-tionism, which is another complex topic, abundantly discussed in the specia-lized literature, see Schröder (1998), Wimsatt (1972), and Primas (1985, 1993, 1998).

Generally, reductionism and emergence are presented as two opposite fronts: emergence deals with the onset of novel properties of the whole, which are not present in the basic components, and as such has an upwards direction. Reductionism generally looks down from a certain level of complexity, trying to explain each level on the basis of the components. There are, of course, different positions within reductionism, including cases of radical reductionism. For example, the more radical views are those of Oppenheim and Putnam (1958), Nagel (1961), and Reichenbach (1978), all cited by Atmanspacher and Bishop (2002). These views are basically grounded on the idea that chemistry and all natural science can be reduced to physics, something which is strongly opposed by Primas (1993). The strongest form of reductionism maintains (according to Ayala, 1983; and cited by Primas, 1998):

... that organisms are ultimately made up of the same atoms that make up inorganic matter, and of nothing else.

Many people, including many scientists, would tend to agree with such a statement. However, just for this reason, it is important to make an important clarification: *the difference between structure and properties*. There is nothing wrong, in principle, with reductionism if it stops at the level of structure: we can all agree that water consists of hydrogen and oxygen; and that living cells are constituted by molecules, which in turn are constituted by atoms. The problem with reductionism is not at this level, but with the claim that the *properties* of water can be reduced to the properties of hydrogen and oxygen.

In other words, the previous assertion by Ayala that all living systems consist of the same atoms that make up the inorganic matter, is something with which we can agree. However, what is missing in this view is the appreciation of the emergent properties – from atoms to molecules, from molecules to genes and enzymes, from these to cells, and from cells to organs; whereby each time, at each level of complexity, there is the onset of novel properties that are not present at the lower level – and, importantly, that cannot be explained on the basis of the constituents. Thus, a cell can be said to be constituted by atoms and molecules, but the essence of a living cell, or of a human being, cannot be satisfactorily expressed in terms of atoms and molecules. The affirmation that instead it may be so, corresponds to the very reductionist paradigm that is very difficult to defend; and this brings us to the next point.

9.4 Deducibility and predictability

Can the emergent properties be deduced from the properties of the components?

This relation between emergent properties and properties of the basic components has been much debated in the literature. One school of thought claims that the properties of the higher hierarchic level are *in principle* not deducible from the components of the lower level. This is the so-called "strong emergence" or radical emergence, that demands, as formulated by Schröder (1998) that:

the [. . .] relation between an emergent property of a whole and the properties of its parts is [. . .] one of non-explanatoriness.

This idea is an old one, as it was put forward by the British emergentists such as Mill (1872), Alexander (1920), Broad (1925), and to some extent by Morgan (1923), and has also been discussed in the following literature (see, for example, Wimsatt, 1972, and McLaughlin, 1992). In other words, the emergent property of the whole is inexplicable, i.e., non-deducible from the properties of the parts.

Opposite to "strong emergence" is "weak emergence," an epistemic point of view that more pragmatically asserts that the relationship between the whole and the parts may not be established because of technical difficulties, such as the lack of computational power or insufficient progress of our skills. Atmanspacher and Bishop (2002) discuss this point at length.

The ontological view that emergent properties of molecules cannot be explained as a matter of principle on the basis of the components is opposed by several scientists, who argue that this is tantamount to assuming that a mysterious force of some undefined nature is at work. In fact, strong emergence may sound like a kind of vitalistic principle – and this is something we do not necessarily want to reintegrate in science. For example, Bedau (1997) writes:

. . . to judge from the available evidence, strong emergence is one mystery we don't need.

The discrimination between a matter of principle (strong emergence), and a matter of practical difficulty (weak emergence), takes us again to the introduction of this chapter, where I cited Primas on the relation between ontic and epistemic views of things. The epistemic view generally contains elements of the ontology, and in this sense I believe that the above discrimination between strong and weak emergence is not always possible – and perhaps it does not always make sense. Take the case of myoglobin, with its 143 amino acid residues. Can the properties of myoglobin be predicted on the basis of the properties of the 20 amino acids?

There are 20^{143} possibilities of making a chain with 143 residues. In principle, the folding and then the binding properties of each of these different sequences could be calculated and eventually predicted, but this would take millions of years – aside from the fact that we really do not know yet how to make these calculations. Therefore, it is not simply a matter of time, but also a matter of our intellective power. At this point, is the impossibility of predicting the properties of myoglobin a matter of principle (strong emergence), or is instead a matter of practical impossibility (weak emergence)?

I would maintain that it is not possible to make such a discrimination. Thus, more often than not, weak emergence and strong emergence cannot be sharply divided from one another.

At any rate, is the main consideration about emergence valid, namely that the novel properties are not deducible from – and are not reducible to – those of the components? It should be clear that this does not imply mysterious forces, but simply reflects the limits of our capability.

Thus, going back to our cell: at each level of increasing complexity in going from atoms/molecules to biological complexes, to organelles and to the entire organism, there arise novel properties that cannot be described in terms of the lower constituents, and therefore the properties of a cell, or any other living organism, cannot be interpreted on the basis of atoms and molecules. This is so, regardless of whether it is a matter of principle or a practical difficulty. Emergence really makes the difference between the reductionist interpretation and a more holistic view of reality.

9.5 Downward causation

It is generally accepted that the development of emergent properties, which is an upward (or bottom-up) causality, is attended by a downward – or top-down – causality stream. This means that the higher hierarchic level affects the properties of the lower components, as reflected by Schröder (1998):

Downward causation is the influence the relatedness of the parts of a system has on the behavior of the parts [. . .] it is not the influence of a macro-property itself, but of that which gives rise to the macro-property, viz., the new relatedness of the parts.

There is an ongoing discussion in philosophical literature on the relation between emergence and downward causation – also called macro-determinism – see for example Bedau (1997), Schröder (1998), Thompson and Varela (2001) and Thompson (2007). It is generally assumed that emergence and downward causation take place simultaneously; in particular, Thompson and Varela (2001) and Thompson (2007) like to combine the occurrence of upward causation (emergence) with downward causation by using the notion of cyclic causality. Evan Thompson (2007) now likes to use the term *reciprocal causation*, thus eliminating the uni-directionality of the word *downward*. (See also the conversation with Evan Thompson in Part II of the book.)

Generally, the point can be made that molecular sciences, and chemistry in particular, offer very clear examples of downward causation, as defined above. In the trivial chemical example of the formation of water from the two gaseous components oxygen and hydrogen, this formation profoundly affects the properties of both hydrogen and oxygen (due to the formation of binding orbitals). Likewise, the electronic orbitals of carbon atoms, and those of oxygen and hydrogen, are changed when the molecule of benzene is formed. In chemistry, relatedness means interaction, and any form of chemical interaction modifies the properties of the components.

It is important to emphasize that this kind of chemical relatedness does not need to assume any special effect other than the normal laws of chemistry and physics. As already mentioned, special effects are often invoked in the literature – particularly in the old "British emergentism." This has been one of the reasons why emergentism in general has been criticized, and once special forces are eliminated from the picture, then this criticism immediately loses validity, as pointed out by Schröder (1998) in his criticism of McLaughlin (1992).

Of course, molecular-science examples are not the only ones to show the effect of downward causation. This is so for all aspects of emergence. Consider the progression of hierarchic levels that go from the individuals to the family to the tribe to the nation: it is clear that once the individuals have a family, the rules of the family affect and change the properties of the individuals, and so on.

9.6 Emergence and dynamic systems

We go now, as anticipated, to dynamic, non-equilibrium systems. In the previous chapter, it was already shown that self-organization has a very strong component in the domain of physics of non-linear systems, and actually some examples of self-organization in non-equilibrium systems have already been given, such as the beehive structure in silicon oil subjected to thermal gradient. This aspect will now be considered again, with a stronger emphasis on the emergent properties.

As already mentioned in the previous chapter on self-organization, the theoretical background of this aspect of emergence can be traced back to the dissipative structures of Prigogine. A dissipative structure in these terms is an open system that is far from equilibrium, and yet is still maintaining a form of stability. In a pendulum, dissipation is caused by friction, which decreases the speed of the pendulum and eventually brings it to a standstill. However, if energy is continuously provided to the system, the oscillating pattern is maintained through a flow of energy. In more complex systems, depending upon the initial conditions and fluctuations of the energy flow, the system in its dynamic behavior may encounter a point of instability – the bifurcation point – at which it can branch off with the emergence of new forms of structure and properties. According to Fritjof Capra (2002):

The spontaneous emergence of order at critical points of instability is one of the most important concepts of the new understanding of life [...] It has been recognized as the dynamic origin of development, learning and evolution.

This statement makes clear the synergy between emergence and self-organization. Capra then goes on to discuss the implementation of this kind of emergence not only in biological structures, but also in social systems (management, information, language), as well as in psychology. Again, the emphasis is on the intrinsic instability of the system, and as such it is also useful to recall a citation by Prigogine (1997):

Once instability is included, the meaning of the laws of nature [...] change radically, for they now express possibilities of probabilities.

Because of this, as Christidis puts it (2002), matter acquires new properties when estranged from equilibrium, namely when fluctuations and instabilities are dominant. For a series of examples in biology, chemistry, and engineering, see Strogatz (1994 and 2001).

How do we go from these theoretical concepts to living structures? Certainly living cells are open systems; and in fact, in this case, the link with the above thermodynamic relations is particularly interesting: the second law of thermodynamics states that the entropy of a closed system can only increase, or remain constant, in which case the system has reached a state of equilibrium. The cell does not proceed towards equilibrium just because the system is not closed; rather, it is an open system with a constant flux of energy during its life cycle. Autopoiesis, as we have seen, can also be viewed as an expression of the same concept. To maintain this flux, dissipation is required, and we are thus back to the notion of dissipative structures. Actually, the term *dissipative* is often referred to structures that emerge as a result of self-organization and use dissipation to retain their organization. This should not be confused with static, equilibrium systems. They are just two different aspects of the physical world.

We should mention again Bak's work on self-organized criticality, illustrated in the previous chapter, as in this case the notion of emergence concerns large-scale phenomena of geological and astronomical dimensions (Bak *et al.*, 1988).

The link between biological systems and non-linear dynamic systems is also an important issue in this field. This is stressed well in the contribution by Stuart Kauffman (Side Box 9.1). This box is a good introduction to the next section, which concerns emergence in some more complex biological systems.

Side Box 9.1

The sciences of complexity

Stuart A. Kauffman

Stuart Alan Kauffman is a medical doctor, a theoretical biologist, and complex systems researcher. He is an emeritus professor of biochemistry at the University of Pennsylvania, and a seminal member and an external professor of the Santa Fe Institute. Kauffman has published three major books, among them is *At Home in the Universe: The Search for the Laws of Self-Organization and Complexity* (1995).

Kauffman is the pioneer and founding father of biocomplexity research. He was able to look at gene regulatory networks from a new and different point of view, and he started working on biocomplexity, showing that the behavior of genetic networks depends critically on the level at which the genes are connected. At the end of the 1960s, Kauffman proposed applying models of random Boolean networks to simplified genetic circuits. These were very early models of large genetic regulatory networks, proposing that cell types are dynamical attractors of such networks and that cell differentiation steps are transitions between attractors. Recent evidence strongly suggests that cell types in humans and other organisms are indeed attractors. In 1971, he suggested that the zygote may not access all the cell type attractors in the repertoire of the genetic network's dynamics, hence some of the unused cell types might be cancers. This suggested the possibility of "cancer differentiation therapy," for which some evidence may now be accumulating. In the same year, Kauffman proposed the self-organized emergence of collectively autocatalytic sets of polymers, specifically peptides, for the origin of molecular reproduction. Reproducing peptide, DNA, and RNA collectively, autocatalytic sets have now been made experimentally.

He is best known for arguing that the complexity of biological systems and organisms might result as much from self-organization and far-from-equilibrium dynamics as from Darwinian natural selection. His hypotheses stating that cell types are attractors of such networks, and that genetic regulatory networks are "critical," have found experimental support.

The FidiPro program brings Kauffman to Tampere University of Technology. The subject of the program is stochastic modeling of gene regulatory networks. The project will focus on modeling the gene regulatory networks using gene expression data.

The sciences of complexity appear to afford a countermeasure to the dominance of reductionistic science, which has held away for over 300 years. It is not that Newton was unaware of the need to integrate the parts of a system into an understanding of the whole. Indeed, such an integration of his three laws of motion and law of universal gravitation gave us classical physics. Rather, in many of the fields mentioned above, studying the collective behaviors of heterogeneous systems of many interacting parts is central to their intellectual mission, is typically beyond the scope of current analytic techniques, and is gradually coming to the fore.

In physics, the late Danish physicist, Per Bak, and his colleagues Tang and Wiesenfeld startled the field in the late 1980s by producing a widely quoted paper on self-organized criticality. Bak and others applied this model widely – to the size distribution of earthquakes and the distribution of clusters of matter in the universe, to the size distribution of extinction events in the biological record. Self-organized criticality has been tested and succeeds in some cases, not in others. More important is the spirit of the effort: find collective variables and predict their collective behaviors, largely independent of the details of the individual parts.

In chemistry, one area that has received outstanding attention is that of RNA folding, shape, and evolution. Peter Schuster, Walter Fontana, Peter Stadler, and their colleagues have made major contributions. Among the concepts here are "energy landscapes" for computer folded models of RNA molecules, the evolution of model RNA sequences over these landscapes in sequence space, the folded shapes of model RNA sequences, and the existence of connected "neutral" pathways across sequence space among model RNA molecules that fold to the same shape.

Biology is emerging as a core field for the sciences of complexity. We have entered the post-genomic era. For 50 years, molecular biologists have focused brilliantly on the examination of specific genes, their "upstream" regulatory "cis" sites, the transcription factors binding such sites and the regulation of the activities of specific genes. It is now becoming abundantly clear that understanding the integrated holistic behavior of such networks is the next overwhelming challenge. One approach, begun by this author, is an ensemble approach in which one learns all the constraints present in real genetic networks, such as the number of molecular inputs that regulate genes and the distribution of that number across genes, the number of genes regulated by given genes, the control rules governing the activities of specific genes as a function of their inputs, the distribution of sizes of feedback loops, and so on.

Other approaches to genetic networks include the study of small circuits with either differential equations or stochastic differential equations. The use of stochastic equations emphasizes the point that noise is a central factor in the dynamics. This is of conceptual importance as well as practical importance. In all the families of models studied, the non-linear dynamical systems typically exhibit a number of dynamical attractors. These are sub regions of the system's state space to which the system flows and in which it thereafter remains. A plausible interpretation is that these attractors correspond to the cell types of the organism. But in the presence of noise, attractors can be destabilized.

Progress in understanding genetic networks also includes attempts to solve the inverse problem. It is now possible to study the level of expression of some 10,000 genes simultaneously by obtaining RNA sequences from a set of cells. If the population of cells is differentiating, the patterns of gene activities change over time. Thus a "movie" can be constructed showing the waxing and waning of abundances of RNA sequences using gene chip arrays. An industry is growing that attempts to derive features of cells and circuits from such arrays. In one line of work, the inverse problem consists in viewing the movie and deducing the circuitry and logic driving the behavior of the genes.

In a variety of ways, the sciences of complexity are coming to bear on economics. For example, an agent-based model of a stock market has been constructed and shown that both fundamental pricing behavior and speculative bubbles can occur. Another just emerging area of the commingling of the sciences of complexity and standard economic theory is in the area of economic growth. Models in complexity theory have shown evidence that the rate of growth of an economy can be linked to the total diversity of goods and services in the economy, which are able to recombine combinatorially to create novel goods and services.

In computer science, advances have been made on several forefronts. These include the study by Crutchfield and colleagues of the emergence of pseudo-particles in cellular automata, where the particle interactions can carry out computations.

In management science, use of rugged landscape models has been brought to bear on questions concerning the adaptability of organizations, optimal organizational structure, the co-evolution of organizations, and further topics.

In summary, the sciences of complexity study the emergent collective behaviors of systems with many heterogeneous interacting parts. Those collective behaviors can now be studied with the computers that have become available. Computers will become more powerful. In a variety of scientific areas, the need to study the integrated behaviors of complex systems is becoming utterly apparent. This new body of science is science, not fad. It will grow, like all science, by stumbling forward, making mistakes, but with moments of real progress. That progress will complement the power of reductionism in leading us to a deeper understanding of myriad phenomena. Some of us, myself included, hope that general laws will emerge. That tale remains to be told.

9.7 Life as an emergent property

Biology offers innumerable examples where the increase of complexity is attended by the onset of sophisticated emergent properties. In the previous chapter, the beehive-like structure of silicon fluids was mentioned; if we now consider a real honeycomb, each bee appears to behave as an independent element, acting apparently on its own account, but the whole population of bees gives a highly sophisticated collective emergent structure. The same can be said for an anthill and for other social animals – in the previous chapter, two patterns of flying birds were given (Figure 8.14), among other examples of "swarm intelligence."

Another very interesting phenomenon, which belongs to self-organization of "simpler" living systems, is the "quorum sensing." This is a cell density-dependent signaling mechanism used by many species of bacteria (Funqua *et al.*, 2001). Above a certain threshold of cell concentration, new collective properties appear in the colony, which are not present at lower cell density. Quorum sensing controls several important functions in different bacterial species, including the production of virulence factors and biofilm formation, as well as the capability of colonizing higher organisms, for example, in *Pseudomonas aeruginosa*, as well as in the appearance of bioluminescence in *Vibrio fischeri* (Miller and Basler, 2001; Smith and Iglewski, 2003). It is basically an intercellular signaling, based on the production of special molecules that activate genes, which in turn produce special proteins. There are already studies referring to the semantics and syntaxes of such a language (Ben Jacob *et al.*, 2004), as well as to inter-kingdom signaling (Shiner *et al.*, 2005). I find particularly fascinating the onset of bioluminescence in *V. fischeri* above a certain threshold of cell density, which is used as a recognition mechanism for members of the same species, and favors grouping and pairing. Strangely enough, this phenomenon is not usually considered in the field of emergence or complexity; however, it is again the onset of collective properties due – in this case – to a clear chemical signal.

In all these cases, we come to a concept that we have already mentioned, namely that regular patterns of self-organization are realized without any externally imposed rules. There are complex self-organization patterns, but localized organization centers are not present. In this regard, I would like to cite Francisco Varela in one of his last interviews (Poerksen, 2004):

Consider, for example, a colony of ants. It is perfectly clear that the local rules manifest themselves in the interaction of innumerable individual ants. At the same time, it is equally clear that the whole anthill, on a global level, has an identity of its own [...] We can now ask ourselves where this insect colony is located. Where is it? If you stick your hand into the anthill, you will only be able to grasp a number of ants, i.e., the incorporation of local rules. Furthermore, you will realize that a central control unit cannot be localized anywhere because it does not have an independent identity but a relational one. The ants exist as such but their mutual relations produce an emergent entity that is quite real and amenable to direct experience. This mode of existence was unknown before: on the one hand, we perceive a compact identity, on the other, we recognize that it has no determinable substance, no localizable core [...]

Actually, this concept of self-organization and emergent properties as a collective ensemble, without an organized localized center, is nowadays under scrutiny in cognitive sciences: several scientists in the area would now agree that the very notion of "I" is an emergent property arising from the simultaneous juxtaposition of feelings, memory, thoughts, remembrances, and so on, so that the "I" is not localized somewhere, but it is rather an organized pattern without a center. In order to make clear the analogy between this last concept and the organization/emergence in the case of social insects, let's again cite Varela, in the same reference as before:

This is one of the key ideas, and a stroke of genius in today's cognitive science. There are the different functions and components that combine and together produce a transient, non-localizable, relationally formed self, which nevertheless manifests itself as a perceivable entity [...] we will never discover a neuron, a soul, or some core essence that constitutes the emergent self of Francisco Varela or some other person.

For more on cognitive science regarding this and analogous concepts, the reader is referred to the books by Damasio (1999), Varela (1999), and le Doux (2002).

9.8 Self-organization and finality

We have already addressed the question of finality in Chapter 3, concluding with Lucretius about the primacy of structure, in the sense that nature does not proceed with the aim of constructing functions; these are the consequence of the structure, which is built in large part by the vagaries of contingency. However, at the level of the smaller design, one can still ask the question, whether the beehive is there at the aim of providing a house for the bees, or one can ask whether the peacock displays his colorful wheel at the aim of seducing the beautiful female in his neighborhood. And we will now reconsider this argument here, also on the basis of the opinion of well-known authors, such as Ernst Mayr.

The relation between self-organization of the living and finality is an old issue in the philosophy of science, particularly in biology, something that goes back to Aristotle and Plato, and remains in the present with Jacques Monod, Stephen Jay Gould, and Ernst Mayr. Finality (a term of Latin origin) here is used to mean a purposely planned direction, or goal-directed pursuit. Some authors prefer the term *finalism*, for example, Philippe Dailleur, in a 2009 article in *Interdisciplinary Encyclopedia of Religion and Science* (which also includes a rich bibliography on the subject). In the philosophy of science and scientific literature, the term *teleology* (from the greek *télos*, end or purpose; and *lógos*, principle, law) is used, often juxtaposed with the term *teleonomy*, and this dichotomy is actually the main subject of this section.

The starting observation is that indeed so many things in life appear to have a goal: everything that is born strives to live as much as possible; bees seem to have as their main goal the construction of their beehive; and the same is true for ants, termites, and so on. Almost all birds aim at building a nest, beavers at building dams, and apparently all living organisms aim at having a progeny. A seed planted in the ground gives rise to a tree: is this a plan, a finality to be seen here? And isn't this so in the development of an organism, from the baby to the adult? Evolution appears to be full of designs, somebody says that the finality of evolution is to create mankind. And for some, the big bang and the expansion of the universe may reflect an apparent finality, so that there might be even a justification to the notion of cosmic teleology. Actually, the idea of cosmic teleology, and teleology in general, was in fact very strong throughout the nineteenth century, being postulated also by important intellectuals like Leibniz and Lamarck, and later by Taillard de Chardin.

According to Ernst Mayr, and nowadays to almost all of modern science, teleology in nature can be explained with natural causes. This term, *teleology* – Mayr observes – has been applied to quite different kinds of processes or phenomena. (See his essay, *Teleological and Teleonomic: A New Analysis* (Mayr, 1974a), where he distinguishes several aspects of the question; see also Mayr, 1988, 1992; and Pross, 2005).

The most common, which he called *teleomatic*, is when the apparent purpose is due to the natural laws, like the falling of a stone, or the flowing of the river towards the ocean – phenomena due to the law of gravity. The first principle of thermodynamics, which makes the heat flow from a warm to a colder body, or the second one, which makes a chemical process go towards the most stable compound, gives also a series of apparently teleogical (in fact, teleomatic) phenomena. According then to Mayr, the apparent cosmic finality, from the big bang onwards, is due to a series of teleomatic processes (with the addition of stochastic perturbations). The birth of biosphere is non-deducible from these principles. And therefore, simply, there is no reason to invoke a cosmic teleology – indeed, a notion that modern science should reject.

The same – Mayr argues (1974a, 1974b) – is true for biological evolution: also in this case the apparent teleology can be explained and interpreted using natural laws. The biological evolution from the initial prokariots to eukariots, then to multicellular organisms, and to more complex plants and animals, can be fully explained in terms of natural selection (i.e., selective forces generated by the competition of living organisms and by the adaptation to the changing environment). This is the Darwinian view, which culminates in the consideration that the behavior of a living organism, be it an amoeba, an oak tree, or a mammal, is due to its genetic program. In this case, we talk about *teleonomy*, processes due to the existence of genetic programs, which give a kind of mechanistic explanation to teleological phenomena.

In conclusion, teleology is concerned with design, purpose, and directive principle, which is characteristic of the natural or human enterprise. By contrast, teleonomy (a term actually first introduced by the cybernetist

Norbert Wiener (Rosenblueth et al., 1943) is a form of philosophical naturalism, which views nature as lacking in design and purpose. Teleology would argue that humans have eyes because they have been programmed and predetermined, in order to fulfill the function of seeing. Modern evolutionary theories accept instead the principle of contingency, according to which there is a primacy of the structure, whereby the structure is constructed in nature – both in prebiotic and biological evolution – without a predetermined plan.

In biology, once the principles of Darwinian evolution have been accepted, according to which evolution operates without a pre-set goal, it is then the term *teleonomy* that should be used. It has been linked to the genomic imperative, by asserting with Mayr (as reported in his 1974a essay, he proposed the following since 1961):

... it would be useful to restrict the term teleonomic to systems operating on the basis of a program of coded information.

Also Monod (1971) has emphasized in this respect the notion of teleonomy to indicate an activity that is directed towards the realization of a biological program. The most salient program is the genetic one, which implies the species reproduction and evolution. Monod's term *program* is very important in this context, as it does not imply finality towards a future pre-set state (teleology), but rather a series of events which are coherent with and determined by the program itself – whereby the result of natural selection is constantly governed by the selective value of the reached terminal point. However, it should be pointed out that also the notion of "program" has been criticized, for example, Denis Noble suggests that we should not use this term (Noble, 2006) – as there is no programmer.

But here, at this level, on whether namely a genetic program of a living organism responds to a teleological or rather a teleonomic principle, the debate is still ongoing. Also, some philosophers (Lazzara, 2001) like to say that teleonomy is an *escamotage* in order not to talk about teleology. And there is even an old joke in the field (ascribed to Haldane), according to which teleology is the biologist's mistress without whom he cannot live, but with whom he is unwilling to be seen in public.

In fact, this simple dichotomy teleology/teleonomy in the things of nature, still gives rise to a series of doubts and questions. Let's take again human actions, for example. Some of our behavior is determined by genetic constraints (e.g., maternal love, the mating instinct, or the killing-ape instinct). All this can be seen as behavior based on genetic programs; should it then be said that we are witnessing actions of teleonomy? Or perhaps should we say that these actions start as teleonomic, but then, due to the guide of human intelligence and consciousness, may become teleological?

This might be a good compromise, but it does not simplify things, as the guide of *mind* is also present in many animals. There is also the consideration that the notion of finality implies a judgment by the observer, who has to define whether there is finality in a given act. For example, if we look at the amoeba moving in a sugar gradient, our mind will project that this movement has the finality of searching for food. So, there appears to be a hint of subjectivity in the very notion of finality.

These are all questions which can be asked, and make this field interesting and challenging to explore.

Concluding remarks

A meaningful point about emergence is the notion of non-predictability: the fact that novel, unpredicted properties can arise. This means – and is an exciting view – that there might be a vast arsenal of unforeseeable properties arising from the intelligent or serendipitous assemblage of components.

A second, unrelated point is the observation that the main concepts in the field of emergence become clear and simple, when using biological or chemical examples. This represents an important contribution of molecular science to the field of philosophy, a contribution – as already noted – not taken duly into account in the chemical literature and in chemistry textbooks. It would be advisable to introduce the notion of emergence in college textbooks, so as to provide young chemistry and biology students with a broader perspective.

Another important point about emergence is that life itself can be seen as an emergent property. The single-cell components, such as DNA, proteins, sugars, vitamins, lipids, etc., or even the cellular organelles such as vesicles, ribosomes, etc., are *per sé* inanimate substances. From these non-living structures, cellular life arises once the space/time organization is defined. The consideration that life is an emergent property gives the notion of emergence a particular significance. No vitalistic principle, no transcendent force, is invoked to arrive at life – and this, as mentioned already, has two consequences: (1) the mechanism of life, at least in principle, can be explained in terms of molecular components and their interactions; (2) it becomes conceivable to make some simple forms of life in the laboratory.

We have considered previously the cell, and claimed that one would not grasp its essence by saying that it is composed solely of atoms and molecules. We need an emergentist view, a systems approach accepting the emergent properties at each of the various levels of increasing hierarchic structural complexity. Some of the cognitive scientists would add that, even at this level, there is something missing, and this is the "observer" – the one who really gives meaning to *the rose* in terms of history, literature, and poetry. As we have illustrated in the chapters on autopoiesis and cognition, the notion of a rose is different depending on whether the observer is a Westerner educated in romantic literature, or an Eskimo who has never seen a rose. (See, about that, the strong point made by Maturana in our conversation.) Here is where the notion of emergence may become co-emergence between the object and the observer throughout his/her consciousness.

Chapter 10

Self-replication and self-reproduction

Introduction

The combination of self-organization and emergence may give rise to a most important aspect of life: growth and reproduction. This is the property of a given structure to make copies of itself by assembling smaller parts taken from the environment, this assemblage being generally catalyzed by specialized catalysts. Thus, DNA does not reproduce itself alone, of course, what is needed in addition to the monomeric nucleotides is a good number of enzymes to make the process possible. However, it is the structure of DNA that possesses in its design the information to organize the process.

Let's start with a semantic note about the two terms, *self-replication* and *self-reproduction*. Although often treated synonymously, there is in fact a difference. Self-replication (the word comes from the Latin *replica*) means a faithful molecular copy, while self-reproduction refers rather to a statistical process of making very similar things. Thus, cells self-reproduce, while molecules self-replicate. It is not simply a semantic question, as according to Dyson (1985), self-reproduction processes, being less precise, may have started first in the early evolution and preceded self-replication processes, which require more complex control and editing.

Self-reproduction is rightly seen as the main motor for the development of life on Earth. Life unfolds in time by repetition patterns: one generation of dolphins gives rise to a new generation of dolphins; one family of roses gives rise to a new family of roses. This is a subtle, invisible multi-branched thread that links the present beings to all those that have existed before and to all those who will come. Looking back, everything condenses into the original cellular family, the whimsical LUCA (last universal common ancestor), which, by the way, should not be confused with the origin-of-life protocell, a much older event.

10.1 Self-replication and nonlinearity

There are some good reasons why, in the field of the origin of life, the emergence of self-reproduction is of such paramount importance. Before the implementation of self-replication, any interesting structure that might have originated in the prebiotic scenario would have decayed due to degenerative processes, and would have disappeared leaving no trace. Instead, with self-reproduction (as soon as the rate of self-reproduction is larger than the rate of decay), an increase in concentration of this structure would be possible.

Also, if the self-replicating structure were to possess chemical information, and in addition the capability to mutate, then self-reproduction, information, and evolution would occur all in the same system; something already very close to life. It is then clear why the search for self-reproduction mechanisms is the holy grail of the research on the origin of life.

We have mentioned already in Chapter 3 that Shreion Lifson (1997) utilized a nice arithmetic to illustrate the power of the autocatalytic self-replication. He took the example of a normal hetero-catalytic process that, making one molecule of B from A at the rate of one per second, would require 6×10^{23} s to make one mole of B. Instead, an autocatalytic process requires only 79s to make one mole of B (at the same rate of 1 reaction per second, each molecule of B gives rise to 2 molecules B, then 2 B give rise to 4, and 4 to 8, and so on).

It may be recalled that the intellectual history of self-replication can be traced back to von Neumann and his cellular automata (Neumann and Burks, 1966). (It is interesting to note that Alan Turing completed a Ph.D. dissertation in 1934 to 1938 under von Neumann's supervision). Von Neumann was indeed a complex personality, who, aside from working in the Manhattan Project to develop the first atomic weapons, worked in quantum mechanics and devised the architecture used in most non-parallel-processing computers. Of most importance to us here is the fact that he created the field of cellular automata (Neumann and Burks, 1966), constructing theoretically the first examples of self-replicating automata. Besides being a theoretician, von Neumann had a practical mind and his interest in self-replicating devices was also to show that practical operations such as mining might be accomplished successfully through the use of self-replicating machines.

The names of von Neumann and Turing lead into the field of artificial life, a subject that falls outside the scope of this book. Freitas and Merkle (2004) discuss this subject beginning from von Neumann's cellular automation model, and usefully describe a large series of self-replication models presented in the artificial life literature.

Self-replication is one basic concept in the hypotheses regarding the origin of life on Earth. We have already discussed the myth of self-replication in the case of the prebiotic RNA-world. And we have also seen how nice and simple arithmetic is flown when applied to real chemistry.

10.2 Self-replicating, enzyme-free chemical systems

When considering self-replication, one immediately thinks of DNA, and we have already pointed out that DNA replication is only made possible by a large family of enzymes. Likewise, a virus *per sé* is not capable of self-replication; the replication is made by the living host cell. However, the double-strand nature of DNA is the most ingenious device for self-replication, in the sense that each of the two strands contains the information for making a complementary one. Figure 10.1 illustrates

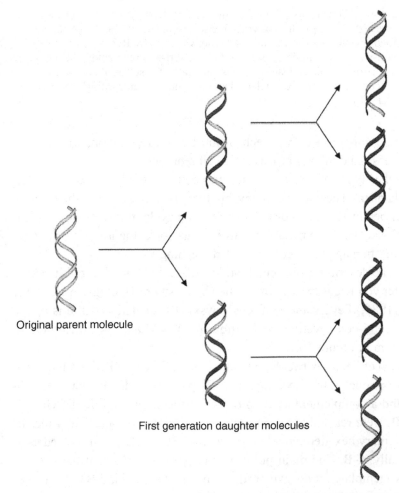

Original parent molecule

First generation daughter molecules

Second generation daughter molecules

Figure 10.1 Schematic diagram of semiconservative replication. Parental DNA is shown in gray and newly synthesized DNA in black.

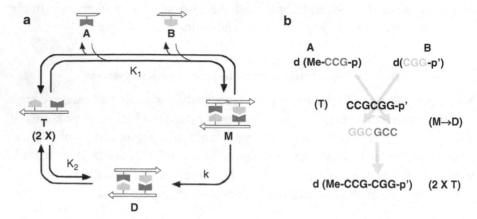

Figure 10.2 General mechanism for a minimal self-replicating system. For more details, see the text. *Panel (a)*: the enzyme-free self-replication scheme. (Adapted from von Kiedrowski, 1986). The same scheme was used by Rebek (1994); see text, with non-nucleotidic molecules. *Panel (b)*: a self-complementary hexanucleotide (**T**) act as template, leading to its autocatalytic self-replication from couples of complementary trinucleotides, **A** and **B**, according to von Kiedrowski (1986).

the well-known semi-conservative mechanism of DNA replication, which is the basis of the meiosis and mitosis processes in cell reproduction.

Are there other types of self-replication in nature, possibly based on a quite different mechanism? There are not many, but there is a famous case: the formose reaction, described in 1861, and based on a reaction cycle of formaldehyde. This reaction has already been mentioned in Part I of this book (Figure 2.5), where we said that this reaction appears to be of limited meaning for life.

The first work describing oligonucleotide self-replication was published in 1986 by Gunter von Kiedrowski, now at the University of Bochum in Germany (Kiedrowski, 1986; Sievers and von Kiedrowski, 1994). The scheme is analogous to the semi-conservative replication of DNA and is illustrated in a schematic form in Figure 10.2a.

This approach utilized a particular hexanucleotide CCGCGG-p' as a template: a 5' terminally protected trideoxynucleotide 3'-phosphate d(Me-CCG-p), indicated as **A**, and a complementary 3'-protected trideoxynucleotide d(CGG-p'), indicated as **B**, were reacted in the presence of a condensing agent to yield the self-complementary hexadeoxynucleotide d(Me-CCG-CGG-p'), indicated as **T**, which is actually **A-B**. The main point of this synthesis was that the product **T** could act as a template for its own production, i.e., it could bind **B** to its moiety **A** and **A** to its moiety **B** [see Figure 10.2 (b)].

The hexamer formation proceeds via the ter-molecular complex **M**, and the proximity between **A** and **B** in this complex facilitates their covalent linkage.

Thus, once the complex **D** is dissociated, two **T** molecules are formed, and the autocatalytic self-replication process can start with the progression described above: two give four, four give eight, eight give sixteen, and so on.

Notice the particular features of this kind of oligonucleotide: the hexameric sequence is said to be self-complementary, since two identical molecules can form a duplex via Watson and Crick bases. Actually, note that the two hexanucleotide sequences are palindromic. It may also be noted that two parallel pathways compete for the formation of the template **T**, namely the template-dependent, autocatalytic pathway, and the template-independent, non-autocatalytic one. This competition is the reason why the initial rate of the autocatalytic synthesis was found to be proportional to the square root of the template concentration – something that von Kiedrowski and colleagues called "the square-root law of autocatalysis." As Burmeister (1998) put it:

... a square root law is expected in the previously described cases, in which most of the template molecules remain in their double-helical complex (duplex) form, which leaves them in an "inactive" state. In other words, a square root law reflects the influence both of auto-catalysis and product inhibition.

Autocatalytic reactions that do not have this competition have a reaction order of one instead of a half (see also Zelinski and Orgel, 1987).

One limit to this beautiful chemistry lies in the requirement of self-complementarity of the self-replicating sequences. Julius Rebek and his group were also active at about the same time with enzyme-free self-replication of chemical structures. Unlike von Kiedrowski's group, he did not use nucleotides, but rather a replicator consisting of an adenosine derivative and a derivative of Kemp's acid (Rotello *et al.*, 1991; Rebek, 1994).

Interesting elaborations of this original scheme were presented later by von Kiedrowski's group (Achilles and von Kiedrowski, 1993; Kiedrowski, 1993; Sievers *et al.*, 1994). Also, their successive developments should be mentioned. By a technique called SPREAD (Surface Promoted Replication and Exponential Amplification) the authors (Luther *et al.*, 1998) combined the original scheme of Figure 10.2 to an insoluble matrix, obtaining a quite complex reproduction scheme.

These last reactions are very elegant and ingenious; however, they have not been regarded as relevant for the origin of life.

A higher degree of complexity is offered by the elegant work of Nicolaou and coworkers (Li and Nicolaou, 1994). In this case, the self-replication of DNA is based on a triple helix.

Of particular originality is the work by Ghadiri and coworkers, at the Scripps Institute, on self-replicating polypeptides (Lee *et al.*, 1996). They showed that a 32-residue alpha-helical peptide based on the leucine-zipper

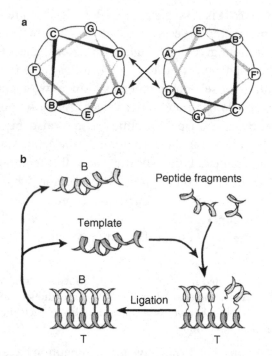

Figure 10.3 Coiled-coil structures and their role in peptide condensation. *Panel (a)*: two peptides containing heptad repeats (**A** to **G** and **A'** to **G'**), both able to form α-helices. Hydrophobic interactions (**A–A'** and **D–D'**) lead to coiled-coil structures. (Adapted from Paul and Joyce, 2004). *Panel (b)*: a full-length peptide (**T**) acts as a template forming a coiled-coil structure with peptide fragments and directing their condensation in other full-length peptides (**B**). (Modified from Ghosh and Chmielewski, 2004.)

domain of the yeast transcription factor GCN4, can act autocatalytically, leading to its own synthesis by accelerating the thioester-promoted amide-bond condensation of 15- and 17-residue fragments in neutral, dilute aqueous solutions.

The original system is based on peptides that contain heptad repeats, where the first and fourth positions of the repeat are hydrophobic amino acids. Such sequences form α-helices, which assemble into coiled-coil structures, as represented in Figure 10.3. The principle is then the same as that used for von Kiedrowski's self-replicating nucleotides (Kiedrowski 1986), in the sense that a full-length peptide template (having in this case 32–35 residues) directs the condensation of the two half-length peptide substrates.

The field of self-replicating peptides, initiated by Ghadiri's group (Lee *et al.*, 1996, 1997) was then elaborated by Chmielewski and coworkers (Yao *et al.*, 1997, 1998; Issac and Chmielewski, 2002; Li and Chmielewski, 2003), as well as by Mihara's group (Matsumura *et al.*, 2003; Takahashi and Mihara, 2004). Further advances from Chmielewski's laboratory have eliminated the initial problems of slow release of the newly formed condensation product and have actually approached exponential amplification. For reviews, see Paul and Joyce (2004), as well as Ghosh and Chmielewski (2004).

Ghadiri's group also made the observation (Severin *et al.*, 1997) that the 33-residue synthetic peptide, based on the coiled-coil structural motif, efficiently catalyzes the condensation of the two shorter peptide fragments. Depending on the substrates used, rate enhancements of tenfold to 4,100-fold over the background were observed. Furthermore, they extended and developed this work to include ecological systems consisting of hypercycles (Lee *et al.*, 1997). Starting with an array of 81 sequences of similar 32-residue coiled-coil peptides, the authors estimated the relative stability difference between all plausible coiled-coil ensembles, and used this information to predict the auto- and cross-catalysis pathways and the resulting plausible network motif and connectivity.

Networks of self-replicating replicators are supposed to be the next step in complexity, and in fact, Ghadiri's and coworkers devised a complex hypercycle network of self-replicating and cross-replicating peptides (Lee *et al.*, 1997; Ashkcnasy *et al.*, 2004). Whether such complexity can teach us anything about the origin of life and/or evolution remains to be seen. One argument in favor of this notion is that such catalytic intercrossing networks may suggest a way by which the world of peptides and the world of nucleotides may have interacted with each other. Ellington and coworkers have taken a step in this direction by devising a system in which there is peptide-templated nucleic acid ligation with a RNA aptamer (Levy and Ellington, 2003). In this general framework, the work by Mihara's group, who have incorporated nucleobase analogues in a self-replicating peptide (Matsumura *et al.*, 2003), should also be noted.

The general question arising in connection with Ghadiri's and Chmielewski's work, aside from its brilliancy and chemical ingenuity, is whether self-replication of peptides is *per sé* relevant for the origin of life. Probably the answer is negative, but the very important message is a general one, the indication namely that autocatalytic self-replication processes are not mysterious, strange chemical pathways, but on the contrary, they enjoy a certain degree of generality in the world of chemistry.

10.3 One more step towards complexity

One of the dreams of present-day researchers is still the construction of a minimal enzyme capable of making copies of oligonucleotides – an RNA polymerase. The ideal operation scheme is presented in Figure 10.4. Here, the enzyme first covalently binds an activated trinucleotide, then a hexanucleotide

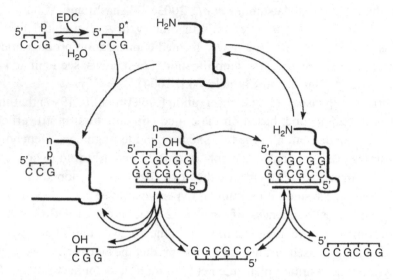

Figure 10.4 Principle of a minimal replicase. (Adapted from Maynard-Smith and Szathmáry, 1995. See also, with comments, Burmeister, 1998.)

can bind by complementarity with the triplet, a second triplet can now be added, and the enzyme links together the two trinucleotides.

Thus, two palindromic hexanucleotides are fabricated, and when their duplex divides, the cycle can repeat itself – and each time, from one hexanucleotide, two are formed. Such an enzyme, or ribozyme, does not exist yet, although Paul and Joyce (2002) came close to it. This is a self-replicating system based on an RNA template that also functions as a ribozyme to catalyze its own replication – considered as the first example of a self-replicating system with a reaction order of unity.

Quite interesting is the next level of complexity, in which more self-replicative cycles interact and cooperate with each other. This gives rise to the notion of *hypercycle*, originally developed by Eigen and collaborators (Eigen, 1971; Eigen and Schuster, 1977 and 1979; Eigen *et al.*, 1981), and schematized in Figure 10.5.

The notion of hypercycle refers to self-replicating informational macromolecules and introduces into the mechanism the basic principle of Darwinian evolution. One important point is the notion of "quasi-species": due to the ongoing production of mutant sequences, selection does not act on single sequences, but on mutational "clouds" of closely related sequences, referred to as quasi-species. The evolutionary success of a particular sequence depends not only on its own replication rate, but also on the replication rates of the mutant sequences it produces, and on the replication rates of the sequences of which it is

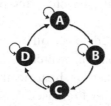

Figure 10.5 A simple graphic rendering of the notion of *hypercycle*. Each of the units **A, B, C** and **D** is a replicator. The rate of replication of each unit is an increasing function of the concentration of the unit immediately preceeding it. (Adapted from Eigen, 1971; see also Eigen and Schuster, 1977, 1979.)

a mutant. As a consequence, the sequence that replicates the fastest may even disappear completely in selection-mutation equilibrium, in favor of more slowly replicating sequences that are part of a quasi-species with a higher average growth rate. However, although the sequences that replicate more slowly cannot self-sustain, they are constantly replenished, as sequences that replicate faster mutate into them. Thus, at equilibrium, removal of slowly replicating sequences due to decay or outflow is balanced, so that even relatively slowly replicating sequences can remain present in finite abundance (Eigen, 1971; Eigen and Schuster, 1979; Schuster and Swetina, 1988). The concentrations of reagents are mutually dependent, and the systems are coupled cyclically so that each unit helps its neighbor to replicate better.

The system is symbiotic, and is considered the basis of co-evolution; as such it has also been investigated by many authors from a theoretical point of view. An in-depth discussion was presented by Kauffman (1993).

10.4 Self-reproducing micelles

So far, only the self-replication mechanisms of linear molecules have been described; it is now time to consider closed spherical structures, such as micelles and vesicles. Here, the term *self-reproduction* will be used rather than self-replication, because the population increase is generally based on statistical processes. The self-reproduction of micelles and vesicles has been briefly considered in this book as part of the discussion on chemical autopoiesis (see Chapter 2, Sections 2.9 through 2.11); we will revisit this topic in Part V of this book, during a discussion on the minimal cell project. (As already stated, a certain degree of repetition is unavoidable in this book.)

The work on self-reproducing micelles and vesicles was initiated in my laboratory in Zürich in the 1980s, at a time in which I was very much involved with reverse micelles – and in fact, reverse micelles were the first spherically closed system that underwent self-reproduction.

We have already discussed the structure and properties of reverse micelles, as small (1–2 nm in diameter), spherical surfactant aggregates built in an apolar solvent (usually referred to as oil), whereby the polar heads form a polar core that can contain water – the so-called water pool. The connection with autopoiesis is historically important, because it was thanks to the collaboration with Francisco Varela that the work started, as described in Section 2.9.

A few micellar systems of this kind were developed (Bachmann, 1991), and this allowed the arising of the field of self-reproducing micelles. The compartmentation in reverse micelles has also been utilized to host the rather complex von Kiedrowski reaction of the self-replication of a hexanucleotide (Böhler *et al.*, 1993). In this case, while the hexanucleotide self-replicates according to the von Kiedrowski mechanism, the reverse micelles also undergo a self-reproduction – thus providing a system in which shell and core replication occur simultaneously (although not coupled to each other). Later on, the procedure for self-reproduction of reverse micelles was developed into a procedure for aqueous micelles and for vesicles.

Let us recall in more detail the case of the micellar aqueous system, as this is actually the most relevant work for the later development in the field of minimal cells. The chemistry is based on fatty acids, which build micelles at higher pH ranges and vesicles at pH *c.* 10.0–10.5 (Hargreaves and Deamer, 1978a). The interest in fatty acids lies also in the fact that they are considered possible candidates for the first prebiotic membranes, as already mentioned.

The experimental apparatus is particularly simple [see Figure 10.6 (a)], also a reminder of a possible prebiotic situation: the water-insoluble ethyl caprylate is overlaid on an aqueous alkaline solution, so that at the macroscopic interphase there is a hydrolysis reaction that produces caprylate ions. The reaction is initially very slow, as shown in Figure 10.6 (b), but eventually the critical micelle concentration (cmc) is reached in solution, and thus the first caprylate micelles are formed.

Aqueous micelles can actually be seen as lipophylic spherical surfaces, to which the lipophylic ethyl caprylate (EC) avidly binds. The efficient molecular dispersion of EC on the micellar surface speeds up its hydrolysis (a kind of physical micellar catalysis), and caprylate ions are rapidly formed. This results in the formation of more micelles [Figure 10.6(c)]. However, more micelles determine more uptake of the water-insoluble EC, with the formation of more and more micelles: a typical autocatalytic behavior.

The increase in micelle population was directly monitored by fluorescence quenching techniques, as already used in the case of the self-reproduction of

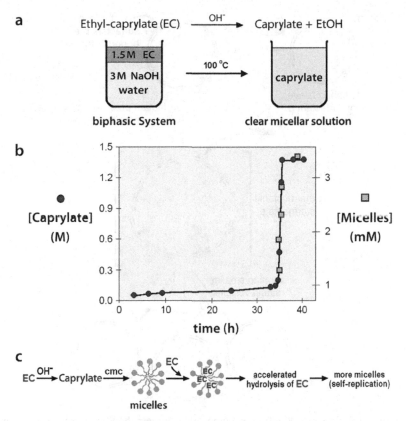

Figure 10.6 Autocatalytic self-reproduction of aqueous caprylate micelles. *Panel (a)*: experimental apparatus. *Panel (b)*: cynetic diagram of the reaction. *Panel (c)*: mechanism of the reaction. More details are in the text. (Adapted, with some modifications, from Bachmann *et al.*, 1992.)

reverse micelles (Bachmann *et al.*, 1992). Micelles could then be converted into vesicles – detected by electron microscopy – by lowering the pH with the addition of CO_2.

This kind of process was reinvestigated later, among others by Micheau and his group (Buhse *et al.*, 1997, 1998).

In the following years, this type of self-reproduction experiment was extended to vesicles. The experimental set up was similar, with a water-insoluble S–S precursor of the S-surfactant binding at the surface of the vesicles, as indicated in Figure 10.7. Again, the large hydrophobic surface of the vesicles accelerates the hydrolysis of S–S. This starts an autocatalytic self-reproduction process, as S molecules remain in the membrane, which grows and eventually divides, giving rise to more aggregates.

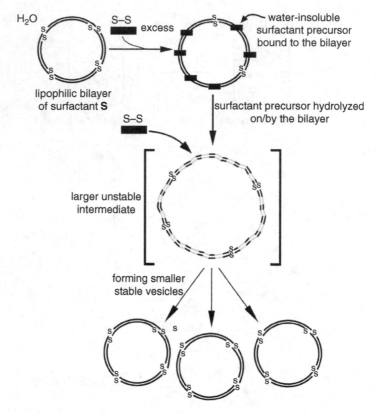

Figure 10.7 Self-reproduction scheme of vesicles; **S–S** represents a water-insoluble precursor that binds to the bilayer and is hydrolyzed *in situ*.

10.5 Self-reproducing vesicles

The time course of a typical experiment is shown in Figure 10.8, reporting the hydrolysis of caprylic anhydride catalyzed by spontaneously formed caprylate vesicles. Note the sigmoid behavior, characteristic of an autocatalytic process.

The lag phase is due to the preliminary formation of vesicles, and in fact the length of the lag phase is shortened when already formed vesicles are pre-added. In this work, an analysis of the number and size distribution of vesicles at the beginning and the end of the reaction was also performed by electron microscopy. Some mechanistic details of these processes will be discussed later on, when talking about synthetic biology.

The sigmoidicity of the curves is strongly dependent on pH as well as on temperature, a phenomenon which has not yet been studied in detail.

Self-reproduction studies were conducted on other fatty acid/anhydride systems, including attempts to correlate self-reproduction with stereospecificity by

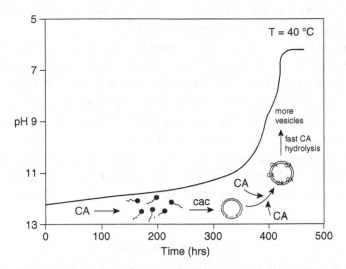

Figure 10.8 Time course of a typical experiment of self-reproduction of caprylate vesicles. **Cac** is the critical aggregation concentration; **CA** is caprylic anhydride.

using chiral fatty acids. This was the case for the vesicles of 2-methyldodecanoic acid (Morigaki *et al.*, 1997).

The rate of hydrolysis of the chiral 2-methyldodecanoic anhydride, catalyzed by chiral vesicles, was studied with the aim of possibly combining exponential autocatalysis and enantioselectivity – i.e., to use the power of exponential growth to discriminate between two diastereomers. No significant effect of this kind was found. However, a marked difference in the behavior of homo-chiral and "racemic vesicles" was observed at 10 °C: the racemic vesicles separated out in gel-like form, whereas the homochiral ones remained stable and continued to self-reproduce during hydrolysis of the anhydride. Also, a significant influence of temperature on the hydrolysis rate and cooperativity was observed (Morigaki *et al.*, 1997).

One should also mention the case of giant vesicles(GV). GV are vesicles with radius dimensions exceeding 10 uM, and as such they have quite particular features (Walde et al., 1994a; 1994b). Due to their dimensions, they can be visualized by optical microscopy, one at a time. Also, with particular micro-injection techniques (similar to those used for cellular work), biochemicals can be introduced directly inside. They are mentioned here because, following the work with vesicles, a procedure has been developed for inducing self-reproduction of giant vesicles in Sugawara's group (see Takakura *et al.*, 2003).

Let us emphasize again that the mechanism of self-reproduction of micelles and vesicles can be considered autopoietic, since growth and eventually division comes from within the structure itself (in vescicles, initial reaction takes place on the bilayer, but the bilayer is part of the structure).

It is important to point out the main message of these experiments. This is that by a very simple set-up, a spontaneous self-reproduction of spherical compartments can be obtained. Since such spherical compartments can be considered as models and/or precursors of biological cells, the hypothesis was put forward (Bachmann *et al.*, 1992), that this autocatalytic self-reproduction process might have been of relevance for the origin of life.

There is another, quite interesting process that leads to self-reproduction of vesicles. This is based on the so-called "matrix effect," which will be considered in the next part of this book, together with other properties of vesicles.

10.6 Nanobacteria?

When we talk about self-reproduction, in meetings or in private sessions, there is often somebody asking about nanobacteria. What are they? Well, the fact is that nobody seems to know, really. The things started with an interesting paper by a Finnish research team lead by Olavi Kajander (Kajander and Çiftçioglu, 1998). Accordingly, these nanobacteria – for which the authors actually used the name *Nanobacterium sanguineum* – are supposed to be 20–200 nm diameter in size, with a unique membrane structure, and they replicate, although slowly.

There were also older reports, for example, by Günther Enderlein, a German zoologist and pathologist; Wilhelm Reich, the famous physician; Royal Rife; and Gaston Naessens, also a pathologist, all claiming the existence of minute and ancient life forms, ubiquitous in nature, and present in humans, actually as a kind of indicator to determine the state of health in humans.

It sounds good and exciting; however, a report by Cisar *et al.* (2000) splashed water on fire, asserting that what Kajander calls nanobacteria are not really bacteria at all, but a new type of self-replicating crystals. However, as we have said at the very start, the notion of nanobacteria at large still lives. Thus, Uwin *et al.* (1998) found novel organisms, which he called "nanobes," as small as 50 nm, in Australian sandstones at a depth of 3–5 Km underground. And Folk (1993) identified life forms with sizes varying from 25 to 200 nm in hot spring carbonates, while Vancanneyt *et al.* (2001) brought evidence of ultra-small marine bacterial forms (most notably *Sphingopyxis alaskensis, c.* 200 nm) in sea water and even in Greenland's 120,000 years-old glacial ice. This brings us to the more recent statement that nanobacteria may be present inside clouds in the atmosphere and may play a role in global infectious diseases (Sommer and Wickramasinghe, 2005). And Samoylov, Vodyanoy, and colleagues (2005, using a modern, potent optical microscopy device) have reported evidence for the existence of "proteons" in blood, which are capable of replication. And there is the work of Lida Mattman (1992), who described variants of partly pleo-morphic, cell-wall-deficient forms included in red blood cells.

It is fair to say that all these reports notwithstanding, the large majority of contemporary life science researchers do not give credibility to the world of nanobacteria. Thus, take this section as historical information, and perhaps as the stimulus for yet another intriguing area of investigation.

Concluding remarks

No longer than 20 years ago, self-replication was one of those mysterious processes considered the monopoly of living matter. The fact that we are now able to achieve it in the laboratory means that we understand self-replication and self-reproduction in terms of simple rules of chemistry. In turn, this means that we have proceeded a step further in the understanding of one of the mechanisms of life. Of course, this is just one step, but it shows that conceptual and experimental progress in the ladder of the transition to life is possible.

We have also learned that self-replication is not a prerogative only of nucleic acids, but it can be shared by different kinds of chemical families; see the formose reaction, the self-replicating peptides, and the self-reproducing micelles and vesicles. The list should include the cellular automata and the other devices of artificial life.

Self-reproduction of vesicles and liposomes is important because it represents a model for cell reproduction. Despite this, in the various reviews on self-reproduction in recent years, practically no mention is made of such micelles or vesicle systems. The reason lies most probably in the bias of old-fashioned biochemists, according to which self-replication is tantamount to nucleic acid only. As I have mentioned, it may take one more generation before a system view of the processes of life takes over on the reductionist, DNA(RNA)-centered view.

I would like to end this chapter with a citation by Dyson (1985), cited in turn by Lifson (1997), as it contains a flavor of criticism towards the emphasis on the importance of self-replication in the evolution of life, and a kind of *ante litteram* stance for systems biology:

I have been trying to imagine a framework for the origin of life, guided by a personal philosophy which considers the primal characteristics of life to be homeostasis rather than replication, diversity rather than uniformity, the flexibility of the genome rather than the tyranny of the gene, the error tolerance of the whole rather than the precision of the parts [...]. I hold the creativity of quasi-random complicated structures to be a more important driving force of evolution than the Darwinian competition of replicating monads.

There are probably nowadays more and more people who would take this stand; and this is probably good.

Questions for the reader

1. In Chapter 8 on self-organization, we met several examples of ordered constructions, like the beehive, the termite or ant nests, the pattern of swarm intelligence, and the bird nests, among others. The common understanding of most of science is that these constructions are not examples of a precise purpose, (what we called *finality*) and that bees, ants, etc., just follow a genetic program blindly – the notion of teleonomy, as opposite to teleology. Do you accept this view? And how much are you ready to stretch it? For example, is the male peacock, when it displays its wings to attract a female partner, also a case of teleonomy? And what about the young man engaged in the courtship of a nice young woman?

2. In Chapter 9, we have discussed emergence, as the principle which gives rise to novel properties. Here, the idea is that we cannot predict the emergent properties, nor can we explain them on the basis of the properties of the smaller components. (Notice the emphasis on "properties" as opposed to structures.) This introduces a kind of blind spot in science, a mysterious camp, which is also a potential source of future creativity and surprise. Do you accept this idea, or do you find it alien to the spirit of modern science?

3. We have seen self-replication/self-reproduction in the last chapter of Part III. For many people working in the area of the origin of life (for example, people in the prebiotic RNA-world camp), self-replication is the way to start. For others, self-replication, being based on precise and interlocked molecular configurations, is the result of earlier mechanistic pathways. Ultimately, we do not know how it came to be. What is your intuition saying about these two perspectives?

Part IV

The world of vesicles

Chapter 11

The various types of surfactant aggregates

Introduction

In the previous two chapters, we have seen the general principles of self-organization and emergence based on examples from the most important classes of biopolymers, proteins, and nucleic acids. We have also seen, rapidly, a quite different class of aggregates, those formed by amphiphilic molecules and micelles and vesicles. In this chapter, we will consider these surfactant aggregates in more detail.

The importance is twofold, in qualitative as well as quantitative terms. On the one hand, biological amphiphiles like lipids and phospholipids are important constituents of cell membranes in the entire living world. If you consider that each of us has c. 100 billion cells, and each cell may contain an average of 10^{9}–10^{10} phospholipid molecules, you get a quick idea of the relevance of such compounds.

On the other hand, in a quite different perspective, surfactant molecules are extremely important for technological applications – from the pharma industry to cosmetics (soaps, cosmetics and the like) and to the heavy industry – in petrol extraction and mechanical washings. In fact, the term *surfactant* indicates a basic property: the capability to reduce the surface tension (or interfacial tension) between two liquids or between a liquid and a solid. They may thus act as detergents, wetting agents, emulsifiers, foaming agents, and dispersants (Falbe, 1987). The world production amounts to several million tons per year, and roughly two-thirds of this is in the form of anionic surfactants, followed by non-ionic surfactants, with a small share of cationic surfactants.

For the present discussion, amphiphiles have to do with one of our main concepts, that of "compartment" and (if this word exists) "compartmentation." This is, of course, very relevant for biology at large, and to start with, it is then proper to compare the various kinds of compartments, such as those formed by micelles, reverse micelles, cubic phases, and vesicles. This will be useful for discussing the biochemical reactions in vesicles, which will be dealt with later on in this chapter.

11.1 General properties of surfactant aggregates

We have already seen, talking about self-organization, how and why amphiphilic molecules tend to form aggregates such as micelles, vesicles, and other organized structures.

Figure 11.1 illustrates the phase diagram obtained for CTAB (cetyl trimethyl ammonium bromide, a surfactant). In this example, the type of aggregate that can be obtained is determined by the relative concentration of the three (in this case) components.

The molecular structure of the surfactant influences the form of the aggregate, and, to predict this, there are some empirical rules (Israelachvili *et al.*, 1977, Israelachvili, 1992) based on the geometrical parameters of the surfactant molecule. In particular (see Figure 11.2), the volume **V** occupied by the

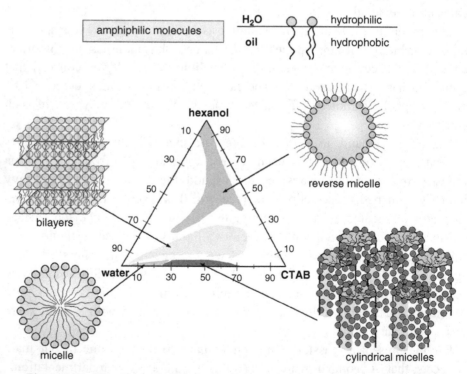

Figure 11.1 The phase diagram in the case of the positively charged surfactant CTAB (cetyl trimethyl ammonium bromide): depending on the relative concentration of CTAB, water, and hexanol, quite different organized structures are formed.

self assembly may be described in terms of the curvature
which exists at the hydrocarbon-water interface

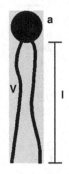

a: head group area

l: length of the hydrocarbon chain(s)

V: volume of the surfactant molecule

the surface packing parameter: V / a x l

Figure 11.2 The *surface packing parameter*, **V** / (**a** × **l**), based on the geometrical parameters of the surfactant molecule, determines to some extent the form of the aggregate. More details are in the text.

surfactant, the head area **a**, and the length of the tails, **l**, are seen as the important determinants. Accordingly, when the ratio **V**/ (**a** × **l**) is around unity, bilayers tend to be formed; when this ratio is less than one, or larger than one, tendentiously spherical aqueous micelles or, respectively, reverse micelles are preferentially formed.

In reality, however, it is not always possible to predict the aggregate form on this basis. Actually, the relation between molecular geometry and thermodynamics continues to be rather elusive; this is also because a series of environmental factors (salt, pH, and temperature) affect the form of aggregate.

Depending upon the nature of the polar head group, surfactants are classified as anionic, cationic, non-ionic, and zwitter ionic. Table 11.1 shows a few examples with the corresponding *cac* (critical aggregate concentration) and the aggregation number, N_x (Pfüller, 1986; Carey and Small, 1972).

The basic common denominator for all these applications is qualitatively well understood: as already mentioned, surfactants and their aggregates permit the mixing between phases or substances that are *per sé* immiscible with each other – mostly oil and water. This is how grease is washed off from our hands when we use soap, the removal being mediated by the oily core of micelles. In turn, micelles and vesicles permit the formation of an extraordinarily efficient interfacial system. Figure 11.3 gives a dramatic demonstration of this, showing that the total surfactant surface of a concentrated soap solution in your sink may well correspond to the surface of a stadium!

Table 11.1 *Some different types of surfactants with their basic properties*

SURFACTANT	cac (mM)	N_x
Anionic		
Sodium dodecylsulfate (SDS)	8.1	62
Sodium dodecanoate	24	56
Sodium cholate	14	3
Cationic		
Hexadecyl trimethyl ammonium bromide (CTAB)	0.92	61
Dodecyl ammonium, chloride	15	55
Non-ionic		
Octyl glucoside	25	84
Dodecyl-(polyethylenglycol(23)-ether (Brij 35)	0.09	40
4-ter-octylphenyl-(polyethylenglycol(9/10)-ether (Triton X−100)	0.31	143
Zwitter ionic		
Sodium taurocholate	3.1	4
N-dodecyl-N,N-dimethylammonium-3-propansulfonate (Sulfobetain SB12)	3.3	55

cac: critical aggregate concentration, above which aggregates (micelles and vesicles) are formed.
N_x: average aggregation number.

Spherical Micelle of caprylate ions

Micellar surface
$A = 19.6 \times 10^{-14}$ cm^2

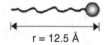

$r = 12.5$ Å

[Miscelles] (mol/lit.)	1.7×10^{-10}	8.5×10^{-8}	8.5×10^{-5}	1.0×10^{-3}
Micellar surface in a litre of solution	19.6 cm^2	1.0 m^2	10^3 m^2	0.012 Km^2
Equals a surface of a	Passport photo	Desk	Swimming pool	Stadium

Figure 11.3 Calculation of the interfacial area formed by an anionic surfactant (caprylate) at different micelle concentrations.

It is no surprise then that surfactants are used so extensively in technical applications; and of course, this large surface area achieved by surfactant aggregates immediately inspires ideas of applications in basic chemistry, too. For example, if the surfactant head has some catalytic properties, these could be extended and developed into an almost incredible dimension.

As far as life sciences are concerned, there are two main reasons of fascination: firstly, above a certain critical concentration, structural order is achieved starting from a chaotic mixture of disordered surfactant molecules. As discussed in Part III with examples and figures, this process is under thermodynamic control, and the increase of order is attended by an increase of entropy and a decrease of free energy.

Secondly, there is the emergence of compartments, with an inside that is physically different from the outside. This may, in turn, permit reactions to take place inside the compartment, which would not take place in the outside medium. The discrimination between inside and outside, typical of compartments, is the first structural prerequisite for the living cell and the living in general.

All these notions are beautifully applied in nature. Life is dominated by water as a background medium; however, living beings contain a considerable amount of lipophilic compounds that are *per sé* incompatible with water – the hydrophobic amino acids and peptides, or the aromatic bases of nucleic acids; and then lipids. How is this compatibility warranted? The compatibility of lipids with water is made possible by the concomitance of two almost opposite effects: a process of de-mixing, by which lipids make their own micro phase, thereby avoiding contact with water. However, in each of these micro phases there is the exposure of the hydrophilic head groups to water, by virtue of which the micro phase becomes soluble – or at least compatible with water. Figure 11.4 shows the well-known

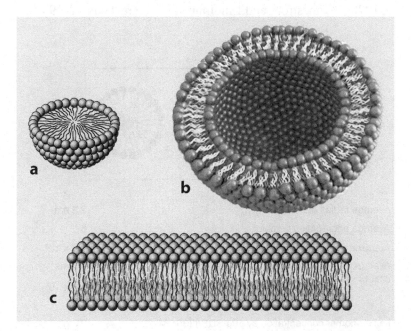

Figure 11.4 Schematic views of three ordered structures that can be formed by mechanically dispersing a suspension of phospholipids in aqueous solution: (**a**) micelle; (**b**) liposome; (**c**) bilayer sheet.

structural analogy between the biological membrane and liposomes, which will be described in the next section.

This principle of separating hydrophobic micro phases and rendering them water compatible by exposing hydrophilic moieties to water can be seen in other structural biological domains, for example in proteins (formation of hydrophobic domains or patches) and with the aromatic bases in DNA (segregated in the interior core of the double helix, avoiding water).

Let us now see some basic features of the various compartments.

11.2 Aqueous micelles

Aqueous micelles have diameters ranging typically from 0.5 to 5 nm, and being so small, do not scatter visible light and give transparent solutions. Figure 11.5 shows some basic parameters for aqueous micelles, relative to the well-known SDS (sodium dodecyl sulfate). Micelles are thermodynamically stable, and this is a significant difference with respect to most large vesicle aggregates.

The thermodynamics of micelle formation has been studied extensively. There is a mass action model (Wenneström and Lindman, 1979) that assumes that micelles can be described by an aggregate M_m with a single aggregation number m. A more complex analysis assumes a multiple equilibrium model, allowing aggregates of different sizes to be in equilibrium with each other (Tanford, 1978; Wenneström and Lindman, 1979; Israelachvili, 1992).

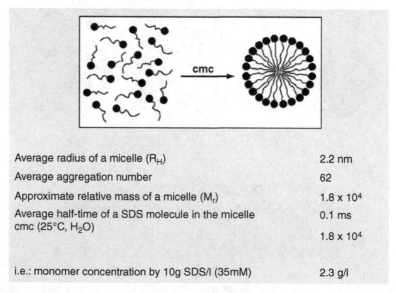

Average radius of a micelle (R_H)	2.2 nm
Average aggregation number	62
Approximate relative mass of a micelle (M_r)	1.8×10^4
Average half-time of a SDS molecule in the micelle	0.1 ms
cmc (25°C, H$_2$O)	1.8×10^4
i.e.: monomer concentration by 10g SDS/l (35mM)	2.3 g/l

Figure 11.5 Basic parameters for aqueous micelles of sodium dodecyl sulfate (SDS).

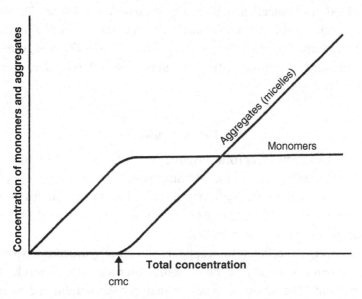

Figure 11.6 Monomer and micelle concentrations as a function of total concentration (schematic). Most single-chained surfactants containing 12–16 carbons per chain have their critical micelle concentration, *cmc*, in the range 10^{-2}–10^{-5} M, while the corresponding double-chained surfactants have much lower *cmc* values due to their greater hydrophobicity. (Adapted from Tanford, 1978.)

In the mass action model, the micellar system can be described by only one parameter, and despite this simplicity, a good qualitative description of the main physical properties is obtained, for example the onset of *cmc* (critical micelle concentration), as shown in Figure 11.6. Notice that the formation of micelles becomes appreciable only at the *cmc*, and after that, by increasing the surfactant concentration further, all added surfactant is transformed directly into micelles, so that the surfactant concentration in solution remains constant at the level of *cmc*.

Hydrophobic substances added to an aqueous micellar solution tend to be entrapped into the "oily" interior, and if the substance contains a fluorophore, its fluorescence properties may change drastically upon entrapment, due to the change in the environment.

Micellar catalysis is a broad field (Fendler and Fendler, 1975; Rathman, 1996; Rispens and Engberts, 2001), and caution is needed when using this term. Often the term *catalysis* is used when there is an increase of the velocity of reaction, but this does not always mean that the velocity constant is increased (namely that there is a decrease of the activation energy). Rather, the velocity effect can be due to a concentration effect operated by the surface of the micelles. This is, in fact, the case for the autocatalytic self-reproduction of micelles, which we discussed in Chapter 10, where the lipophilic precursor of the surfactant is concentrated on the hydrophobic surface of the fatty acid micelles (Bachmann *et al.*, 1992), a feature that has given rise to some controversy (Mavelli and Luisi, 1996; Buhse *et al.*, 1997, 1998; Mavelli, 2004).

11.3 Reverse micelles

Reverse micelles form in aprotic organic solvents, such as hydrocarbons or CCl₄, and can be seen as a core containing water (the water pool) solubilized in an oily environment by the hydrophobic tails. Figure 11.7 shows a scheme of reverse micelle and the structure of AOT (aerosol-octyl or aerosol-OT), the most popular surfactant for reverse micelles.

A typical reverse micellar system appears as a clear hydrocarbon solution, and the water content can usually be varied within an order of magnitude, say from 0.5 to 5 percent. The amount of water that can be solubilized is generally expressed as molar ratio w_o between water and surfactant molarities. This ratio determines the size of the water pool and the stability of the micelles in a given solvent (for details about these and other properties of reverse micelles, see the papers by Pileni, 1981; Boicelli *et al.*, 1982; El Seoud, 1984; Luisi and Straub, 1984; Luisi and Magid, 1986; Levashov *et al.*, 1989).

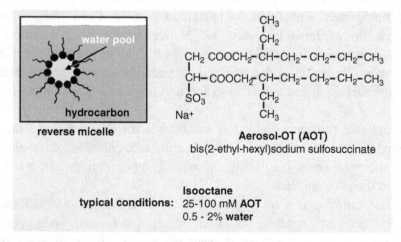

Figure 11.7 Schematic representation of a reverse micelle in cross section. The structure of the most popular surfactant for reverse micelles, the aerosol-octyl (AOT), is shown, as well as the typical conditions to obtain reverse micelles with this surfactant. The molar ratio w_o between water and surfactant concentrations is related to the size of the water pool and to the stability of the micelles in a given solvent.

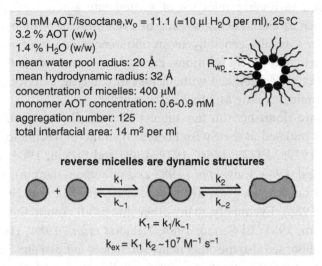

50 mM AOT/isooctane,w_o = 11.1 (=10 μl H_2O per ml), 25 °C
3.2 % AOT (w/w)
1.4 % H_2O (w/w)
mean water pool radius: 20 Å
mean hydrodynamic radius: 32 Å
concentration of micelles: 400 μM
monomer AOT concentration: 0.6-0.9 mM
aggregation number: 125
total interfacial area: 14 m^2 per ml

reverse micelles are dynamic structures

$$K_1 = k_1/k_{-1}$$

$$k_{ex} = K_1 k_2 \sim 10^7 \, M^{-1} \, s^{-1}$$

Figure 11.8 Some structural details and dynamic properties of reverse micelles. (Adapted, with modifications, from Fletcher and Robinson, 1981, and Harada and Schelly, 1982.)

Micelles are thermodynamically stable; at the same time they are highly dynamic, as they break and re-form rapidly, and exchange material by continuous collisions and fusion processes (Fletcher and Robinson, 1981; Harada and Schelly, 1982), although aqueous micelles also present similar highly dynamic behavior (Ulbricht and Hoffmann, 1993). Figure 11.8 gives an illustration of this for reverse micelles. It is also interesting that structures of the reverse micelle type have been observed *in vivo* (de Kruijff *et al.*, 1980) where they seem to absolve specific, albeit not dramatically important, functions.

The fact that micelles are equilibrium systems means that the final state, for example, the average micellar size and all corresponding physical properties, are reached regardless of the pathway used to form the micelles. The final state is thus independent of the mixing order of the components, and does not depend on the previous history of the sample. It is possible to make larger or smaller micelles by changing the environmental parameters (e.g., salt concentration) – and if two different micellar solutions are mixed with each other, the final mixed state will have a unique average dimension and shall acquire its own equilibrium state.

11.4 Entrapment of biopolymers in reverse micelles

The entrapment of water-soluble solutes in reverse micelles has represented from the very beginning a quite active area of inquiry. It is worthwhile to consider this issue in some more detail.

The addition to reverse micelles of a moderate amount of water-soluble substances such as amino acids, sugars, or even proteins, which are insoluble in hydrocarbons, results generally in an entrapment into the water pool, and gives also rise to transparent solutions. Particularly interesting is the process by which enzymes are solubilized without loss of catalytic activity in a micellar solution containing as little as 1 to 2 percent water.

The literature flourished on the subject of "micellar enzymology" in the 1980s, a field initiated at the Swiss Federal Institute of Technology in Zürich (Luisi *et al.*, 1977a, 1977b; Luisi, 1979; Luisi and Straub, 1984; Luisi, 1985; Luisi and Magid, 1986; Luisi *et al.*, 1988 and references therein); as well as by Martinek and Levashov's group (Martinek *et al.*, 1978, 1981, 1986; Martinek and Berezin, 1986). Eventually, many other groups all around the world joined the field (Pileni, 1981; El Seoud, 1984; Hilhorst *et al.*, 1984; Han and Rhee, 1986; Waks, 1986; see also the other many references in Luisi and Straub, 1984). In this way, enzymatic reactions, occasionally even characterized by an enhancement of the turnover number, could be carried out in practically apolar solvents – such as isooctane, decane, and chloroform.

The importance of the particular compartmentation in this field is made apparent by a series of interesting and partly still unexplained effects. For example, when the amount of water is varied in the reverse micellar solution, the maximum enzyme activity – even in the case of hydrolases – is not observed at higher water-content values, but with relatively low amounts of water. In addition, the local pH – due to the constraints of the water pool – is anomalous with respect to the pH value in water (El Seoud, 1984; Luisi and Straub, 1984).

Reverse micelles are the first compartment structures for which the phenomenon of self-reproduction was described (Bachmann *et al.*, 1990; Bachmann, 1991). This experimental work – as already mentioned in Chapter 6 – was a follow-up of a theoretical study by Varela and Luisi (Luisi and Varela, 1990), and is a work that eventually brought to light the self-reproduction of aqueous micelles and vesicles.

Figure 11.9 shows a classic example of compartmentation of reverse micelles: chymotrypsin entrapped in reverse micelles catalyzes the peptide synthesis, and since the product of the reaction is water insoluble – and soluble in the hydrocarbon – it is expelled outside of the micelles, and this drives the chemical equilibrium more completely towards synthesis (Barbaric and Luisi, 1981).

Enzymes are large molecules with respect to the initial size of the water pool, and their inclusion in the reverse micelles brings about a reorganization of the micelles (Zampieri *et al.*, 1986).

Another phenomenon of particular interest is the *phase transfer* of proteins from water to a supernatant reverse micellar solution. In this case, supernatant to

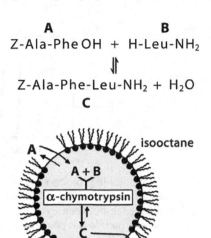

A   **B**
Z-Ala-Phe OH + H-Leu-NH$_2$

$\Updownarrow$

Z-Ala-Phe-Leu-NH$_2$ + H$_2$O
C

Figure 11.9 A case of selective compartmentation in reverse micelles, permitting the synthesis of a peptide by the reverse protease action. The product **C**, produced in the water pool, is expelled into the outside hydrocarbon environment due to its insolubility in water. (Modified from Barbaric and Luisi, 1981.)

the concentrated aqueous protein solution, there is a reverse micellar solution with a small w_o (5–15) – and the surprising thing is that the protein (which is *per sé* water-soluble) leaves the aqueous phase to go up into the micellar phase, and generally in a quantitative way. The extraction into the micellar phase can be operated also starting from the solid state of the protein. In both procedures, a good degree of specificity has been observed, in the sense that, starting from a mixture of two or more proteins, and depending on the environmental conditions (salt, pH, temperature), one protein can be extracted preferentially to the other one(s) – see Leser and Luisi (1989, 1990, and references therein). One example is given in Figure 11.10.

This process poses interesting questions from the thermodynamic point of view on why and how a water-soluble protein should prefer to migrate into the micellar phase. I will not dwell here on all the theoretical studies concerned with this; the reader is referred to the specialized studies on the subject (Caselli *et al.*, 1988; Bianucci *et al.*, 1990; Maestro and Luisi, 1990).

I would like instead to report a relatively old phase transfer experiment, which is really remarkable for its implications on the phenomenon of overcrowding, an issue that we have already mentioned in Part I of the book, and which will be discussed at length in Part V. For this experiment, a colored protein has been used, cytochrome C, so that one can easily follow the progress of the colored protein from one phase to the other.

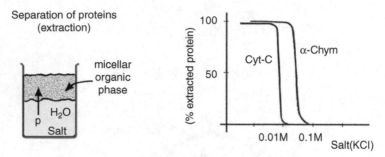

Figure 11.10 The uptake of water-soluble proteins (**p**) from an aqueous solution to an AOT micellar solution as a function of the salt concentration in the water phase. Note the remarkable difference between cytochrome-C (Cyt-C) and α-chymotrypsin (α-Chym), so that in principle it is possible to separate one from the other.

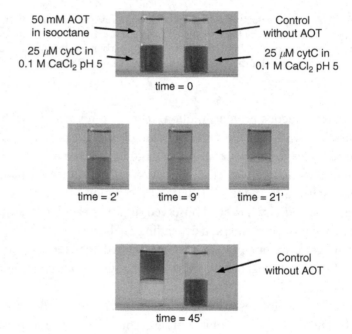

Figure 11.11 Phase transfer of cytochrome C from water (25 μM) to a reverse micellar phase. In 45 minutes, we have the new equilibrium with all the water-soluble protein in the micellar phase.

As the Figure 11.11 shows, results are indeed surprising: in less than one hour, the enzyme is totally transferred into the supernatant micellar solution. Consider now the question of the concentration: the initial concentration of the enzyme in the aqueous phase was 25 μM, and since the percentage of water in the micellar solution was 1 percent, we can calculate the value of 2.5 M for the concentration of the protein in the water phase of the micellar solution!

We have therefore operated a very remarkable overconcentration effect, based on a spontaneous transfer process.

The process of protein transfer in reverse micelles, as interesting as it is, has never reached the stage of a large-scale industrial or semi-industrial separation for proteins, although research on reverse micellar and micellar enzymology has continued into the present day. See, for example, the work by Crans and Levinger (2012) on the old and fascinating question of the pH in the water pool, or the work by Dodevski *et al.* (2014), who were capable of studying the entrapped proteins by NMR.

Even micro-organisms can be solubilized in reverse micellar solution, thus permitting microbiology in overwhelmingly organic solvent (Häring *et al.*, 1985, 1987; Hochköppler and Luisi, 1989, 1991; Hochköppler *et al.*, 1989; Pfammatter *et al.*, 1989, 1992; Famiglietti *et al.*, 1992, 1993). Again, the percentage of water in such systems is very limited, and the reason for and mechanism of this phenomenon is still largely unexplained.

Nucleic acids can also be solubilized in reverse micelles, including ribosomes and plasmids (Imre and Luisi, 1982; Palazzo and Luisi, 1992; Pietrini and Luisi, 2002, 2004; Ousfouri *et al.*, 2005), which also gives rise to a series of interesting structural and thermodynamic questions. In particular, high-molecular-weight DNA appears to acquire a condensed form in reverse micelles, with the characteristic "psi-spectrum" (Imre and Luisi, 1982; Pietrini and Luisi, 2004; Ousfouri *et al.*, 2005). This super-condensation of the DNA macromolecules, and their non-covalent cross-linking to yield the psi-spectrum, is due to the restricted environment. It is also a case of overcrowding, again with strong similarities to biological systems.

Figure 11.12 shows one experiment (Ousfouri *et al.*, 2005) in which DNA has been added to a micellar solution containing an "empty" micellar water pool of only 10 nm diameter; the presence of the large DNA brings about a rearrangement of the micellar structures, producing very large aggregates (up to 1 µm), which are very stable and host the DNA. These can be very large reverse micelles, or cylindrical, tube-like long structures – this has not been clarified yet.

It is also interesting to recall one additional feature of reverse micelles, found by serendipity in the search of conditions to make reverse micelles from lecithin. When traces of water are added to a hydrocarbon solution containing lecithin, a gel is formed (Scartazzini and Luisi, 1988; Luisi *et al.*, 1990), as can be seen in Figure 11.13. This "organogel," as it has been dubbed, can entrap a series of different molecules. In fact, this gel has been utilized for transdermal transport (Willimann and Luisi, 1991) and for an interesting chemistry in the semi-solid state (Fadnavis and Luisi, 1989; Scartazzini and Luisi, 1988), and it has been the subject of interesting relationships between the high viscosity and the structure of the aggregates in apolar solvents (Schurtenberger *et al.*, 1990, 1991).

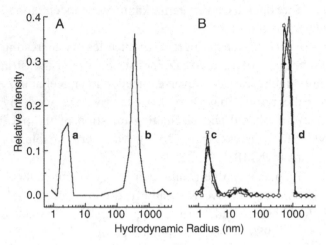

Figure 11.12 Dynamic-light-scattering size distribution (angle 120°) of a C_8PC reverse micellar solution, containing aqueous DNA solution, $w_0 = 5$. *Panel A*: 0.5 mg ml^{-1} DNA. *Panel B*: 4 mg ml^{-1} DNA. In *Panel B*, three size distributions are plotted, referring to: 15 min (—); 1 day (-♦-); 6 days (-○-) from the preparation of the micellar solution; **a** and **c** are empty micelles; **b** and **d** are DNA-containing micelles. (Ousfouri *et al.*, 2005).

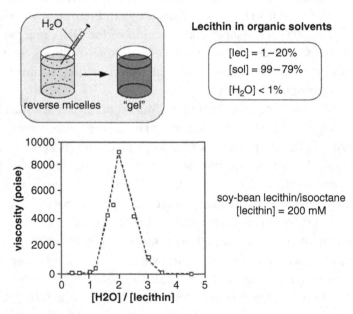

Figure 11.13 The simple addition of a minimal amount of water to a hydrocarbon solution of lecithin (or other phospholipids) brings about the formation of an organogel. (Adapted from Scartazzini and Luisi, 1988.)

As already mentioned, reverse micelles are dynamic systems that rapidly exchange compartmentalized materials. There is one limit to this, however: when the enclosed solutes are macromolecules, this does not happen anymore; if two different populations of reverse micelles are mixed, one, say, with enzymes and the other with nucleic acids, the two macromolecules are not going to interact with each other. This might be due to a stronger interaction of the macromolecule with the interior of the micelle, but there are no systematic studies on this effect.

11.5 Water-in-oil microemulsions

There is another example of water-in-oil compartmentation, which is substantially different from the transparent, thermodynamically stable reverse micelles: water-in-oil microemulsions. These can be prepared by adding to the "oil" a small amount of aqueous surfactant solution, with the formation of more or less spherical aggregates (water bubbles) having dimensions in the range of 20–100 μm in diameter.

These systems are generally not thermodynamically stable, and tend to de-mix with time. However, they can be long-lived enough to permit the observation of chemical reactions with their kinetic time course.

These "bubbles" fuse and mix with each other rather efficiently upon stirring. One example is given in the Figure 11.14, relative to one experiment in which protein expression – the green fluorescent protein – was achieved by mixing two populations of bubbles containing the complementary reagents for protein expression (Pietrini and Luisi, 2004; see also Stano and Fiordemondo, 2006).

There are fascinating reports on the use of microemulsions as compartments for protein expression (see, for example, Martini & Mansy, 2011; Kato *et al.*, 2012; Torre *et al.*, 2014). This subject will be dealt with more specifically in Chapter 15, dealing with synthetic biology.

11.6 Cubic phases

One should also mention another family of non-spherical compartments that are much less popular than micelles or vesicles, but in my view are also very interesting. These are the cubic phases, so called because of their cubic symmetry, although this kind of compartmentation is not so well defined as in the case of micelles or vesicles. In fact, many different types of cubic structures have been described (Mariani *et al.*, 1988; Lindblom and Rilfors, 1989; Fontell, 1990; Seddon, 1990; Seddon *et al.*, 1990; Luzzati *et al.*, 1993).

Many single-chain amphiphiles form cubic phases when added to water in a given composition. Two of the most well known are di-dodecyl-phosphatidyl ethanolamine, and mono-olein. Figure 11.15 shows some idealized bi-continuous cubic structures of the former, including typical inverse

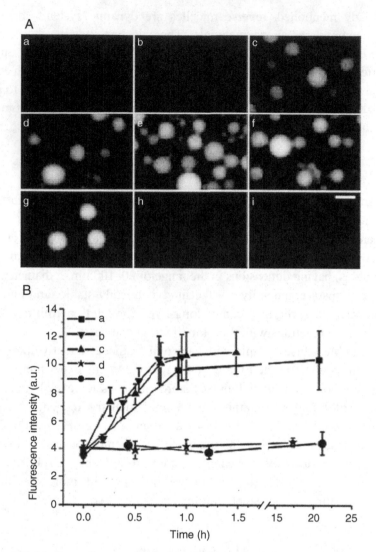

Figure 11.14 Green fluorescent protein (GFP) synthesis in water-in-oil emulsion as visualized by fluorescence microscopy. (Adapted from Pietrini and Luisi, 2004.) In the figure are shown the compartments in which GFP has been expressed (green, in the original micrographs). *Panel A*: Typical micrographs of the cell-free GFP synthesis in Span 80 (0.45% v/v)/Tween 80 (0.05% v/v)/ aqueous solution (0.5% v/v) in mineral oil emulsion droplets, preparation at 4 °C incubation at 37 °C: (**a**) 0 min, (**b**) 11 min, (**c**) 23 min, (**d**) 32 min, (**e**) 44 min, (**f**) 57 min, (**g**) 21 h. Negative control: (**h**) 0 min, (**i**) 21 h. The bar represents 50 μm. *Panel B*: Kinetics of the cell-free GFP synthesis in emulsion droplets, whereby on average 10 droplets with diameters of 30–60 μm are evaluated per time point; **a, b** and **c** are three independent experiments of the cell-free enhanced GFP synthesis in emulsion droplets, and **d** and **e** are two independent experiments of negative control.

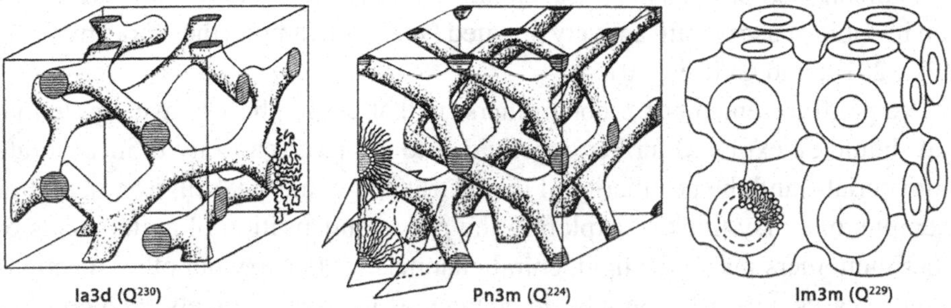
Ia3d (Q^{230})
Pn3m (Q^{224})
Im3m (Q^{229})

Figure 11.15 Idealized structures of hydrated didodecyl-phosphatidyl ethanolamine showing some typical bicontinuous cubic phases. (Adapted from Seddon *et al.*, 1990; see this reference for the indicated crystallographic nomenclature.)

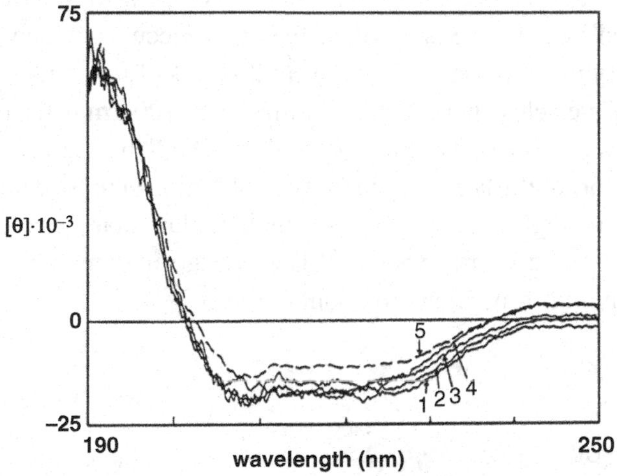
75
[θ]·10^{-3}
0
−25
190
250
wavelength (nm)
5
1 2 3 4

Figure 11.16 Temperature-dependent circular dichroism spectra of 1.2×10^{-4} M melittin in a 43% (w/w) 1-palmitoyl-2-linoleoyl-L-3-phosphatidylcholine (PLPC), cubic phase (10 mM tris-HCl buffer, pH 7.4). Spectra were taken during a heating cycle (**1**, 5 °C; **2**, 15 °C; **3**, 25 °C; **4**, 35 °C; **5**, 45 °C); $[\theta]$ is the mean residue ellipticity. (Adapted from Landau and Luisi, 1993.)

ones. This is also highly viscous and optically transparent as are most of the other cubic phases.

Several years ago, the very interesting observation was made that cubic phases can incorporate proteins up to 50 percent of their weight (Ericsson *et al.*, 1983). Usually cubic phases also remain transparent after incorporation of proteins, and in fact, it has been possible to carry out circular dichroic investigations of enzymes in such systems (Larsson, 1989; Portmann *et al.*, 1991; Landau and Luisi, 1993), as shown in Figure 11.16, and even to follow spectroscopically the course of enzymatic reactions (Portmann *et al.*, 1991).

As mentioned previously, the chemistry and biochemistry of compartmentation in cubic phases is still actively pursued (see, for example, the works by Chen
et al., 2014; and Caffrey, 2015).

The point to remember, among others, is that cubic phases have a potential
that should be explored further, mostly due to the large capacity to incorporate
biomaterials, and the peculiarity of the restricted geometrical environment. For
example, one may think of exploring the poly-condensation of amino acids or
other monomers inside the lipid, chiral channels, or to carry out other reactions
in those cases where it would be advantageous to have relatively high concentrations of reagents in a restricted tubular environment.

11.7 Size and structural properties of vesicles

Let us now look at vesicles. We are dealing in this case with spherical aggregates
that are formed by a double layer of surfactant molecules, which can be lipids.
Their sizes vary in a vast range, going from 20 nm to 1,000 nm diameter for the
"conventional" vesicles, then reaching values up to 100 μm in the case of the so-
called giant vesicles (GV). They can be "unilamellar," that is, formed by a single
double layer, or multi-lamellar, in a typical onion-like structure as shown
schematically in Figure 11.17; SUVs (small unilamellar vesicles) have very
small diameters (< 50 nm), whereas LUVs (large unilamellar vesicles) have
diameters spanning from 50 nm to about 400 nm.

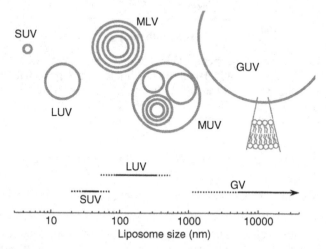

Figure 11.17 Schematic representation and approximate relative size of the
various vesicles. **SUV** stands for small unilamellar vesicles, **LUV** for large
ones, **GUV** for giant unilamellar vesicles. Notice the difference between the
unilamellar (**UVs**) and the multi-lamellar vesicles (**MLV**), and that they can be
contained one in the other (**MVV**).

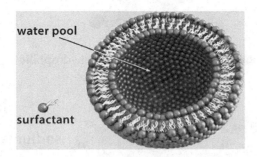

Figure 11.18 Schematic view of a liposome. The double layer, having external hydrophilic head groups, permits the coexistence of these lipid structures in water. The inner "water pool" is shown.

Liposomes can be defined as vesicles made out of lipids, although often the two terms are used synonymously. Figure 11.18 illustrates again a liposome, showing the inner "water pool," and the double layer with the external hydrophilic head groups, which permit the coexistence of these lipid structures in water [see also panel (b) of Figure 11.4].

Unilamellar vesicles are the most useful vesicles since generally their physicochemical properties depend only, or mostly, on their size. Moreover, diffusion of solutes inside unilamellar vesicles is simpler than in other cases. Unfortunately, most of the common preparation methods do not produce unilamellar vesicles. (For details on the preparation of vesicles and liposomes, see the specialized literature, e.g., Angelova and Dimitrov, 1988; Bianucci *et al.*, 1990; Lasic, 1995; Stano *et al.*, 2006; Walde *et al.*, 1994a; Stano and Luisi, 2007).

Two well-known vesicle-forming surfactants, fatty acids and palmitoyl-oleoyl-phosphatidylcholine (POPC), are shown in Figure 11.19. In both cases, the hydrophobic and hydrophilic parts are emphasized.

Since the bilayer shell in a sphere is necessarily curved, the number of amphiphiles constituting the inner layer is smaller than the number of amphiphiles present in the outer layer. For a 100-nm vesicle similar to the structure schematized in Figure 11.18, the calculated number of POPC molecules is 3.74×10^4 in the inner layer, and 4.36×10^4 in the outer layer, assuming a mean head group area of one POPC molecule of 0.72 nm and a bilayer thickness of 3.7 nm (Walde *et al.*, 1994b).

Oleate, along with most long-chain fatty acids, forms vesicles spontaneously, on simple addition of its concentrated aqueous or methanol solution into water, while POPC and other lipids do not form liposomes spontaneously when simply added to water. However, they do so when they are added to water as an alcoholic solution, or by first preparing a lipid film from an organic solution

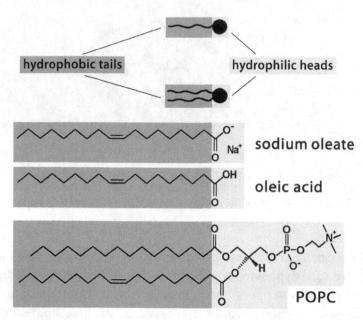

Figure 11.19 Most common vesicle-forming surfactants, showing the hydrophobic tails (dark grey, left) and the hydrophilic heads (light grey, right). The chemical name of oleic acid is *cis*-9-octadecenoic acid. POPC stands for palmitoyl-oleyl phosphatidyl choline.

(by evaporation), then adding water and stirring to induce a vortex. See Figure 11.20 for a qualitative illustration of these methods. Generally, a broad distribution of size is obtained, and by various procedures of extrusion (filtering through polycarbonate filters), relatively monodisperse distributions can be obtained.

One very simple way to prepare small unilamellar vesicles is by the injection method, as shown in Figure 11.20 (a), that is, when vesicles are formed by adding a methanol solution of surfactant into the aqueous solution of the solute to be entrapped (Domazou and Luisi, 2002; Stano *et al.*, 2004). The solvent evaporation method, illustrated in Figure 11.20 (b), is also frequently used. By these methods, eventually followed by extrusion, vesicles of different average size, say 50 nm and 200 nm, can be prepared, and if the two preparations are mixed with each other, a bimodal distribution is observed. In other words, the two species, contrary to the case of micelles, do not fuse and equilibrate with each other. This is because vesicles and liposomes are generally not equilibrium systems; they are kinetically trapped systems – the activation energy to change size is too high. This also means that in all these cases the liposomal system does not reach (does not have?) a state of absolute minimal energy.

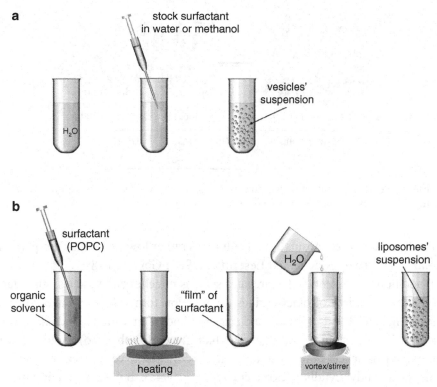

Figure 11.20 Schematic illustration of vesicle preparation. (**a**) Some surfactants spontaneously form vesicles when added to water as an aqueous or methanol solution. (**b**) Liposomes can be obtained from POPC and other water-insoluble phospholipids following a simple sequence: POPC is first dissolved in an organic solvent; then a POPC film is obtained by evaporation. At last, adding water under vigorous stirring, the liposomes are produced.

There is another important point about the thermodynamics, and this is the fact that vesicles are generally metastable systems. Contrary to micelles, which can be indefinitely time-stable, vesicles tend to aggregate and precipitate with time. However, this process can take many hours or days, and does not prevent reliable physical and chemical studies.

Going back to the physical properties, it should be mentioned that vesicles can undergo temperature-induced phase transitions. In fact, liposomes can pass from a more ordered state at low temperature (the so-called solid-like state) to a more disordered state at a higher temperature (liquid-like state, or liquid crystalline phase), as qualitatively illustrated in Figure 11.21. The phase-transition temperature T_m is a function of the length of the acylated chains, and the longer the chain, the higher the transition temperature.

The transition temperature also influences the physical parameters, such as permeability and stability (liposomes can only be prepared at temperatures

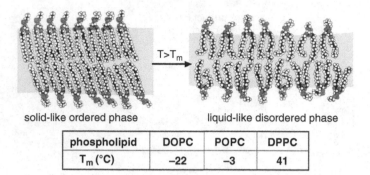

solid-like ordered phase   liquid-like disordered phase

phospholipid	DOPC	POPC	DPPC
T_m (°C)	−22	−3	41

Figure 11.21 Illustration of the phase-transition temperature of liposomes, with the T_m values. (Modified from Robertson, 1983.)

above T_m), and the encapsulation of solutes is rather low below this temperature (Janiak *et al.*, 1976; Machy and Leserman, 1987; Gennis, 1989).

In addition to conventional aggregates such as micelles and vesicles, surfactants can assume a variety of other curious aggregation forms. Ribbons and complex forms of helical strands, tubules, and micro-cylinders of various kinds have been observed. For example, 5'-(1,2-dimiristoyl-*sn*-glycero(3) phospho)cytidine (DMP-cytidine) forms helical strands (Itojima *et al.*, 1992). Stable vesicles can be prepared from phosphatidyl-cytidine (Bonaccio *et al.*, 1994a), but small environmental changes produce the formation of helical patterns and other complex, stable, geometrical figures (Bonaccio *et al.*, 1994b; see also Bonaccio *et al.*, 1996).

In general, the variety of architectures built by surfactants and lipids is extremely high, and small variations in the surfactant chemical structure can bring about significant changes in the supramolecular structure of the aggregates. Due to this fine-tuning, the relation between surfactant structure and supramolecular structure presents considerable difficulties, and is thus very difficult to predict the aggregate form, particularly for a novel surfactant.

11.7.1 The water pool of vesicles

The stiffness of liposomes brings about a series of problems in the chemistry of compartmentation; for example, as already mentioned, liposomes do not fuse with each other and therefore do not exchange material, contrary to micelles. Also, vesicles are generally characterized by a very poor permeability; compounds swimming in the outside bulk water are not easily promoted inside. Table 11.2 gives some values of the permeability of some common substances into lecithin liposomes. Even water and glycerol do not permeate easily in and out the double layer.

The restricted permeability of liposomes may allow for a significant concentration gradient across the bilayer, for example, with small vesicles a pH

Table 11.2 *Permeability of the liposome-membrane lecithin (T > T$_m$, pH = 7)*

	Permeability coefficient (cm · s^{-1})
Water	$4 \cdot 10^{-3}$
Glycerin	$5 \cdot 10^{-6}$
Urea	$4 \cdot 10^{-6}$
Tryptophan	$4 \cdot 10^{-10}$
Glucose	$\sim 10^{-11}$
Cl$^-$	$7 \cdot 10^{-12}$
Lysine	$5 \cdot 10^{-12}$
Na$^+$	$1 \cdot 10^{-12}$

Generally, the permeability for polar, charged molecules and for molecules with a high molecular weight is small.
Maximal permeability: T = T$_m$
Data from: Brunner *et al.*, 1980; Cevc and Marsh, 1987; and Chakrabarti *et al.*, 1994.

difference of 5 units can be maintained (Swairjo *et al.*, 1994); and a comparably large gradient of phosphate ions (0.5 M inside and 0.05 M outside) is observed. Conversely, once something has been entrapped, the leaking out is not an efficient process. Partial leaking can, however, be a disturbing feature.

The relative rigidity and poor permeability of liposomes is not inconsistent with an intense dynamics of the surfactant molecules composing them. Particularly important are the flip-flop movements, which permit an exchange of the surfactant molecules from one layer to the other. The rate depends on the structure; for example, the rate is faster in oleate vesicles than in POPC liposomes. The other important dynamic aspect concerns the dissociation and reassociation of the surfactant from the bilayer.

These in-and-out movements of the lipid molecules from the bilayer should not be confused with a chemical equilibrium between free surfactant and aggregate: it is not that the liposomes are being continuously destroyed and re-formed (as in the case of micelles and reverse micelles), but rather that localized changes – around a given stable, kinetically trapped, supramolecular structure – are taking place.

Having considered the properties of the liposome membrane, it is now useful to look inside the water pool. We can start by calculating the internal aqueous volume of the liposomes, as this will be the region where reactions will take place, particularly when considering liposomes as biological cell models.

Table 11.3 shows some of these values for POPC liposomes of various radii, calculated (assuming mono-lamellarity).

Table 11.3 *Geometrical properties of unilamellar POPC liposomes (from Walde, 2000, personal communication)*

Radius outer, inner (nm)	Internal aqueous volume of one liposome (10^{-20} l)	Surface of one liposome: outer, inner, total (10^3 nm^2)	Number of POPC molecules: outer, inner, total ($n \times 10^3$)	Liposome molarity ($\times 10^{-7}$) for a 10 mM POPC solution	Total internal aqueous volume ml mmol^{-1} POPC
20, 16.3	1.81	5.03, 3.34, 8.37	6.98, 4.64, 11.6	8.61	0.94
40, 36.3	20	20.1, 16.6, 36.7	27.9, 23.0, 50.9	1.96	2.37
100, 96.3	374	126, 117, 243	175, 162, 336	0.297	6.69
500, 496.3	51200	31400, 31000, 62400	4360, 4300, 8660	0.0115	35.59
5000, 4996.3	5.22×10^7	3.14×10^5, 3.14×10^5, 6.28×10^5	4.36×10^5, 4.36×10^5, 8.72×10^5	1.15×10^{-4}	360.67

What is the physical meaning of these data? The internal radius is 3.7 nm smaller than the external one, a difference which corresponds to the thickness of the bilayer. Liposomes with an external radius of 100 nm are made up of 336,000 lipid molecules, and in this case each compartment has an internal volume of 3.74×10^{-12} µl. This means that the total internal aqueous volume in 1 liter of a 10 mM POPC solution is 66.9 ml, namely 6.69 percent of the total volume.

If the vesicle radius is now increased to 500 nm, the internal aqueous volume becomes 512×10^{-12} µl, the total internal volume of all liposomes in a 10 mM POPC solution becomes 355.9 ml, namely 35.59 percent of the total volume.

About the comparison of vesicles of different sizes in Table 11.3, note that the total surfactant concentration remains constant, and therefore the total surface remains constant (this is given by the number of vesicles multiplied by the total surface of one vesicle, namely the product of the third and fifth column). The total internal volume calculated in Table 11.3 represents then the volume that we have at our disposal when using compartments for biological or chemical reactions.

11.7.2 The case of oleate vesicles

Of all mentioned prebiotic membranogenic molecules, the ones that have gained more attention are long-chain fatty acids. In addition to their prebiotic relevance, these compounds are relatively simple from the structural point of view, and most of them are easily available. In fact, the first investigations on self-reproducing aqueous micelles and vesicles were carried out with caprylate (Bachmann *et al.*, 1992) and most of the recent studies on vesicles involve vesicles from oleic acid/oleate (for simplicity we will refer to them as oleate vesicles).

Very important in this field are the pioneering studies of David Deamer and his group. Figure 11.22 shows the behavior as a function of pH, illustrating in particular how oleate makes micelles at higher pH values, where the carboxylate is completely ionized, while vesicles are formed at lower pH values.

Quite interesting is the increase of pK_a, from a typical value for carboxylic acids of 4.5 to 8.5: this large pK shift is due to the mutual proximity of carboxylate groups, which makes the dissociation of a proton more difficult. This is due to the collective behavior of the aggregate, a nice example of an emergent property. In fact, most of the studies on oleate vesicles have been performed in the pH range around 8.0–8.5. In this regard, one should mention the work by Haines (1983) relative to the proton-conducting pathway along the membrane surface, operated by anionic lipid head groups.

As for most long-chain fatty acids, there is a spontaneous vesiculation, with a size distribution that depends on the preparation procedure – initial

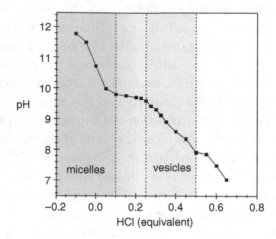

Figure 11.22 The behavior of oleate surfactants as a function of pH: equilibrium titration curve of sodium oleate at 25 °C. Note the micelles at higher pH, and the vesicles at lower pH. (Modified from Cistola *et al.*, 1988.)

concentration, velocity of injection, pH, etc. The *cac* (critical aggregate concentration) values for oleate is in the millimolar range, which means that at the operational concentration of 10–50 mM there will be a significant concentration of monomer in equilibrium with the aggregate. This consideration allows us to go back to the question of whether vesicles are chemical equilibrium systems. Oleate vesicles cannot be considered proper chemical equilibrium systems, however they behave in a mixed way, with some features that are typical of micelles in equilibrium (Luisi, 2001).

As apparent from Figure 11.22, there is a gradual transition from the oleate micelle region to the vesicle region, and in fact, a simple way to obtain vesicles is to inject a few microliters of a high-alkaline sodium oleate solution into a pH 8–8.5 aqueous buffer.

11.8 Local versus overall concentration

The existence of compartments in solution (case of micelles) or in a suspension (case of vesicles) brings up the interesting question regarding the definition of concentration for solutes, which are entrapped in them – in particular the question of local versus overall concentration.

Consider Figure 11.23, relative to reverse micelles – in which we have 1 mM of the solute **A** in 1 liter of solvent, so that the overall concentration (i.e., relative to the total volume) will be 1 mM. Now, if this solute **A** is only soluble in water, it will be localized only in the water pool; and if we have 1 percent water, then its

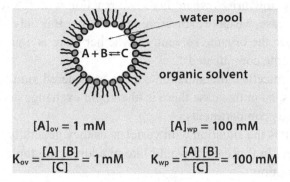

Figure 11.23 Overall (**ov**) versus local (**wp**) concentration in reverse micelles.

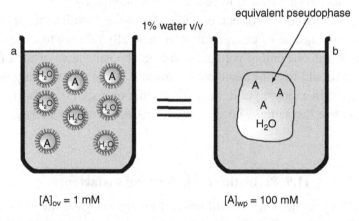

Figure 11.24 Schematic representation of the "equivalent pseudophase" in the case of reverse micelles. Due to the fast exchanging rate among micelles, one can view the system in terms of equivalent pseudophase, represented in b. Given a reagent **B** in the medium (not shown in the figure), if the exchange rate is fast in respect to the reaction we consider (then for this reaction between **A** and **B**) we must use local concentration (**wp**); if the exchange rate is slow (as in case of protein-containing micelles), then we must use overall concentration (**ov**). In general, the concentrations are modified by a parameter which takes into account the intermicellar exchange rate.

local concentration will be 100 mM. We have, in principle, two different sets of concentration, depending on how we look at the system, and this is then true for all properties that have the dimension of concentration, for example, the equilibrium constant, as in the numerical case shown in the figure.

In the case of reverse micelles, at normal conditions, we have a very rapid exchange among all of them; they continuously fuse with each other, so that the situation can be depicted as in Figure 11.24: we have an aqueous internal pseudophase that contains all solute, and its water pool concentration is then 100 mM.

The difficulty now arises when there is a reaction with the reagent **B** in the medium. In this case, we have to discriminate whether **B** is only soluble in water, or only soluble in the organic solvent, or whether there is a partition, whereby things may become complicated.

For aqueous micelles, the situation can be considered similar to that of the reverse micelles, as in this case there is also rapid exchange and fusion among the small aqueous compartments.

More difficult is the situation with vesicles, as they generally do not fuse and do not exchange. In this case, I would like to address the topic of local versus overall concentration.

Let us take one single enzyme molecule inside one vesicle of one picoliter volume. One molecule in one picoliter (10^{-12} l), does it mean 10^{12} molecules in one liter? This enzyme should have a concentration of 10^{-11} M?

This is mathematically correct, but does this make chemical sense? Probably not, because the notion of concentration is generally relative to many independent molecules in one unit of volume, so that in choosing any portion of the total volume, we should find the same concentration. When there is only one molecule entrapped in a single point, be it a liposome or a spot in a matrix, it seems difficult to state the same. Those are concepts not properly considered in the literature.

11.9 Prebiotic vesicle-forming surfactants

Phospholipids are the main constituents of most biological membranes, being produced in modern cells by complex enzymatic processes.

However, there have been attempts to show that they could have been formed also non-enzymatically, *via* prebiotic chemistry. Hargreaves and Deamer were trying to accomplish this kind of prebiotic chemistry already at the end of the 1970s starting from glycerol, fatty acids, and orthophosphates (Hargreaves *et al.*, 1977; Hargreaves and Deamer, 1978a, 1978b). The synthesis of phosphatidylethanolamine was attempted also in Oró's group (Rao *et al.*, 1987). Later on, the question of ancestral lipid biosynthesis and early membrane evolution has been re-examined (Pereto *et al.*, 2004). However, this question cannot be considered satisfactorily solved as of yet.

A different approach was proposed by the late Guy Ourisson and his group. They started from the consideration that the amphiphilic molecules in primitive membranes must have been very different from the modern eukaryotic ones (Ourisson and Nakatani, 1994). They argued that simple poly(prenyl) or di(polyprenyl) phosphates satisfy all conditions for being really primitive. In fact, these compounds form vesicles, provided that they contain at least 15 carbon atoms (Pozzi *et al.*, 1996). Ourisson and Nakatani (1999) also argued that in principle it is possible to synthesize these compounds starting from C_1-C_4 molecules, by reactions involving no enzymes and only acidic catalysts. The starting molecules are formaldehyde and phosphoric acid, which are present in a prebiotic environment (Miller and Parris, 1964; Oró, 1994), and isobutene, which appears to be present in comets, volcanic solfataric gases, and fluid inclusions in Archaean geological formation (see Ourisson and Nakatani, 1999, and references therein).

Evidence that is more convincing was obtained for the synthesis of simpler compounds, such as straight-chain fatty acids, which we have mentioned before, adding that these compounds form spontaneously stable vesicles. Nooner *et al.* (1976) reported prebiotic synthesis of these compounds. Later on, monocarboxylic acids have been observed from a spark discharge synthesis (Yuen *et al.*, 1981) and from a Fischer-Tropsch type of reaction (McCollom *et al.*, 1999; Rushdi and Simoneit, 2001). The evidence for a possible prebiotic synthesis of such compounds has come from astrobiology: thus, Dworkin *et al.* (2001) reported their synthesis in simulated interstellar/pre-cometary ices. Amphiphilic components were observed in the Murchison carbonaceous chondrite by Deamer and his group (Deamer, 1985; Deamer and Pashley, 1989). In this regard, we have seen already, in Part I of the book, Deamer's pictorial evidence for the prebiotic existence of membrane-forming surfactants.

11.10 Giant vesicles

The so-called giant vesicles (GV) can be often observed as by-products in the normal preparation of vesicles, but they can be obtained by specific methods, for example by electro-formation (Angelova and Dimitrov, 1988).

Giant vesicles have been the subject of several international meetings and specialized literature (see Luisi and Walde, 2000; Fischer *et al.*, 2000). There are several reasons for this interest. One is that, because of their size, they can be evidenced by normal optical microscopy. Figure 11.25 shows, as an example, the transformations brought about by the addition of a water-insoluble precursor (oleic anhydride) to oleic acid giant vesicles (Wick *et al.*, 1995).

Oleate GV, under optical microscopy, display size and form transformation upon addition of oleic anhydride. In both left- and right-hand panels, microphotographs show that there is a kind of duplication, with the production of a novel GV, and particularly interesting is the experiment

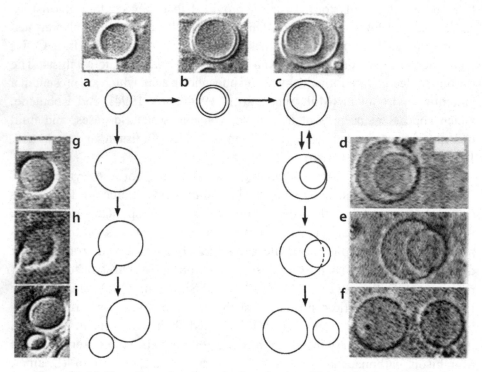

Figure 11.25 Transformations brought about by the addition of oleic anhydride (a water-insoluble precursor) to oleic acid giant vesicles. (Adapted from Wick *et al.*, 1995).

on the right-hand panel: here, one smaller vesicle is formed inside the larger one, then it "jumps out" and has about the same dimension as the "parent" GV.

A second reason of interest is that, by special micro-manipulation (similar to that used in cell biology), it is possible to inject chemicals directly and quantitatively inside the GV compartment. An example of the effect of an enzymatic reaction inside GV is shown in the Figure 11.26.

By working with giant vesicles, the chemist acquires the working habits of a cell biologist, suffering, however, from being obliged to work with only one compartment at a time – at variance with his preference to work with an Avogadro's number of particles. A nice review on the methods of preparation of GV is to be found in a work by Walde and colleagues (2010).

We will see then in Chapters 14 and 15, devoted to synthetic biology, that GVs have acquired a particular meaning as bioreactors for complex biochemical transformations.

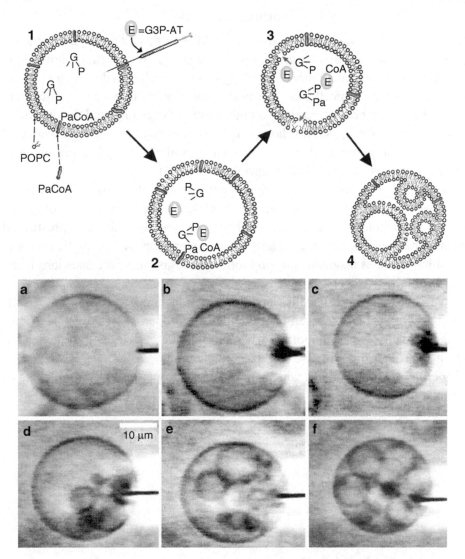

Figure 11.26 Effects of a micro-injection of *sn*-glycerol-3-phosphate acyltrans-ferase (G3P-AT) into POPC/Palmytoyl CoA (PaCoA) giant vesicles. As shown in the scheme of the upper panel, after the injection (**1**), G3P-AT interacts with G3P and PaCoA, determining a partial hydrolysis of PaCoA from internal membrane wall (**2**). The depletion of PaCoA (**3**, small arrows) produces shrinkage of the vesicle, followed by the formation, on its inner surface, of smaller compartments (**4**). The lower panel shows a series of phase-contrast micrographs showing the transformations of a single giant POPC/PaCoA vesicle, induced by micro-injection of 180 fl of G3P-AT solution (250 µg ml^{-1}). (**a-f**): Phase-contrast micrographs, taken 0, 10, 20, 50, 140 and 300 s, respectively, after the injection of G3P-AT into the vesicle. (Adapted, with some modifications, from Wick et al., 1996).

Concluding remarks

We have seen how varied and rich is the world of surfactant aggregates, and it is proper to repeat here that these compounds have a great importance both in the technical world and in nature. These aggregates are all based on non-covalent bonds, but whereas micelles are thermodynamically stable systems, vesicles are not. Micelles are characterized by a high degree of exchange and fusion, which is not the case for liposomes; however, liposomes can reach a high degree of stability – for example, they can be heated until 90 °C, as in certain PCR experiments, maintaining their integrity (see Part V).

Vesicles display in general no significant permeability. This is a problem in the chemistry and synthetic biology of vesicles, as we will see in the last part of this book. In biological cells, permeability of metabolites is due to sophisticated membrane proteins, which are a later evolution product, and prebiotic routes to the flow of solutes inside/outside prebiotic vesicles is one of the questions in the origin of life.

We have seen how micelles and vesicles can undergo processes of self-reproduction, generally based on the chemistry/properties of fatty acids. This, understandably, is very important when considering vesicles as cell models. In the next chapter, we will also see how vesicles and liposomes can change their shape, and grow and divide, which are important properties in the mimicking of cellular life.

Chapter 12

Vesicle reactivity and transformations

Introduction

Vesicles are good models for the outer shell of biological cells. This is due to the bilayer spherical structure, which is also present in most biological cells, and to the fact that they can incorporate biopolymers and host biological reactions. Vesicles also have the capability of self-reproduction, an autocatalytic reaction already illustrated in previous chapters. Some additional aspects of this process will be considered here, together with some particular properties of their growth – the so-called matrix effect.

We will begin with some theoretical, geometrical considerations, and then proceed to the experimental aspects.

12.1 Growth and division of vesicles: some geometrical relationships

The processes of changes of form and dimension in vesicles are generally linked to the relation between surface and volume. In order to consider this point more in detail, let us examine Figures 12.1, 12.2, and 12.3 (I am indebted to Pasquale Stano for planning these three figures and calculations therein; see also Stano *et al.*, 2006).

Figure 12.1 illustrates the geometrical parameters characterizing division in the simplifying assumption of spherical vesicles, when division takes place with constant volume, or with constant surface, or with constant radius.

When the total volume remains constant, growth and division proceed at the expenses of added lipids, and finally the volume of each particle is one-half of the initial one. For the surface to remain constant, as in the second case, water has to leave in order to permit vesicle division, the surface of each obtained particle is then one-half of the initial one, and the final radius and volume are in a simple mathematical relation with the initial parameters.

Finally, if the division occurs with constant radius, the surface and the volume of each particle will be the same as the initial one, which means that both the final total volume and the final total surface will be twice as much as the initial one. Of course, in this case both water and lipids have to be added.

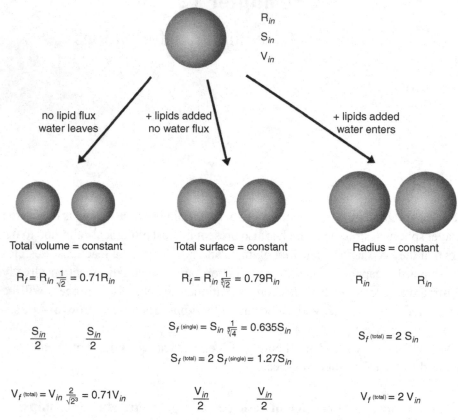

Figure 12.1 Schematic representation of the geometrical parameters in the process of vesicles division, under the assumption of sphericity, and showing the three limit cases of a constant volume, constant surface, and constant radius.

The Figure 12.2 illustrates some simple geometrical relationships in the case of surface growth by a generic factor α, as induced by lipid and water uptake, which brings about an increase of volume and surface.

Here, V_{mid}, S_{mid}, and R_{mid} indicate the geometrical parameters of the intermediate vesicle after uptake of water. The surface of each single vesicle after growth-division increases by a factor $\alpha/2$, and the volume by a more complex function of α. In the particular case of $\alpha = 2$, $R_f = R_{in}$ and we find the same situation described in the correspondent case (constant radius) of previous Figure 12.1.

As already mentioned, all this is valid under the approximation of spherical vesicles. However, also under those limiting conditions, the relation between surface and volume during growth and division is not trivial, as already indicated by Figures 12.1 and 12.2.

Things are more difficult in the case of a non-spherical growth, as schematized in Figure 12.3. And, just to complicate the picture, it is also conceivable

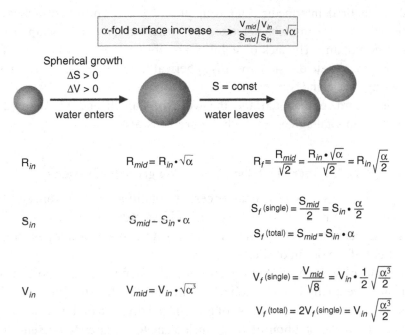

Figure 12.2 Schematic representation of the spherical growth of vesicles by a factor α, with the corresponding geometrical parameters.

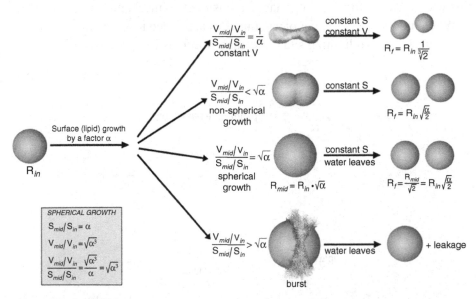

Figure 12.3 Surface vesicle growth as induced by the addition of lipids (and water) by a factor α, eventually followed by division process, and related geometrical parameters. Several cases are illustrated: surface growth at constant volume, increase of volume with non-spherical growth, increase of volume with spherical growth, and a very large increase of volume leading to vesicle burst. The small grey box recalls main geometrical parameters of spherical growth.

that the structural intermediate upon uptake of water and/or fresh lipid, can assume irregular forms, characterized by a non-trivial relationship between surface and volume. The lack of stability possibly associated with these forms can actually facilitate division towards spherical final forms. See in this respect the work by Bozic and Svetina (2004).

In this regard, one should add the work of Fanelli and McKane (2008), although not in total agreement with the previous one.

12.2 Experimental studies on the growth of vesicles

In Chapter 10, we have seen that under certain conditions, vesicles are capable of undergoing an autocatalytic process of self-reproduction. This is a novel, dynamic aspect of the reactivity of such aggregates, which clearly has relevance for the field of the origin of life.

This kind of vesicle self-reproduction can be defined as autopoietic, since growth and eventually reproduction comes from within the structure. We wish to discuss now a more general process of growth and division of fatty acid vesicles: this consists in the addition of fresh surfactant to an already present vesicle suspension, whereby the fresh surfactant is added for example as a micellar solution at high alkaline pH. Many experiments in the literature are carried out in this way, and therefore it is important to consider this mechanism more in detail.

When monomeric oleate or oleate micelles are added to a solution containing oleate vesicles, two limiting situations may occur, as illustrated in Figure 12.4:

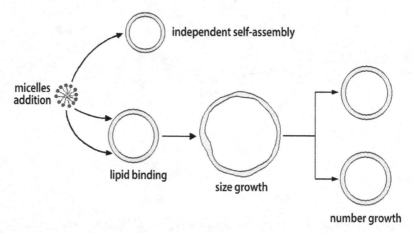

Figure 12.4 Addition of fresh surfactant (or a micellar solution thereof) to a solution containing pre-existing vesicles can follow two alternative (not exclusive) pathways: either formation of new vesicles, by independent self-assembly, or binding to the existing vesicles, which may bring about growth and division phenomena (a highly idealized case is shown, with vesicles dividing in two).

either the added substrate molecules self-assemble by their own course, ignoring the pre-existing vesicles; or, instead, they bind to the pre-existing vesicles. In this latter case, the corresponding uptake can induce growth and eventually division.

It is clear from Figure 12.4, that we can have a competition between the two processes, and the process proceeds depending upon the two relative velocities – the rate of assembly formation and the rate of binding, respectively. If the rate of vesicle formation is *per sé* very high, there is no possibility for binding and for the eventual growth and reproduction of the pre-existing vesicles.

For example, if fresh POPC from a methanol solution is added to a POPC liposome solution, there will be an immediate formation of fresh liposomes, independent from the presence of previous liposomes. In fact, the *cmc* in this case is of the order of 10^{-10} M in favor of the aggregates over the free monomers, and the rate of formation is extremely high. Thus, we are in the presence of an independent self-assembly mechanism.

In contrast to this, we can have the addition of oleate surfactant – in the form of micelles or free monomer – to oleate or to POPC vesicles. In this case, the ratio of the two competitive rates is such that a considerable binding of the added fresh surfactant to the pre-existing vesicles takes place. The efficient uptake of oleate molecules by POPC liposomes (Lonchin *et al.*, 1999) as well as to oleate vesicles (Blöchliger *et al.*, 1998) is well documented in the literature.

How can we investigate and clarify the mechanism?

One way is to label the pre-existing vesicles, and then follow the destiny of the label in the vesicle size distribution. The concept is made clear in Figure 12.5.

The label that has been used to this aim is ferritin (Berclaz *et al.*, 2001a, 2001b), which has been entrapped into vesicles. Ferritin is an iron-storage protein in plants and mammals, and consists of a hollow protein shell of *c.* 12 nm, containing in its center an iron core of *c.* 7.8 nm. This very dense iron core gives rise to a strong scattering contrast, which facilitates detection in electron microscopy.

Once ferritin is entrapped in vesicles, vesicle suspensions can be frozen (vitrified) as thin aqueous layers and examined at low temperature by transmission electron microscopy (TEM). This technique also permits to count the number of ferritin molecules in each vesicle, and in addition to determine the vesicle size and lamellarity (Böttcher *et al.*, 1995). See Figure 12.6.

The distribution of sizes and ferritin content before and after addition of the fresh surfactant can be determined by TEM. If the fresh surfactant does its own thing and does not interact with the pre-existing ferritin-containing vesicles, the

A) *de novo* vesicle formation

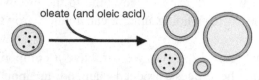

B) growth of the preformed vesicle,
eventually followed by a fission process

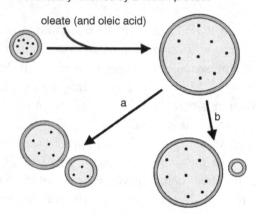

Figure 12.5 Schematic representation of the possible vesicle formation and transformation processes when oleate (and oleic acid) are added to preformed vesicles, which have been labeled with a water-soluble marker (ferritin). *Panel A* shows the situation if only *de novo* vesicle formation occurs. In this case, the added fresh surfactant does not interact with the already present vesicles, due to the very fast formation of its own vesicles. *Panel B* illustrated the case in which the added fresh surfactant interacts and binds to the already present vesicles. Therefore, here a growth in size of preformed and labeled vesicles occurs. This may lead to a fission process, either yielding vesicles that all contain marker molecules (*case a*, a statistical redistribution of the ferritin molecules is thus obtained) or yielding vesicles that do not all contain markers (*case b*). See text for details. (Adapted from Berclaz *et al.*, 2001a, 2001b.)

same initial distribution of ferritin-containing vesicles will be found at the end of the experiment. If instead the added surfactant interacts with the pre-existing ferritin-containing vesicles, inducing their transformation, the distribution of ferritin-containing vesicles will be changed in a way that reflects the mechanism of growth and/or division.

These considerations have given rise to an intensive TEM investigation (Berclaz *et al.*, 2001a, 2001b). Compare now Figures 12.4 and 12.6; and then, Figure 12.7, which shows the size distribution of "empty" (no ferritin) and filled vesicles at the start; it also shows the size distribution of empty and filled

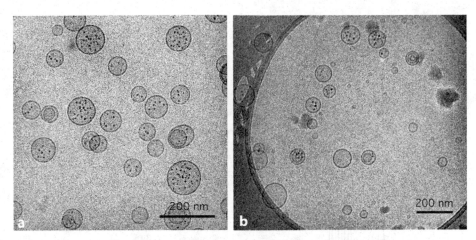

Figure 12.6 Cryo-TEM micrographs of (**a**) ferritin-containing POPC liposomes prepared using the reverse evaporation method, followed by a sizing down by extrusion through polycarbonate membranes with 100 nm pore diameters ([POPC] = 6.1 mM) and of (**b**) the vesicle suspension obtained after addition of oleate to preformed POPC liposomes ([POPC] = 3 mM, [oleic acid + oleate] = 3 mM). (Adapted from Berclaz *et al.*, 2001a, 2001b.)

vesicles after oleate addition; and finally the comparison between filled vesicles before and after addition.

It is clear that there is a significant growth of the vesicles under the given conditions: in particular, a population of vesicles with dimensions larger than 200 nm, and extending up to 500 nm, has been formed anew. This is direct evidence of the growth mechanism depicted in Figure 12.5.

However, this is not all: as shown in Figure 12.8, under certain conditions there is a small but significant concentration of small ferritin-containing vesicles of sizes that were not present before the addition of oleate. This clearly shows the process of division, as illustrated in Figure 12.5, *Panel B, case a*.

In conclusion then, there is direct evidence that vesicles grow in size when fresh surfactant is added; and direct evidence of division processes. All this is, of course, very important when vesicles are regarded as models for biological cells – the argument that will come next.

The growth of fatty acid vesicles has been reinvestigated by Chen and Szostak (2004) who, by the use of stopped-flow and fluorescence resonance energy transfer (FRET) techniques, have provided interesting insights into the kinetics of this process. We should also cite at this point the work of the Sugawara group on growth and division of synthetic surfactants with GV (Takakura *et al.*, 2003, 2014).

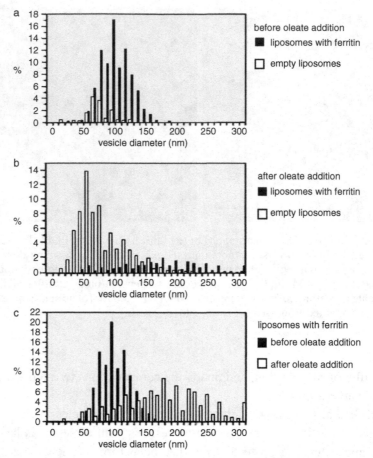

Figure 12.7 Number-weighted size distributions as obtained by cryo-TEM (adapted from Berclaz *et al.*, 2001a, 2001b). (**a**) Distribution for the pre-formed POPC vesicles ([POPC] = 1.9 mM). (**b**) Distribution for the vesicle suspension obtained upon addition of oleate to pre-formed ferritin-containing POPC vesicles ([POPC] = 0.2 mM; [oleic acid + oleate] = 5 mM). Empty (□) and ferritin-containing (■) vesicles are represented individually in the histogram. (**c**) Direct comparison of the number-weighted size distribution of the pre-formed POPC vesicles, which contained at least one ferritin molecule (■) with the number-weighted size distribution of the ferritin-containing vesicles obtained after oleate addition to pre-formed POPC vesicles (□). Note that the total of all ferritin-containing vesicles was set to 100 percent.

12.3 The matrix effect

Closer investigation of the kinetics of vesicle growth and reproduction caused by addition of fresh surfactant to a solution of pre-existing vesicles has given additional surprising results.

Consider the experiment illustrated in Figure 12.9, which shows the rate of oleate vesicle formation when an aliquot of concentrated surfactant is added to

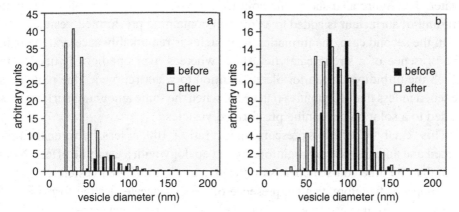

Figure 12.8 Experimental demonstration of the process of vesicle division upon addition of fresh oleate surfactant to ferritin-labeled pre-existing POPC liposomes. (Adapted from Berclaz *et al.*, 2001a, 2001b). Comparison of the "absolute" number-weighted size distribution (**a**) of the empty and (**b**) filled pre-formed POPC liposomes ([POPC] = 6.1 mM; ■) with the vesicles obtained after addition of oleate ([POPC] = 3 mM, [oleic acid + oleate] = 3 mM; □).

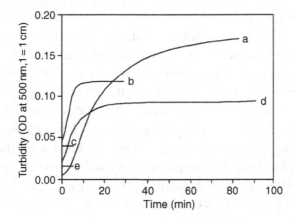

Figure 12.9 Effect of pre-added vesicles on the formation of oleic acid/oleate vesicles. Turbidity, measured at 500 nm (1 cm path length), is plotted as a function of time. T = 25 °C. Curve (**a**): 62 μl of 80 mM aqueous sodium oleate was added to 2.438 ml of 0.2 M bicine buffer, pH 8.8 ([oleic acid/oleate] = 2 mM). Curve (**b**): 62 μl of 80 mM aqueous sodium oleate was added to 2.438 ml of a 2 mM oleic acid/oleate "100 nm vesicle" suspension (0.2 M bicine buffer, pH 8.8 [oleic acid/oleate] = 4 mM). Curve (**c**): turbidity of 2 mM oleic acid/oleate "100 nm vesicles." Curve (**d**): the same as (**b**), but using a "50 nm vesicle" suspension. Curve (**e**): turbidity of 2 mM oleic acid/oleate "50 nm vesicles." (Modified from Blöchliger *et al.*, 1998.)

water. The figure also shows the comparison with the case in which the same amount of surfactant is added to a solution containing pre-formed vesicles.

In the second case, the formation of vesicles is remarkably accelerated, as in the presence of a strong catalytic effect: whereas over one hour is needed to reach the turbidity plateau for oleate addition to water (curve **a**), the plateau is reached in less than 10 minutes (curve **b**), when the same amount of surfactant is added to a solution containing pre-formed vesicles.

This "catalytic" effect is present also in a ratio 1:100, or less, between the pre-added and added surfactant, reinforcing the analogy with a catalytic effect. Note also that the time course depends on the size of the pre-added vesicles.

It appears, therefore, that the presence of vesicles accelerates the formation of "new" vesicles. It is not easy to rationalize how and why. What comes to mind is a general observation from the field of surfactants, that pre-organization makes the organization of further material easier. For example, there is no spontaneous vesiculation when POPC is simply added to water, as previously described in Chapter 11, and no significant amount of vesicles is formed by "vortexing" (namely, stirring to produce a vortex). However, when there is a pre-existing film of POPC in the flask (namely, a pre-organized surface), simple vortexing of the water suspension is enough to produce a large amount of POPC liposomes (see Figure 11.20).

There is another important observation arising from Figure 12.9: this is the height of the turbidity plateau. Note that in the experiment described by curve **b**, the final concentration of surfactant is double compared to the experiment represented by curve **a**; hence, one would have expected a turbidity plateau of roughly double intensity. The contrary is true: the turbidity plateau is lower in the solution with more surfactant. How can this be explained?

To this aim, we should recognize first that turbidity is sensitive not only to concentration, but also to the size of the particles. A lower plateau may indicate that we are dealing with smaller particles. Thus, the most likely interpretation of Figure 12.9 is that curve **b** corresponds to a much narrower size distribution. This is in fact what studies of dynamic light scattering show. This is shown in Figure 12.10, while an actual experiment, carried out with dynamic light scattering, is shown in Figure 12.11.

The surprising result from the Figure 12.11, is that the size distribution of the newly formed vesicles is extremely close to that of the initially present pre-formed ones. This has been confirmed using different sizes of the starting monodisperse solutions of vesicles.

It is as if there was a kind of stamp, that makes the new vesicles have the same size as those that are already present in the solution (the term "template effect" – rather than matrix effect – might also have been appropriate, except that this term is generally used in connection with macromolecular primary sequences with an informational content).

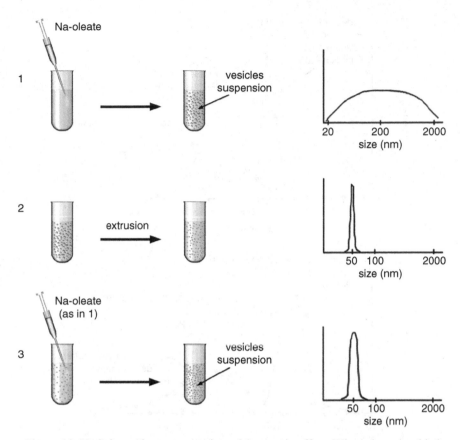

Figure 12.10 Schematic representation of the matrix effect. When oleate is added to water, vesicles form spontaneously, in a very broad distribution of sizes (1). By particular extrusion techniques, one can prepare a vesicle solution with a very narrow size distribution, shown in (2). When to this solution the same amount of oleate is added as in (1), the result is not a broad distribution of sizes, as in (1), but a narrow distribution (3) very close to the one initially present.

Note that the data in Figure 12.11 report the light-scattering intensity, which is a weight average. In terms of number average, the number of particles after addition is larger than the initial one (since the concentration has doubled and the size of the particles is more or less the same).

These numbers have been considered in detail (Rasi *et al.*, 2003), and in particular it was calculated that in the case of the oleate addition to POPC (where the radius actually shifts slightly towards smaller values), the number of particles becomes more than double. More details on this phenomenology, in terms of the effects of relative concentration of the reagents, of the size of the pre-added vesicles, and of the methods of addition, are given in the original literature (Lonchin *et al.*, 1999; Berclaz *et al.*, 2001a, 2001b; Rasi *et al.*, 2003, 2004; Luisi *et al.*, 2004; Stano *et al.*, 2006).

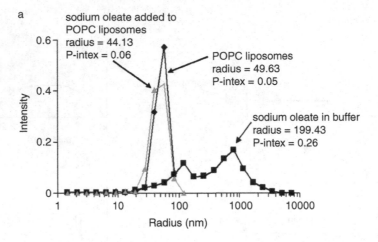

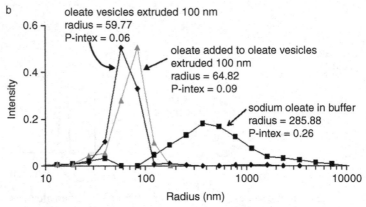

Figure 12.11 Matrix effect for sodium oleate addition to pre-formed POPC liposomes (**a**) and to pre-formed oleate vesicles (**b**). Note the close similarity between the initial size distributions (the pre-formed liposomes or vesicles) and the final size distributions (after addition of oleate). By contrast, the size distributions of vesicles formed in the control experiments (no pre-existing liposomes nor vesicles) are, in both **a** and **b** cases, very broad. (Adapted, with modifications, from Rasi *et al.*, 2003.)

It has also been shown that the matrix effect is not limited to a first addition of fresh surfactant, but can be repeated several times, thus increasing the number of particles of the same size distribution up to an order of magnitude (Rasi *et al.*, 2003; Stano *et al.*, 2006). The addition of fresh surfactant can also be carried out with a continuous reactor.

The effect is not limited to oleate, as DDAB (dimethyl-didodecyl-ammonium bromide) also shows: when added to itself or to POPC liposomes, DDAB is characterized by a strong matrix effect (Thomas and Luisi, 2004). The matrix effect was reinvestigated by Szostak's group, who confirmed the basic findings

(Hanczyc *et al.*, 2003); and also by Ueno's group (Chungcharoenwattana and Ueno, 2004, 2005a), and independently by Rogerson (Rogerson *et al.*, 2006), providing new and interesting insights into the matrix effect.

The matrix effect is thus a way to reproduce a vesicle population of a given size distribution. In an origin of life scenario, the constancy of size during self-reproduction is probably important, as it would have ensured a constancy of physicochemical and biological properties over various generations.

In the case of the matrix effect, contrary to the autopoietic experiments described earlier, there is no need of water-insoluble precursors. It is the very addition of the same surfactant to an already existing family of vesicles that brings about the multiplication of the same size distribution. All that is needed is an initial narrow distribution of vesicles, and a continuous addition of fresh surfactant. Methods to obtain narrow size distributions in the case of sponta-neous vesiculation have been described (Domazou and Luisi, 2002; Stano *et al.*, 2006). In fact, a prebiotic scenario may be conceived, where the fresh surfactant is continuously synthesized *in situ* and, thanks to the matrix effect, the same vesicle size is propagated over and over again. Of course, there is no way to demonstrate that this is what really happened in prebiotic times; it is fair, however, to claim that, given the simplicity of the process, there is a reasonable probability that a process of this sort may have occurred (Luisi *et al.*, 2004).

12.4 Fusion of vesicles

The process opposite to vesicle division is fusion, when two or more vesicles come together and merge with each other, yielding a larger vesicle. As outlined in the previous chapter, vesicle fusion is generally not a spontaneous process. If two populations of POPC liposomes with different average dimensions are mixed with each other, they do not fuse to produce a most stable intermediate structure; they stay in the same solution as two stable, distinct species. This is connected to the notion of kinetic traps, as discussed previously, and is sup-ported by theoretical and experimental data from the literature (for example, Olsson and Wennerstrom, 2002; Silin *et al.*, 2002, Bozic and Svetina, 2004).

This situation has recently been reinvestigated on the basis of dynamic light scattering, and Figure 12.12 shows the coexistence of two POPC liposome species obtained by extrusion. The two species coexist for days in this way without interacting with each other; and the same finding is obtained with other sizes (Cheng and Luisi, 2003).

There have been many reports in the literature of attempts to induce fusion, but results are not very clear; this is partly because the term *fusion* is used in a rather indiscriminate sense. I believe that this term should be restricted to those

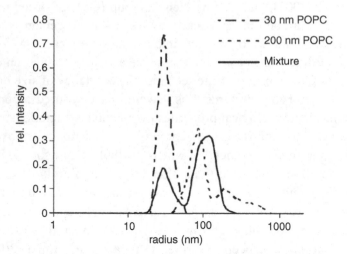

Figure 12.12 Intensity-weighted size distribution of POPC in 0.2 M bicine buffer solution, pH 8.5; 0.5 ml solution of 1 mM 30 nm extruded POPC vesicles mixed with 0.5 ml solution of 1 mM 200 nm extruded POPC vesicles. The measuring angle is 90°. (Adapted from Cheng and Luisi, 2003.)

processes where there is exchange and mixing of the water pools and formation of a new vesicle species. Instead, the word *fusion* is often used to indicate the formation of a complex between vesicles, without making clear what happens afterwards. Afterwards, there can be partial exchange of solutes or not; and often the two vesicles may depart from each other and things return to more or less the initial situation. This cannot be defined as fusion.

In the literature, the use of Ca^{2+} is often used to induce "fusion" – but also in this case it is not always clear what really happens. One illustration of the addition of calcium as a non-fusion process is given in Figure 12.13, showing the effect of Ca^{2+} on negatively charged oleate vesicles. One might expect that in this case the metal cations favor fusion. At first sight, this is true: as shown in Figure 12.13, there is an increase of turbidity when calcium is added to the vesicles, suggesting an increase of the molecular mass. However, when EDTA is added to this solution, the turbidity value is reduced to the initial value, showing the reversibility of the aggregation phenomenon: no fusion had taken place. To confirm that real fusion does not take place in this experiment (and similar ones), fluorescence techniques have been employed, with the typical terbium/dipicolinate (DPA) assay (Wilschut *et al.*, 1980). The Tb^{3+} ion is encapsulated in one vesicle species, and DPA in the other. The mixing would result in a large increase of fluorescence (observed in the control without liposomes) if the two chemicals came into contact. Nothing of this sort is observed, confirming that the two vesicle species aggregate without exchanging their water-pool material.

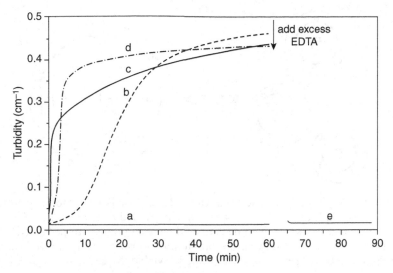

Figure 12.13 Effect of Ca^{2+} on the turbidity change upon mixing oleate vesicle solutions: 0.25 ml 1 mM 60 nm radius extruded oleic acid vesicles + 0.25 ml 1 mM 200 nm radius extruded oleic acid vesicles + 1.5 ml bicine buffer. Calcium ion concentration: (a) 0 mM; (b) 1 mM; (c) 2.5 mM; (d) 5 mM; (e) added excess EDTA to (d). (Adapted from Cheng and Luisi, 2003.)

If fusion of conventional vesicles is generally difficult, it seems to go better with GVs. In this regard, it is proper to cite the work of Yomo's group: the authors have been able to induce fusion between two populations of GVs, one containing DNA, the other containing a kit for protein synthesis and no DNA, obtaining protein expression only after fusion (Sunami *et al.*, 2010; Caschera *et al.*, 2011).

Going back to conventional vesicles, one efficient way to bring them to fusion is by the use of opposite charges in the membrane, as already observed by a number of authors (Uster and Deamer, 1981; Kaler *et al.*, 1989; Kondo *et al.*, 1995; Yaroslavov *et al.*, 1997; Marques *et al.*, 1998; Pantazatos and McDonald, 1999). This is schematized in Figure 12.14.

This kind of fusion process is interesting in several respects. It is a way to induce reactivity between the solutes entrapped in two different vesicle species. Fusion between vesicles is also a way to increase the molecular complexity of the incorporated species: for example, one can bring together enzymes and nucleic acids, or more enzyme species in order to induce, in principle, a meta-bolic cycle, and so on.

Consider also that this kind of increase in molecular complexity due to fusion corresponds loosely to a kind of symbiogenesis, which is reminiscent of the prebiotic scenario suggested by Dyson for bringing together the world of nucleic

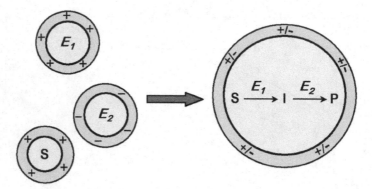

Figure 12.14 Fusion of vesicles as a way to foster reactivity and to increase the molecular complexity of the water-pool content: E_1 and E_2 are enzymes; **S, I** and **P** are enzymatic substrate, intermediate, and reaction product, respectively. This is also a method to circumvent the problem of substrate permeability in lipo-somes. It can be seen as a model of synthetic symbiogenesis.

acids and the world of proteins (Dyson, 1985). This is, in turn, reminiscent of the classic ideas of Lynn Margulis for symbiogenesis at the cellular level (Margulis, 1993).

All this represents work in progress, and to date the thermodynamic and kinetic aspects of this fusion process have not yet been clarified. It is also difficult to make predictions in this field, as unexpected results are often obtained.

One such surprise has been observed recently in a series of experiments in which negatively charged oleate vesicles were mixed with the positively charged DDAB vesicles. Both these surfactants display spontaneous vesicula-tion, each forming a broad distribution of sizes ranging from *c.* 20 to 1000 nm.

Now, when these two species are added to each other, in a given relative concentration, a new species appears with a much *narrower* size distribution. This is shown in Figure 12.15, where the P-index (a measure of the poly-dispersity) is plotted against the molar fraction of DDAB. The P-index drops from the initial value of 0.20 (a very broad distribution) to 0.04, a very narrow distribution (stable for months), at a relative percentage of 0.4 DDAB to 0.6 oleate (Thomas and Luisi, 2004). Between DDAB molar fractions of 0.41 and 0.60, flocculation occurs, which indicates a thermodynamic instability, in agree-ment with other cationic systems (Kaler *et al.*, 1989; Marques *et al.*, 1998; Kondo *et al.*, 1995).

The very peculiar molar ratio 0.4 DDAB to 0.6 oleate, which gives rise to the narrow size distribution, is really noteworthy. This molar ratio corresponds closely to electro-neutrality (this is not at 50:50 molarity, due to the relatively high pK of oleate carboxylate in the bilayer) and suggests that small mixed

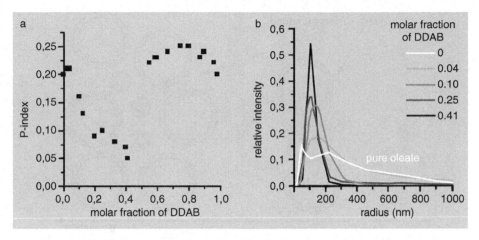

Figure 12.15 Dynamic light scattering (DLS) of vesicle mixtures. (**a**): P-index phase diagram; (**b**): size distributions (from DLS) for DDAB-oleate mixtures, total concentration 1 mM in 0.2 M borate buffer at pH 8.5, 25.0 °C, scattering angle 90°. (Adapted from Thomas and Luisi, 2004.)

vesicles with an approximately equal number of positive and negative charges may enjoy particular stability.

All this indicates the richness of the unexplored in the field of vesicles – and of course, the need for much more study.

12.5 Size competition of vesicles – and interaction with RNA

The uptake of oleate by pre-added vesicles, and in particular the matrix effect, permits regulation of the growth of the size and the number of particles, and in this way it is possible to tackle a series of novel questions. One such question is: *Is there a difference in the rate of uptake of fresh surfactant between two vesicles of different sizes?*

This question is interesting because the relative rate of growth and self-reproduction may simulate a competition between two different kinds of organisms coexisting in the same medium, whereby the added surfactant can be seen as the nutrient for which the two organisms are in competition.

Experiments have been set up in this direction (Cheng and Luisi, 2003) utilizing extruded oleate vesicles having radii of 31.6 and 64.1 nm. For the same surfactant concentration, the total surface area is the same for the two families of vesicles (neglecting the differences arising in aggregation number due to the differences between the inner and outer leaflets). This also means that for the same total concentration, the number of vesicles is much larger in the case of the family with the smaller radius. Generally, in these competition

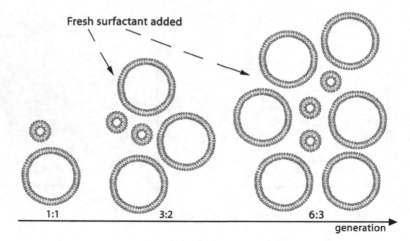

Figure 12.16 Larger vesicles grow faster than smaller ones: a starting ratio 1:1 between larger and smaller vesicles becomes 3:2 and then 6:3 in the following generations (idealized). (Modified from Cheng and Luisi, 2003.)

experiments, care should be taken to work either with a constant number of particles, or with a constant surface area. This has been done in the cited paper, and experiments were carried out under conditions where the vesicle number concentration is roughly the same for the 31.6 and the 64.1 nm families, having shown that the larger vesicles grow faster (by roughly a factor of three).

The greater reactivity of the larger vesicles can be visualized in Figure 12.16, which illustrates qualitatively the outcome of competition during generations of self-reproduction.

Since larger and smaller vesicles are present in the same number, the larger reactivity of the larger vesicles could be explained considering that their total surface and therefore the uptake area is larger. This is the easy explanation. On the other hand, one might have expected that the smaller vesicles, characterized by a greater curvature radius, might have relaxed more eagerly into a relaxed state by absorbing more rapidly the added material.

The issue of the competition between vesicles of different sizes has been examined in Szostak's group, as well, in the very interesting context of the incorporation of RNA (Chen *et al.*, 2004). These authors argue that RNA encapsulated in fatty acid vesicles exerts an osmotic pressure on the vesicle membrane that drives the uptake of additional membrane components, leading to membrane growth at the expense of relaxed vesicles, which shrink. Although the effects are rather small, the authors argue that this difference in growth may have implications on the Darwinian evolution of early cells (Chen *et al.*, 2004). Interesting here is the argument that there is a relation between physical factors of the membrane and evolution.

In a different kind of experiment, CTAB vesicles of different sizes were used to study the binding interaction with t-RNA (Thomas and Luisi, 2005). The work was prompted by the idea of investigating whether RNA might present some kind of specific interaction with phospholipid bilayers. To this aim, vesicles from POPC containing 3.5 percent molar fraction of the positively charged single-chained surfactant CTAB and the t-RNA mixture were prepared. Two different populations of narrowly sized extruded vesicles having an average radius of 40 and 80 nm, respectively, were used to study the possible influence of size on the interaction.

The RNA was then added to the vesicle solution. The size distribution of the vesicles before and after the addition of t-RNA was investigated by dynamic light scattering and by direct ultraviolet (UV) optical density observation.

Surprisingly, it was observed that the aggregation behavior of t-RNA to the charged vesicles was strongly dependent on the vesicle size – although in this experiment the size of the two vesicles differs only by a factor of two. In particular, the larger vesicles aggregated rapidly upon RNA addition, whereas the smaller ones did not, although they too were equally capable of binding RNA. The aggregation of the larger vesicles was completely reversible: as soon as RNAase was added to the aggregates, the initial size distribution was obtained again, which showed that the aggregation process occurred without significantly affecting the vesicular properties. A nice experiment is shown in Figure 12.17: to the mixture of the two populations (light-scattering resolution is not sufficient to show two peaks in this case) RNA was added. One of the two components shifts towards larger radii, while the smaller component is left behind. After addition of RNAse, the original situation is again obtained.

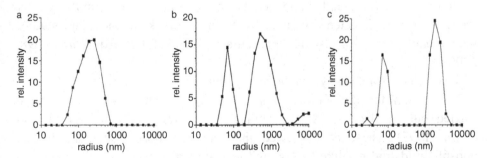

Figure 12.17 Vesicle size selection by RNA added to a mixture of small (~40 nm radius) and large vesicles (~80 nm radius) from 0.5 mM POPC – 3.5% CTAB in 20 mM sodium phosphate buffer (pH 7.0). (**a**) The initial size distribution of the 1:1 mixture; (**b**) the size distribution 4 min after RNA addition, during the selection process; (**c**) the stable final size distribution after 15 min, with one peak for small diameters (~45 nm radius) and a second peak indicating aggregates of large vesicles (> 500 nm). As mentioned in the text, this process is reversible upon addition of RNAase. (Thomas and Luisi, 2005.)

In conclusion, larger vesicles from POPC/CTAB readily aggregated in the presence of RNA in a completely reversible way; smaller vesicles – with the same chemical composition and only half as small – did not. The question arising from these data is whether the inherent ability of RNA to discriminate between different vesicle sizes with such a fine-tuning might have been important with regard to early cell evolution. This connects with the illustration in Figure 12.16, showing that different sizes can grow and reproduce with different efficiency, as well as with the previously mentioned observation of Chen *et al.* (2004); see also further studies by Szostak's group (for example, Chen *et al.*, 2005).

I would like to mention a quite different field of research where there is an interaction between the world of nucleic acids and the world of vesicles. This is at the level of the phosphatidyl nucleotides. This is a new family of amphiphiles, where the phospholipid group is covalently linked to a nucleobase. These compounds have been shown to be capable of forming liposomes (Bonaccio *et al.*, 1994a, 1994b, 1996). See Figure 12.18.

The idea behind this research can easily be understood: to prepare "complementary liposomes," namely liposomes with adenine groups attached to the phosphatidyl moiety; and a distinct family of liposomes with guanine (or guanine with thymine), mix them with each other, and see whether the base recognition also works at the level of liposomes. When this project was started, our research group was rather excited at the possibility of having the two complementary families fuse with each other and form liposomes with two complementary leaflets, held together by Crick-Watson base pairs. Nucleic-acid recognition at the level of vesicles?

Nothing of this sort really happened. Or rather, some kind of complementary recognition was detected, but in a frustratingly weak manner; and studies with monolayers were also not particularly successful (Berti *et al.*, 1998, 2000).

Research does not always go the way you want...

The reason for this failure was evidenced by an NMR investigation carried out by the late Anna Laura Segre's group in Rome (Bonaccio *et al.*, 1997). These data showed that the nucleobases attached to the phospholipids did not like to "swing" in water; they were mostly tacked in the lipophilic bilayer and were not available for "talking to" the complementary nucleobases.

Some other interesting results did emerge from these studies, for example concerning the morphology of these amphiphilic aggregates (Bonaccio *et al.*, 1996).

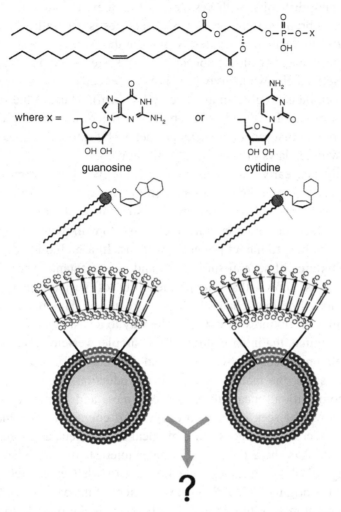

Figure 12.18 A schematic representation of phosphatidyl nucleoside-liposomes, showing the bases attached to the phosphatidyl unit of the lipid. More details are in the text.

Concluding remarks

We have described experiments with vesicles, which gave surprising results. Completely unexpected was the matrix effect or the influence of vesicle size on RNA interaction, or the formation of a sharper size distribution on mixing two much broader size distributions. The fact that, working with liposomes, new effects like that are always being discovered, is rewarding and exciting; however, it also reveals how little we know in the field of liposomes, and in particular with regard to their thermodynamic and kinetic properties.

One inherent difficulty, with vesicles, is that they, unlike micelles, are not chemical equilibrium systems; each vesicle species is generally its own kinetic trap. On the other hand, this peculiarity gives at the same time a strong analogy with biological cells, as cells are not equilibrium systems – and cannot be if they have to preserve their own identity (i.e., horse liver cells and *E. coli* cells do not mix with each other to give an average cell system). Thus, at the level of the double layer of vesicles, the physical characteristics of life are already encountered (membranes that cannot be easily permeated by external solutes and refuse to comply with the laws of chemical equilibrium).

It is likely that early cells were more permissive about permeability, and perhaps an early step in the transition to life is the transition from permeable, simple protocells to hard and impermeable structures, like our present POPC liposomes. In fact, the common stand of chemists to work with pure compounds may not be the best to model prebiotic systems. In a prebiotic scenario, most probably, mixtures of several surfactants and co-surfactants were dominating the scene. It is known that the permeability of vesicles increases when co-surfactants – like long-chain alcohols – are added. This observation about the importance of mixtures would open the way to a vast area of research.

In the meantime, the intense study of the simpler vesicle systems has unraveled novel, unsuspected physicochemical aspects – for example, growth, fusion and fission, the matrix effect, self-reproduction, the effect of osmotic pressure, competition, encapsulation of enzymes, and complex biochemical reactions, as will be seen in the next chapter. Of course, the fact that vesicles are viewed under the perspective of biological cell models, as expressed in the next pages, renders these findings of greater interest. In particular, one tends immediately to ask the question, whether and to what extent they might be relevant for the origin of life and the development of the early cells. In fact, the basic studies outlined in this chapter can be seen as the prelude to the use of vesicles as cell models, an aspect that we will consider in more detail in the next chapter.

Chapter 13

Biochemistry and molecular biology in vesicles

Introduction

We have seen in the previous chapter that liposomes are suitable models for biological cells, or, more precisely, for the shell of biological cells. In order to bear some resemblance with biological cells, biochemical compounds must be present in the interior. This is the large field of entrapment of solutes in vesicles, focusing particularly on biopolymers such as proteins and nucleic acids. In the previous chapter, we have described the entrapment of molecules in micelles, reverse and/or aqueous, and in this case, the entrapment is made possible by the dynamics of micelles, as they are open systems, which break and reform continuously. We know already that the case of vesicles is very different, as vesicles are generally not very permeable. Let us then see, first, how the entrapment can be carried out.

13.1 The entrapment of solutes in vesicles

Only occasionally, depending upon the chemical nature of the solute and/or the lipid surfactant, some restricted and selective permeability is observed. In the experiment illustrated in Figure 13.1, an apolar molecule is capable of permeating inside, where it reacts with phosphate ions by the opening of the ring: the product, being now polar, is trapped inside. In this case, although the permeability is very low, the irreversible chemical transformation of the reagent inside the liposomes drives the incorporation process.

Since permeation does not work for enzymes and other macromolecules, these have to be incorporated during the formation of vesicles. A schematic illustration is shown in Figure 13.2, which shows a classic way to incorporate a macromolecule, an enzyme in this case, inside the liposomes: one starts from a film of the water-insoluble surfactant, e.g. oleate or a phospholipid like POPC, while the solution contains the enzyme to be entrapped. The next step is a robust vortexing, which sends out lamellae of the lipid in water, which then form the

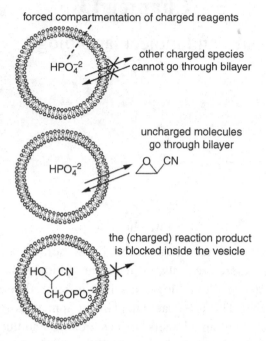

forced compartmentation of charged reagents

HPO_4^{-2}

other charged species
cannot go through bilayer

HPO_4^{-2}

uncharged molecules
go through bilayer

the (charged) reaction product
is blocked inside the vesicle

HO   CN

$CH_2OPO_{3}^{-2}$

Figure 13.1 Reaction features of surfactant aggregates, showing an example of permeability of non-charged molecular species, and forced compartmentation of charged species. (From Luisi *et al.*, unpublished data).

spherically closed liposomes. In the process of closure, the liposome entraps mechanically part of the solute molecules. In this particular experiment, the enzyme is capable of polymerizing ADP into Poly-A, and ADP is left permeate inside the liposomes in a very slow and not very efficient process – but eventually enough ADP permeates inside to bring about the synthesis of the polynucleotides. In this case, as in the previous example, the macromolecule cannot leak out, so it accumulates inside.

Note from the figure a couple of important operational details: only part of the enzyme is being entrapped by this method, and it is then necessary to eliminate by chromatography the enzyme molecules remaining in solution. Then we obtain a suspension of the liposomes, which can be analyzed by sampling out. Later on in this chapter, more details on this system will be given (see Figure 13.6).

There are several other ways to entrap solutes inside the liposomes, and the entrapping efficiency depends on the structure of liposomes (small unilamellar, large unilamellar, multi-lamellar, vesicles, etc.) and from the technique for liposome preparation (Roseman *et al.*, 1978; Cullis *et al.*, 1987; Walde and Ishikawa, 2001).

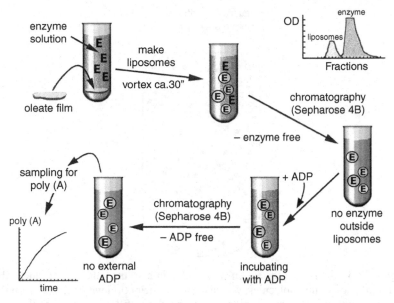

Figure 13.2 One of the procedures for entrapping an enzyme – in the example, polynucleotide phosphorylase (PNPase), inside liposomes. Liposomes are produced by vortexing a POPC film, and when they form and close, they mechanically entrap the solute. The non-entrapped enzyme is eliminated by size exclusion chromatography, then the substrate (ADP in this case) is left to permeate throughout, the ADP excess is again eliminated chromatographically, and the reaction kinetics – due to the internalized enzyme reaction – measured. (From Walde *et al.*, 1994b).

The method indicated in Figure 13.2 is commonly used with phospholipids, but there is a simpler method when the surfactant spontaneously forms vesicles. This is the injection method, as we have seen in the previous chapter, when we have described the entrapment of ferritin in oleate vesicles.

In fact, when such a surfactant is being added to a solution containing the solute, be it an enzyme or any other water-soluble molecule, the *in situ* formation of vesicles is attended automatically by the entrapment of such a solute. We have seen an application of this already in Part I of this book (see Figure 4.6), talking about overcrowding, and we will come back to this in Part V, talking about synthetic biology.

Rationalizing the distribution of solute in the liposomes is not an easy matter. One may expect a kind of Poisson distribution, one characterized by discrete events all independent from each other. A qualitative illustration of this is given in Figure 13.3, which gives the idea that different liposomes may have a different content of the solute. The stochastic nature of this distribution – together with the lack of equilibrium exchange among the liposomes – makes it so the liposomes may have a different reactivity from each other.

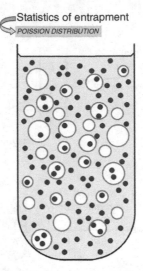

Figure 13.3 The *in situ* formation of vesicles in a solute solution, produces the entrapment of solute molecules, with a certain distribution which depends on the initial conditions. This distribution should produce "filled" vesicles with a different number of solute molecules.

Entrapment of enzymes in liposomes is an active research field. Most of the work presented in the literature describes the entrapment of one enzyme at a time. The review by Walde and Ishikawa (2001) provides a rich account and discussion of the various techniques used to incorporate enzymes inside liposomes, and their possible applications in chemistry, medicine, and industry. Table 13.1 is a modification and simplification of one of their tables (Walde and Ishikawa, 2001, Table 6). This table also describes in a nutshell the corresponding experiments and operational goals, and displays a large variety of enzymes and potential applications.

The aim of entrapping enzymes in vesicles is on the one hand to study enzymatic reactions in a restricted medium, and on the other hand to develop models for cellular reactions. If aiming at models for metabolic pathways, the encapsulation of one enzyme is not enough, and actually, the far-reaching goal would be the entrapment of an entire enzymatic cycle. It is fair to say that there are no significant examples of this yet in the literature, although some experiments along this line have been published. For example, there has been an attempt to entrap in one single POPC liposome the four enzymes responsible for the synthesis of lecithin starting from glycerol-3-phosphate (Schmidli *et al.*, 1991). The idea behind this project was to construct a minimal cell capable of producing its own membrane from within – an idea related to autopoiesis. (We will go into more detail on this later on in this chapter.)

Table 13.1 *Selected examples for enzyme-catalyzed reactions inside lipid vesicles**

Enzyme	Type of study, remarks	References
D-Amino acid oxidase	Investigation of the activity of the entrapped enzyme against externally added D-amino acids (egg PC-based MLV).	Naoi *et al.*, 1977.
Ascorbate oxidase	Entrapped (and partially adsorbed enzyme) was active against externally added ascorbate and O_2.	Mossa *et al.*, 1989; Annesini *et al.*, 1992; Ramundo-Orlando *et al.*, 1993.
Carbonic anhydrase	The entrapped enzyme mainly adsorbed onto the positively charged lipid bilayer.	Annesini *et al.*, 1993.
	Low frequency, low amplitude magnetic fields increase the permeability of enzyme-containing lipid vesicles.	Annesini *et al.*, 1994.
	Activity measured against externally added CO_2.	Ramundo-Orlando *et al.*, 2000.
α-Chymotrypsin	Activity and stability measurements of the enzyme entrapped inside POPC vesicles against externally added Bz-Tyr-pNA. No activity against the larger substrate Suc-Ala-Ala-Pro-Phe-pNA or casein. Inhibition of externally present enzyme by an inhibitor protein.	Walde and Mazzetta, 1998.
	Detailed kinetic measurements and kinetic analysis (by dynamic modeling) of chymotrypsin-containing POPC vesicles.	Blocher *et al.*, 1999.
	Fusion of chymotrypsin-containing vesicles with vesicles containing the substrate Suc-Ala-Ala-Pro-Phe-pNA by addition of partially denatured Cytochrome C.	Yoshimoto *et al.*, 1999.
DNAase I/DNA	Entrapment of the enzyme together with the substrate (DNA) in dipalmitoyl-phosphatidylcholine (DPPC) lipid vesicles.	Baeza *et al.*, 1994, 1990.
β-Galactosidase	Investigation of the change in the permeability of enzyme-containing vesicles against externally added substrate.	Annesini *et al.*, 1994.
β-Glucosidase	Entrapment of the enzyme inside the lipid vesicles led to a stabilization of the enzyme against inhibition by externally added Cu^{2+} ions, as measured with *p*-nitrophenyl-D; β-D-glucopyranoside as substrate.	Sada *et al.*, 1988, 1990.

(Continued)

Table 13.1 (*cont.*)

Enzyme	Type of study, remarks	References
Glucose oxidase	A few measurements on the activity of entrapped glucose oxidase after external addition of glucose in the presence of variable amounts of added insulin.	Solomon and Miller, 1976.
	Activity measurements in egg PC-based enzyme-containing vesicles which had deoxycholate in the membrane against externally added glucose.	Ambartsumian *et al.*, 1992.
Glucose oxidase, in combination with peroxidase or lactoperoxidase	Investigation of the activity of the two enzyme-containing anionic or cationic lipid vesicles against externally added D-glucose.	Kaszuba and Jones, 1999.
Polynucleotide phosphorylase	Experimental demonstration that the enzyme inside dimyristoylphosphatidylcholine (DMPC) vesicles was active against externally added ADP, yielding lipid vesicle trapped poly (A).	Chakrabarti *et al.*, 1994.
	Simultaneous oleic anhydride hydrolysis resulting in a self-reproducing vesicle system.	Walde *et al.*, 1994a.
Trypsin	Activity and stability measurements of the enzyme entrapped inside POPC vesicles against externally added Bz-Arg-pNA and Z-Phe-Val-Arg-pNA.	Graf *et al.*, 2001.
Tyrosinase	Kinetic investigations against different externally added mono- and diphenols.	Miranda *et al.*, 1988.
Urease	Kinetic investigation towards externally added urea. Increased stability of the entrapped enzyme.	Madeira, 1977.
Urate oxidase	Activity measurements in egg PC-based enzyme-containing vesicles which had deoxycholate in the membrane against externally added uric acid.	Ambartsumian *et al.*, 1992.

* Adapted, with shortening and slight modifications, from Table 6 of Walde and Ishikawa, 2001, with the authors' permission.

13.2 On the surface of liposomes

A quite different application of liposomes as reactive compartments is in drug delivery. It was found several years ago that liposomes, because of their hydrophobic nature, strongly interact with the membranes of biological cells and can actually be incorporated inside them by endocytosis or other mechanisms, e.g., fusion (Allison and Gregoriadis, 1974; Gregoriadis, 1976a, 1976b, 1988, 1995; Papahadjopoulos *et al.*, 1989).

It is not always necessary to incorporate the drug inside the liposomes, particularly when it is lipophilic. In this case, the drug can be bound to the liposome bilayer by hydrophobic interactions. An example is given by the binding of camptothecin, an anticancer drug extracted from Chinese plants, to POPC liposomes. In particular, to this aim very small liposomes were obtained by the injection method, which have the advantage over the very large ones of a longer circulation in vivo and a greater stability (Stano *et al.*, 2004).

A drug bound in this way can then be delivered inside the cell. In the cell, phosphatidylcholine (PC) liposomes can be then digested by phospholipases and thus the sequestrated drug is made free. The therapeutic applications are somewhat restricted by the process of elimination of liposomes in vivo by the cellular phagocytic system. This problem can be reduced by making covalent adducts with polyethylene glycol, the so-called "PEGilation," the process that gives rise to the "stealth liposomes" (Woodle and Lasic, 1992). There is a vast amount of literature and intense patent activity in the field, and the work by Gregoriadis is again particularly noteworthy. Several liposome preparations are already commercial or under study (Gregoriadis, 1995), including trans-dermal therapy (Raab, 1988; Foldvari *et al.*, 1990; Lasch *et al.*, 1991; Lasic, 1995), and cosmetics (see for example, Raab, 1988; Cevc and Blume, 1992; Lasic, 1995).

Liposomes can also be utilized for gene transport and corresponding transfection (Gao and Huang, 1995; Zhu, Zhang, and Reszka, 1996; Zhu *et al.*, 1996; Reszka, 1998; Kikuchi *et al.*, 1999) – also a very active field of inquiry.

In this case, the liposomes are generally dubbed with positively charged co-surfactants, such as DDAB (dimethyl-didodecyl-ammonium bromide) or CTAB (cetyl-trimethyl-ammonium bromide), to induce an effective binding with DNA or RNA via electrostatic interactions. There is then the possibility of three different liposome/DNA species, since DNA can then be localized on the surface, inside the water pool, or in either place. It is not yet ascertained which species is the most effective from the point of view of gene transfer, but usually the preferred operational procedure is by surface binding, as this is operationally by far the simplest one.

The surface of liposomes can also be utilized for facilitating chemical reactions. One example is the liposome-aided synthesis of peptides (Blocher *et al.*, 2000, 2001). In this case, the binding of hydrophobic NCA-activated amino acids (NCA = *N*-carboxy-anhydride) to the lipophilic surface of palmitoyl-oleyl-phosphatidyl-choline (POPC) liposomes, has permitted the condensation of Trp-oligomers up to a polymerization degree of 18, which is considerable, given that in water the synthesis by way of the same reaction is limited to oligomers of 5 to 6 due to their insolubility.

13.3 The road map to the minimal cell: complex biochemical reactions in vesicles

The next part of this book (Part V) will consider in detail the construction of the minimal cells in the laboratory, using the liposomes we have been describing until now. Such a research project is a complex enterprise and it may be useful to divide the "road map" to the minimal cell into different milestones of increasing complexity.

The first one, which is already under control in several laboratories, is to carry out and optimize complex enzymatic reactions in liposomes – such as the polymerase chain reaction, the biosynthesis of RNA and DNA, the condensation of amino acids, and so on. I believe that it can be said that this kind of research was pioneered at the ETHZ, the Swiss Federal Institute of Technology, in the early 1990s.

Probably one of the very first examples of enzymatic reactions carried out in liposomes with the aim of building a minimal cell is the work by Schmidli *et al.* (1991), as already mentioned (see Figure 13.4).

The basic idea is to have inside the liposomes the series of reactions that, starting from a relatively simple product (G3P, glycerol-3-phosphate), leads to the phospholipid membrane. In Figure 13.5, the pathway for phosphatidylcholine synthesis can be seen, with the four enzymes G3P-AT (*sn*-glycerol-3-phosphate acyltransferase), LPA-AT (1-acyl-*sn*-glycerol-3-phosphate acyltransferase), PA-P (phosphatidate phosphatase), and CDPC-PT (cytidine-diphospho-choline phosphocholine-transferase).

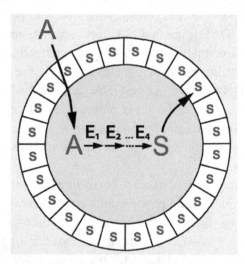

Figure 13.4 A liposome that builds its own membrane with the help of entrapped enzymes: the prototype of the simplest autopoietic minimal cell. In the experiment by Schmidli *et al.* (1991), four different enzymes were entrapped in one single lipsome, with the idea of synthesizing lecithin.

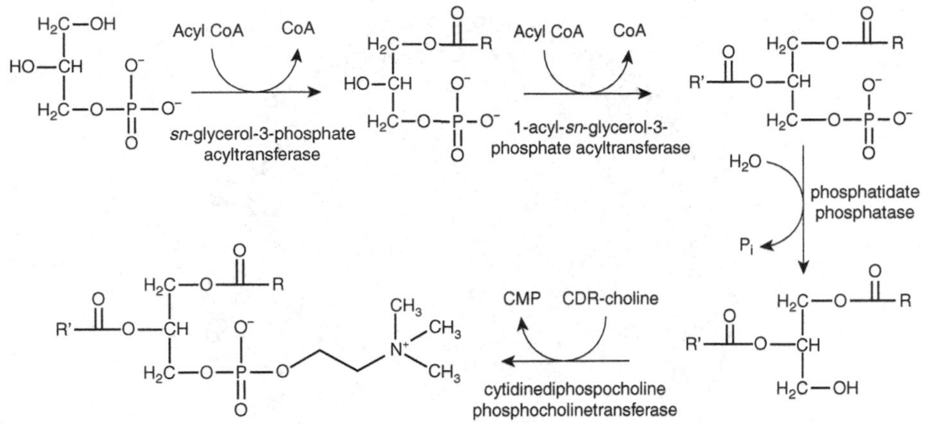

Figure 13.5 Salvage pathway for phosphatidylcholine synthesis. (Modified from Schmidli *et al.*, 1991.)

The enzymes were active when associated with liposomes, with an activity corresponding to that in the original microsomes. It was shown that phosphatidylcholine (PC) was synthesized in the enzyme-containing liposomes, and that its yield was about the same when analyzed as a function of the radioactive *sn*-G3P as substrate or when cytidine-diphosphocholine is the radioactive-labeled substrate. The other substrates used for the reaction were palmitoyl and oleoyl coenzyme A (CoA), so that palmitoyl-oleyl-phosphatidylcholine (POPC) was the synthetized product. In order to visualize better the effect of this synthesis on the physical state of the proteo-liposomes, shorter lecithin chains were also synthesized using hexanoyl CoA and dihexanoyl phosphatidylcholine. The yield was only around 10 percent, and the effect on the liposome size, as examined by light scattering, was in the expected direction, but rather small.

This kind of work, although rather significant, has not been pursued, partly because of the difficulty of obtaining the four enzymes with a sufficient degree of purity. Some work followed with the idea of using liposomes formed simply by phosphatidic acid (the product of the first two enzymes shown in Figure 13.5; see also Luci, 2003).

A different example of enzymatic synthesis in vesicles was also mentioned above with the poly-condensation of adenine diphosphate (ADP) into poly(A) by encapsulated enzymes (Chakrabarti *et al.*, 1994; Walde *et al.*, 1994a), as already shown in Figure 13.2. In the case of Walde *et al.* (1994a), there was a core-and-shell reproduction, as the synthesis of poly(A) – a prototype of RNA – was occurring simultaneously with the self-reproduction of the vesicle shells. It is a reaction that was investigated originally by Oparin in his aggregate systems,

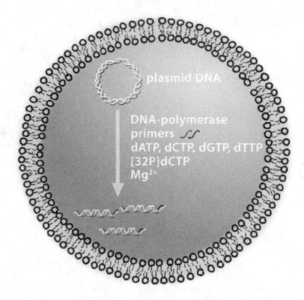

Figure 13.6 The PCR reaction in liposomes (Oberholzer *et al.*, 1995a). Two types of liposomes, obtained from POPC or from a 9:1 (w/w) mixture of POPC and PS, were used in the original experiments. For both POPC and POPC/PS liposomes, all components were added to a lipid film in a buffered solution: 25 nM plasmid DNA (linearized or nonlinearized); 10 nM DNA polymerase; 4.7 pM oligonucleotide primers A3 and A5; 1 mM of dATP, dCTP, dGTP and dTTP; and 0.2 mM $[^{32}P]$dCTP (for the POPC liposomes) or 0.32 mM $[^{32}P]$dCTP (for the POPC/PS liposomes). Mg^{2+} ions were also present.

and appropriately enough, the cited article (Walde *et al.*, 1994a) is titled "Oparin's reactions revisited."

Another complex biochemical reaction is the polymerase chain reaction (PCR). It has also been implemented already a long time ago in liposomes (Oberholzer *et al.*, 1995a). This reaction was interesting from the point of view of vesicle chemistry because the liposomal system had to endure the extreme PCR conditions, with several temperature cycles up to 90 °C (liposomes were practically unchanged at the end of the reaction). Furthermore, nine different chemicals had to be encapsulated in an individual liposome for the reaction to occur (see Figure 13.6). This was carried out by mechanical entrapment from a solution that contained all components: only a minimal number of the *in situ* formed vesicles could entrap all nine components. These odds notwithstanding, there was a significant synthesis. We will come back to this point of the statistics of entrapment later on in this section, when talking about protein synthesis.

A suggestive example of core and shell reproduction was provided shortly after (Oberholzer *et al.*, 1995b) with the use of the famous Spiegelmann and

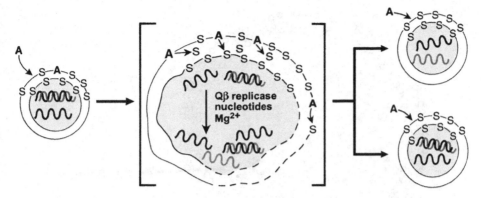

Figure 13.7 Replication of RNA in self-reproducing vesicles. The initial vesicles contained the enzyme Qβ replicase and the four ribonucleotides in excess, as well as the RNA template (the MDV-1 template). The division of vesicles is induced by the addition of oleic acid anhydride and the duplication of the figure is idealized, as in reality division occurs on a statistical basis. (Adapted from Oberholzer *et al.*, 1995b.)

Eigen's enzyme, the Qβ replicase. As illustrated in Figure 13.7, while the enzyme was replicating RNA, the oleate vesicles were multiplying on their own accord. The vesicle self-reproduction was induced by the binding of the water-insoluble oleic acid anhydride, as described earlier. The hydrolysis of the anhydride was followed spectroscopically by FTIR (Fourier transform infrared), and the kinetics of the reaction, as well as the vesicle size distribution, were studied by freeze-fracture electron microscopy. The vesicles, in addition to the enzyme and RNA MDV-1 template, contained the triphosphates of adenine, cytosine, guanine, and uracil, i.e., ATP, CTP, GTP, UTP, (^{35}S ATP), as well as Mg^{2+} ions and buffer. This experiment was operated under excess of Qβ replicase/RNA template, so that the replication of RNA could proceed for a few generations.

At first sight, Figure 13.7 already appears to correspond to a living cell. Still, since reproduction of vesicles occurs statistically, after a few "self-reproduction" steps, the large majority of new vesicles will not contain either enzyme or template, and therefore the system will undergo "death by dilution." Death by dilution is typical of cell systems where there is no regeneration of the macromolecular components from within. A further shortcoming of the system is given by the fact that the two self-reproduction processes (shell and core) are not coupled with each other.

A significant step forward in this field has been provided by the entrapment of an entire ribosomal system in POPC vesicles (Oberholzer *et al.*, 1999). In this first experiment – see Figure 13.8 – poly (Phe) was synthesized, using poly(U) as

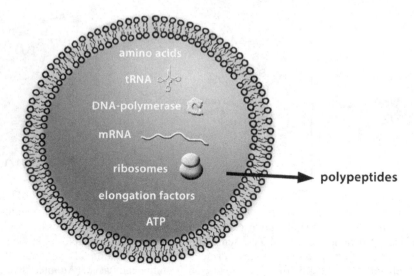

Figure 13.8 The first ribosomal polypeptide synthesis in liposomes. In a first step, a solution without enzymes but with all other components was added to a POPC film, and vortexed in order to induce liposome formation. The resulting suspension was then subjected to seven freeze/thaw cycles in order to increase the entrapment yields. In the second step, to this liposomal suspension, a solution containing all enzymes required for protein biosynthesis was added, followed again by a series of freeze and thaw. All entrapment operations were executed at −5 °C in order to avoid protein biosynthesis in the bulk. For more details, refer to the text and to the original paper by Oberholzer *et al.* (1999).

the m-RNA. Although this work is now obsolete, it contains several elements of interest. Firstly, several components must be entrapped within the same liposome in order for the polypeptide synthesis to occur: the entire ribosomal structure, namely the complex between the subunit 50S and 30S; the messenger RNA (poly(U) in this case); t-RNA^{phe}; the elongation factors EF-Tu, EF-G, EF-Ts; and the substrate Phe (labeled in this case). The strategy of choice was again to form the liposomes in a solution containing all components, and the entrapment yield increased with several freeze and thaw cycles. All procedures for the preparation of liposomes had to be carried out at 5 °C in order to avoid poly(Phe) synthesis outside the liposomes; furthermore, after the extrusion step, ethylene-diamine-triacetic acid (EDTA) in excess was added to the external solution in order to inhibit the poly(Phe) expression outside the liposomes. The overall yield was 5% with respect to the experiment in water without liposomes. The authors argue (Oberholzer *et al.*, 1999) that this yield is actually surprisingly high, considering that the liposomes occupy only a very small fraction of the total volume of solution; and that only a very few of them would contain all ingredients by the statistical entrapment.

This experiment signed the beginning of a new area. In the years immediately following, in fact, this procedure was optimized in the same group, as well as several other groups, to express real proteins, the first being – for obvious detection reasons – the green fluorescence protein (GFP). The "road map," in fact, continues with protein expression, which is the subject of the next part of the book.

Concluding remarks

The area of biochemical reactions inside vesicles represents a very active and exciting field, linking bioorganic chemistry with cellular chemistry. What is generally outstanding from the few examples given in this chapter is the high stability and versatility of liposomes as hosts for this kind of reaction: the fact that liposomes can withstand cycles of 90 °C, and are made to contain dozens of different biochemicals, is now taken for granted, but in no way is this a trivial feature. It permits the first step of the "road map" to the minimal cell, and we will see shortly the next step, describing the synthesis of proteins inside liposomes. This kind of work was pioneered at the ETH in Zurich, and is now the heritage of a large number of researchers in the world.

Actually, in the last chapter of Part IV, we have seen that liposomes are good models to host very complex biochemical reactions, and we have seen a possible way to do the first step of a road map to the protocell. The second step will be one in which the entire ribosomal apparatus would be entrapped in the water pool of liposomes and operate the first complex cellular reactions, for example the synthesis of proteins.

This can be seen as a particular aspect of synthetic biology, and this is what we will describe in Part V – also giving a picture of the field of synthetic biology at large.

Questions – and research proposals – for the reader

1. The fusion of vesicles with opposite charges appears to be an efficient method to increase the molecular complexity. Suppose you distribute all enzymes of the Krebs cycle into five or six different charged vesicle families (some positively, some negatively charged) – then mix. Would the Krebs cycle be reconstituted?
2. Fusion among vesicles may correspond to a primitive form of prebiotic symbiogenesis. Can you conceive an experiment in this light – e.g., fusing mitochondria-containing vesicles with DNA containing vesicles?
3. Vesicles can self-reproduce, however, it is argued that there is no information content passing from one generation to the next. What do you think: is information at the level of membrane shell really so important for self-reproduction of

early protocells? And, importantly: what do you mean by information in a prebiotic scenario?

4. We have seen experiments of growth of vesicles, whereby fresh surfactant binds to pre-formed vesicles. How would you devise an experiment, so that the growth rate is determined – or influenced significantly – not by the shell, but by the vesicle content?

5. The common stand of chemists to study liposomes by working with pure compounds may not be the best to model prebiotic systems. In a prebiotic scenario, most probably, mixtures of several surfactants and co-surfactants were dominating the scene, and probably vesicles were then leakier and more permeable. Studies in this direction may indeed represent an important research project in the origin of life.

Part V

Towards the synthetic biology of minimal cells

Chapter 14

A panoramic view of synthetic biology

Introduction

As we have already mentioned, this last part of the book is devoted to the question of the minimal cell, namely, whether and to what extent by using the laboratory tools of synthetic biology (SB) we will be able to approach the construction of a living cell. Let us begin then with an illustration of SB at large, to focus later on the particular question of the minimal cell.

The brochure of the last synthetic biology congress in London (2014) opens up with a sentence stating that SB is predicted to be worth 10.8 billion dollars by 2016. Not bad for a relatively new branch of life sciences. Although, for some, SB is not really a novel branch of science, but actually a new dress for bioengineering within the broader field of biotechnology. It is indeed an offspring of the union between bioengineering and molecular biology, and in this vast field, it claimed from the very beginning a very ambitious program: that of creating new forms of life, alternative to the extant ones, which would permit us to tackle the energy problems of our world successfully, and produce new drugs cheaply. This ambitious statement becomes less high-sounding if one considers that, in reality, the new forms of life are restricted to bacterial life; and that SB aims at synthesizing simpler biological structures, not necessarily living, but still alternative to those found in nature.

One definition given in one important meeting in Zurich in 2007 (SB3) defines SB as a science intended to "engineer and study biological systems for achieving better understanding of life processes, generating and assembling functional modular components, or develop novel applications of processes."

You see in this sentence two important points: one is the declared interest for the basic aspects of life science (not only the applied ones); and the notion of modularity, which, as we will see, is indeed very important for the development of bio-engineering SB.

Actually, the term *synthetic biology* was coined back in 1912 by the French chemist Leduc in 1912, but only recently has entered with force into the broad field of biotechnology and bioengineering, with a high number of yearly international meetings, several books and review articles, as well as new journals devoted to this area. Much of its success is due to the interest that SB induced in the pharma industry, which is now investing considerable funds in recently organized research units. This pivots on the perspective of new drugs; but SB is under study also for the perspectives of making fine chemicals, as well as for making biofuels, as we will see.

Generally, all that is potentially deriving from genetically modified bacteria. In fact, genetic manipulation is the usual operational way for applied SB. However, we will see soon there is also an important branch of SB that can do without genetic manipulation.

14.1 Main strategies and perspectives of synthetic biology

The field of SB has several working directions now, and for example, O'Malley *et al.* (2007) mention, in a thoughtful paper, that the high-profile areas of research in SB can be understood as encompassing three approaches: DNA-device construction, genome-driven cell engineering, and protocell creation. Benner and Sismour (2005) state instead that synthetic biologists come in two classes: those who reproduce emergent behaviors from natural biology, with the goal of creating artificial life; and those who seek inter-changeable parts from natural biology to assemble into systems that function unnaturally. In this framework, the thought of Endy (see Chan et al., 2005) is also very suggestive. Another view of SB is presented in the conversation with Paul Freemont, here below.

I will make use of a different classification; for me, the more fundamental difference is between the genetic engineering approach (genetic manipulation) and the kind of SB that is based on chemical synthesis or chemical procedures without interfering with the extant genomes. We will come back to this in the next sections. Now, let us consider one point of general interest, also for non-scientists.

This is that SB has evoked strong reactions from the mass media, mostly because of the fear of engineered new, unknown, and potentially dangerous forms of life: bacteria, for example, that can bring about some new epidemic disease. More generally, mass media and lay people have been impressed by the synthesis of fully working viruses and by the cloning experiments with Dolly. There are claims that synthetic biology is being developed without proper attention to possible health risks, environmental problems, and human rights implications.

I believe, as many in the field do, that such fears are out of place; and that, at the same time, bioethical questioning must always be present in this kind of work. This is actually the case in this field; in all SB international meetings I have been visiting, there has been always a section devoted to bioethics. And the question is also amply debated in the literature; see for example, Cho *et al.*, 1999; Maurer *et al.*, 2006; Boyle, 2007; Parens *et al.*, 2008.

The opposition to SB is not relenting. The ETC Group, an international civil societal organization based in Canada, is in the process of challenging Venter's patent application.[1] The problem is not an easy one, and the controversy is going to accompany us for a long time. The question then of the patent application for "new forms of life" is a particularly difficult one; see the conversation in this chapter with an expert in the field, Sarah Lau.

In closing this section, I would like to spend a couple of words on a quite different item, the relation between SB and artificial life (AL) – also because this is a term occasionally used in the SB field, and one that generates some confusion.

My main point is the following: that aside from the general directive of dealing with living forms that are not extant in our nature, AL and SB are different sciences. SB is basically experimental genetic engineering, or experimental chemical syntheses. AL generally is based on computation, computer simulation, cellular automata, and robotics. AL generally is neither concerned with experimental molecular biology and genetics, nor with synthetic chemistry, and in fact in the classic books of AL as originally edited by C. G. Langton (1988, 1993, 1995), a computer scientist, one does not find any experimental genetic engineering work; rather, keywords such as *cell automata, chaos theory,* and *phase transition* are used. If we construct in the laboratory a new bacterium by synthesizing its genes and other biochemical parts, this will be a work of SB and not of AL. If we try to do so using silicium instead of carbon atoms, or if we want to power chemical reactions with batteries instead of ATP, this will be AL. Of course, this is all a matter of definitions, and there may be grey zones where the two overlap, but it is important to keep the two branches conceptually clear from each other. Particularly in recent years, there has been a lot of confusion, which does not help anybody. This chapter deals with SB and not with AL.

---

[1] For data and literature from the ETC group, see www.etcgroup.org/en/materials/publications.html?pub id=8

Conversation with Paul Freemont

Professor **Paul Freemont** is co-director and co-founder of the EPSRC Centre for Synthetic Biology and Innovation (since 2009) and the National UK Innovation and Knowledge Centre for Synthetic Biology (SynbiCITE; since 2013) at Imperial College London. He is also currently the head of a new section of Structural Biology in the Department of Medicine at Imperial (since 2014). He was previously the Head of the Division of Molecular Biosciences (2005–2012), and Head of the Imperial College Centre for Structural Biology (2000–2005), having joined Imperial from the Cancer Research UK London Research Institute. His research interests span from understanding the molecular mechanisms of human diseases to the development of synthetic biology platform technologies and biosensors. He is the author of over 170 scientific publications. He also holds a number of external positions including chair of the Diamond Light Source Scientific Advisory Committee and member of MRC Molecular and Cellular Medicines Board. He has previously held positions including membership of UK Synthetic Biology Roadmap Coordination Group (2011–2012), Royal Academy of Engineering enquiry into Synthetic Biology (2006–2008); member of Cancer Research UK Biological Sciences funding panel (2006–2011); and member of Wellcome Trust Genes Molecules and Cells funding panel (2002–2005). Professor Freemont is a co-founder of the spin-off companies Equinox Pharma, Ltd., and LabGenius, Ltd., sits on the Scientific Advisory Board of Netscientific, Ltd., and has held consultancies with a number of other companies and organizations including Scottish Enterprise, Syngenta, and Rio Tinto. He has also been active in a number of public engagement activities including co-curator of the "Grow Your Own" synthetic biology exhibition at Dublin Science Gallery (2013), co-organizer of the Imperial Fringe Event "Life as we know it" in London (2013), participant in the RCUK Public Dialogue on Synthetic Biology (2010), and contributor to the Royal Academy of Engineering public engagement on Synthetic Biology (2008). He has appeared regularly on radio and television broadcasts on the subject of synthetic biology and has successfully co-supervised Imperial undergraduate iGEM teams since 2006 and acted as an iGEM judge.

PLL: In the beginning of SB, there was the tendency of discriminating SB out of the large field of bioengineering. Now, instead, all became SB, all kinds of genomic and DNA manipulations, even GMO, in fact, all kinds of applied molecular biology. This diffuse broadening brings about a lack of proper character. Is this a problem?

PAUL FREEMONT: I don't see SB as diffuse, although many commentators and researchers confuse SB with the rebranding of other activities like metabolic engineering, chemical and molecular biology, or protein engineering. These fields are among many existing and established areas of research that underpin SB. A widely accepted definition of SB is: "Synthetic biology aims to design and engineer biologically based parts, novel devices and systems as well as redesigning existing, natural biological systems."[2, 3] Note the words 'design' and 'engineering.' What really differentiates SB from molecular biology or other areas of biotechnology is the concept of engineering design and systematic processes fused with human practice and societal considerations,[4] which have never been applied to the repurposing of biological systems in any concerted way before. The field is currently developing these foundational tools and work practices.

PLL: Would you say that the discriminating character of SB is to apply engineering design? With modularity?

PAUL FREEMONT: Yes this is the unique aspect. SB is focused on the biological design cycle where project applications are considered within a systematic design framework. For example, specifications are defined, e.g., to build a metabolic pathway to produce a complex chiral chemical with optimal yields or to construct a whole cell biosensor that detects arsenic from drinking water or detects and neutralizes human pathogens in infected wounds. Biological literature and data are mined to elucidate the genetic components that can be designed to fulfill the specifications (Design stage). These are then assembled (Build stage) and tested in the appropriate host cell (Test stage), and the outcome measured and analyzed to inform and iterate further biological designs (Learn stage). The cycle is underpinned by modeling, simulation, and measurement (where possible) in order to develop an optimal design process to achieve the desired functioning biological system. As this cycle becomes common place and researchers share information and computational tools through databases and open-source web archives, we will be able to learn design rules for building complex genetic circuits that function in different host cells in a more predictable and robust way that has not been possible to date. This design process can be scaled such that lab-based designs and tests can be quickly translated into industrial processes and protocols and one could in part consider SB as an enabling field that allows the rapid industrial translation of biological based solutions to many societal and global challenges.

PLL: UK has invested much money in SB programs, probably more than any other European country. And you have built companies, and foundry... Can you give some figures, and explain how you could convince your government to do so?

PAUL FREEMONT: Currently, the UK research councils' investment in Synthetic Biology is around £180M, including investments made through the Synthetic Biology

---

[2] Royal Academy of Engineering Report: Synthetic Biology: scope, applications and implications, 2008 (http://www.raeng.org.uk/publications/reports/synthetic-biology-report).
[3] Synthetic Biology A Primer, ICL press (2012 – *ISBN*: 978-1-84816-862-6).
[4] A Synthetic Biology Roadmap for the UK: http://www.rcuk.ac.uk/RCUK-prod/assets/documents/publications//SyntheticBiologyRoadmap.pdf

for Growth programme. To lever this investment, my colleague Prof. Richard Kitney at Imperial and I initiated in 2006 a vigorous lobbying activity aimed at funders, policy-makers, and government departments pointing out the likely emergence of synthetic biology as a major transformative bioengineering field with significant economic potential. In 2008, we led a report on synthetic biology sponsored by the UK's Royal Academy of Engineering on the scope and applications of synthetic biology.[5] The report was very well received by many stakeholders and was influential in persuading funding agencies to develop strategies and programs around synthetic biology research. The report also led to the establishment of the Synthetic Biology Leadership Council co-chaired by the Rt Hon David Willetts MP, the then UK Minister of State for Universities and Science which then led to a UK roadmap for synthetic biology.[6] One major strategic development has been the establishment of the National UK Industrial Translation Centre for SB at Imperial College in 2013 called SynbiCITE.[7] The center's mission is to facilitate and accelerate UK academic SB research into industry including new start-ups and SMEs by providing investment, facilities, and expertise. We are currently (December 2014) working with 17 UK universities, 45 SMEs, 8 multi-nationals and have spun-out a number of micro-companies or start-ups. We have established a professional and semi-automated DNA synthesis and assembly foundry with standardized workflows, measurement systems driven by a novel computational platform that will allow researchers and companies to apply the design-build-test-learn cycle on their specific problems. Part of the foundry comprises an automated part-characterization platform, which automatically measures the characteristics of specific biological parts like promoters, repressors, or activators, to automatically produce interactive datasheets stored in our new Synthetic Biology Information System (SynBIS) being developed by my co-director Prof. Richard Kitney.

PLL: Some people, also within your country, see SB as an alternative to oil-based energy. Is this realistic, or still just a slogan?

PAUL FREEMONT: As discussed above, SB provides an engineering-inspired framework that allows the development of foundational tools and standardized laboratory proto-cols/processes that can be applied across a wide range of different applications. One application is expand and extend the design of unicellular organisms to manufacture a variety of commodity and specialized chemicals including biofuels. These applications are already happening with engineered yeast producing bio-diesel at scale.[8] Other chemical products produced by SB approaches include perfumes and fragrances, food additives, silk, cellulose, and pharmaceuticals. However, there is much debate on the use of land-based feedstocks to fuel these bio-manufacturing processes resulting in a reconsideration of this issue. There is much current activity within SB to

[5] Royal Academy of Engineering Report: Synthetic Biology: scope, applications and implications, 2008 (http://www.raeng.org.uk/publications/reports/synthetic-biology-report).
[6] A Synthetic Biology Roadmap for the UK: http://www.rcuk.ac.uk/RCUK-rod/assets/documents/publications/ / SyntheticBiologyRoadmap.pdf
[7] SynbiCITE – the Industrial Translation Engine to accelerate innovation and commercialization of Synthetic Biology at Imperial College London (www.synbicite.com).
[8] Amyris Biotechnologies: http://www.amyris.com/

utilize waste by-products including municipal waste, CO_2, and methane to act as energy sources for cellular production systems such that sustainable SB processes can be established in the medium and long-term.[9]

PLL: What are the more ambitious SB programs you have in mind?

PAUL FREEMONT: The beauty of SB is that the systematic design and engineering approach allows modularity and interoperability and communication between the different foundational tools including standards and processes, information systems, host/chassis cells, and bio parts functionality. Thus I would argue there is no limit to the ambition of SB, other than the limits imposed by the societal acceptability and responsible innovation of the technology and resultant products/solutions. In terms of the more ambitious programs, there is huge interest in developing novel therapeutic interventions using SB including systematic re-programming of stem cells, novel drug delivery systems using biological materials including living cells, 3D printing of human organoids, and *in vivo* disease monitoring systems. Another major and exciting project at the genome engineering scale is the complete synthesis of the yeast genome from *Saccharomyces cerevisiae* or Sc2.0[10]. This international consortium plans to refactor all 16 chromosome of yeast, which can then be scrambled and recombined to generate novel chromosome combinations and cellular phenotypes. This project will establish the first synthetically controlled eukaryotic cell and in my opinion will represent a landmark in the field of SB.

Conversation with Sarah Lau

Sarah Lau is a Partner and Patent Attorney at Kilburn & Strode LLP in London. She works for clients ranging from academic institutes, startups, and SMEs to large multinationals and provides advice on all aspects of the patent procedure. Sarah's practice focuses on the field of biotechnology, including the areas of synthetic biology, next-generation DNA-sequencing technologies, proteomics, medical devices, and antibody technology. She is particularly interested in technologies where the biotechnology field intersects with other fields. Synthetic biology is an area of special interest for Sarah, and she regularly speaks at and attends synthetic biology events.

[9] LanzaTech: http://www.lanzatech.com/   [10] Synthetic Yeast 2.0: http://syntheticyeast.org/sc2-0/

PLL: There has been a burst of activity in synthetic biology (SB) in the last years. Is there a corresponding burst of patents in the field? Can you give some figures (not only UK)?

SARAH LAU: It is difficult to definitively correlate a burst in activity in synthetic biology to a burst in patent filings in synthetic biology. This is due to the lack of an exact definition of synthetic biology, and a corresponding difficulty in identifying suitable keywords to search for synthetic biology patents and patent applications. This is acknowledged by van Doren *et al.* in a paper entitled "The development of synthetic biology: a patent analysis" (*Syst Synth Biol*, 2013, **7**: 209–220). The authors carried out a detailed analysis of international patent applications, which involved searching for international patent applications in databases based on certain keywords (such as riboswitch and microfluidics), keyword-strings (such as synthetic biology, synthetic genome, biological parts, DNA assembly, rational protein design, DNA origami, and genome engineering) and keyword-combinations (such as molecular machine AND protein and multiplex AND genome) and using certain international patent classification (IPC) classes and groups. The authors then analyzed the patent applications retrieved using this search strategy. Based on the methodology used by the authors, a total of 1,195 patents were retrieved. The analysis showed a clear increase in international patent applications over the 20 years from 1990 to 2010. According to the analysis carried out by the authors, in 1990, only 13 international patent applications relating to synthetic biology were filed, whereas in 2010 a total of 86 international patent applications relating to synthetic biology were filed. These numbers actually seem rather low, but the data are of course only based on one study and other methods could be used to quantify the number of synthetic biology patent applications filed. However, in general it does seem that the rise of synthetic biology does correlate with an increase in the number of synthetic biology patent applications.

PLL: Is there a difference, at the level of patents, between the field of SB, and the more general one of bioengineering?

SARAH LAU: There is no real difference between the field of synthetic biology and the more general one of bioengineering from a patent point of view. In terms of how a patent application is examined by a patent office, this is done on a case-by-case basis, in view of the claims of the patent application. The claims are the part of the patent application that defines the monopoly that the patent applicant wishes to seek. Patent law is applied in each country in which a patent application is examined based on the wording of the claims. The field of the patent application, for example whether a patent application is a synthetic biology patent application or a bioengineering patent application, will not affect how the claims are treated by each patent office. In terms of classifying patents and patent applications in terms of technical subject matter for search purposes, the IPC system is used. However, as acknowledged by van Doren *et al.* in the paper mentioned above, emerging technologies often do not have a dedicated location within this system, and there is no specific class within the IPC for designating patents and patent applications relating to synthetic biology or bioengineering. There is therefore likely to be no real difference between the field of synthetic biology and bioengineering in how the subject matter is classified for patent searching purposes either.

PLL: If, say, I modify, or cut 10 percent of the genome of an organism, is the modified organism patentable as a new form of life? And, if yes, let me modify only one out of a million bases of the genome: is this also a patentable new form of life? (Or shall I first show that something like that does not exist in nature?) And generally: is it so easy to patent new forms of life?

Sarah Lau: The issue of patentability of certain types of subject matter is one that is dealt with differently by the patent laws of different countries. Different countries have implemented different laws in terms of the types of subject matter that is fundamentally unpatentable. This can lead to the situation where the patents granted in different countries based on a single international application will have claims of quite different scope. In addition, the test of whether the subject matter of a patent application is eligible for patent protection is only the first hurdle that needs to be overcome by a patent applicant before obtaining a granted patent. The claims of the patent application will also need to comply with the other requirements of patentability according to the laws of that country, for example novelty and inventive step.

In Europe, it would in theory be possible to patent a modified organism having, say, 10 percent of its genome modified or deleted compared to that of the naturally occurring genome, as long as the organism is not a human. The claimed subject matter would also need to comply with the requirements of novelty, inventive step, sufficiency of disclosure, etc.

In the US, there is no fundamental prohibition on the patenting of modified organisms, but recent US case law has expanded the definition of patent ineligible subject matter that has a direct effect on nature-based products. The most recent guidance from the US Patent and Trademark Office (USPTO) on this subject (The "2014 Interim Guidance on Patent Subject Matter Eligibility" issued on December 16, 2014) states that in order to be patent-eligible, a nature-based product must have markedly different characteristics based on structure, function, and/or properties compared to its naturally occurring counterpart. Accordingly, as long as it can be proven that the modified organism has such markedly different characteristics compared to its naturally occurring counterpart, it would in theory be possible to obtain a patent in the US to the modified organism. It can therefore be seen that there may be a difference in patent eligibility in the US between a modified organism having 10 percent of its genome modified or deleted compared to that of the naturally occurring genome and one in which only one out of a million bases of the genome is modified. For any US patent application, the usual patentability requirements of novelty, non-obviousness, enablement, and written description will need to be complied with in order for a patent to be granted.

Another point to note is that, for subject matter such as this, in some cases the biggest hurdle to obtaining a granted patent may be establishing inventive step (non-obviousness), rather than patent eligibility. This, to some extent, will depend on how close the modified organism is in its sequence to its naturally occurring counterpart.

PLL: What about time and cost of a patent in SB?

Sarah Lau: The time and cost involved in obtaining a patent in relation to a synthetic biology invention is no different to that of any other biotechnology related subject matter, which can be extremely variable. The time and cost involved will depend on the number of countries in which it is desired to obtain patent protection. For example, the time involved in drafting, filing, and prosecuting a European patent application up to grant for biotechnology related subject matter is typically around 4 to 6 years, but can often be longer. The monetary cost of this process can vary widely, for example based on the length of prosecution (number of rounds of examination entered into with the European Patent Office) and the countries in which the European patent is eventually put into effect. The total amount could be from around £10,000 to over £100,000. If protection is desired in countries outside Europe, a filing strategy based on an international (PCT) application is usually taken, and in that case costs can be even more variable depending on the number of countries in which subsequent patent applications are filed.

PLL: What is (are) some of the most interesting cases – from the patent attorney point of view – in SB?

Sarah Lau: An interesting synthetic biology patent application is international patent application no. PCT/US2006/039047. This application was filed by The J. Craig Venter Institute and the original claims were directed, amongst other things, to a set of protein-coding genes that provides the information required for growth and replication of a free-living organism under axenic conditions in a rich bacterial culture medium, a free-living organism that can grow under axenic conditions in a rich bacterial culture medium comprising this set of genes, and a method for producing hydrogen or ethanol by growing the organism in a suitable medium. In essence, the patent application relates to an engineered microorganism, which has a minimal genome that may be further engineered and used as a chassis for the production of secondary metabolic products such as hydrogen and ethanol.

The corresponding US patent application no. 11/546,364 is currently being examined by the USPTO. The claims being examined relate to a plurality of bacterial genes comprised on one or more isolated nucleic acid molecules. This patent application is interesting because the USPTO Examiner considers the claims of this application to be directed to a product of nature, specifically naturally occurring genes, and has objected to the claims on this basis. The Examiner has cited the "2014 Interim Guidance on Patent Subject Matter Eligibility" in support of this objection. Therefore, even though the inventors went to great lengths to identify which genes within the organism are essential for life, the USPTO Examiner currently does not consider this invention to be patentable in view of her opinion that the combination of genes are not different in structure or function to those found in the natural state. At the time of writing, this US patent application is still pending and it will be interesting to see how prosecution progresses.

References

van Doren, D., Koenigstein, S., and Reiss, T. (2013). The development of synthetic biology: a patent analysis. *Systems and Synthetic Biology*, 7, (4), 209–220.

14.2 The case of engineering SB

The statement seen above from SB3 in Zurich emphasizes, as already mentioned, two different aspects of SB. One has to do with applied, engineering aspects; the other has to do with inquiry into basic science. Although the two aspects can be, of course, complementary, they correspond to two different ways of conceiving and operating in SB.

Let us deal first with the engineering aspect of SB. Here, I plan to give an overview of the field, reporting the titles of interesting papers; and this list of course cannot be exhaustive. I apologize with all colleagues whom I will not adequately cite.

Engineering SB is the most popular branch of SB among the researchers and mass media. As we have already said, this mostly involves genetic engineering, as the operations are generally relative to genes, genomes, or stretches of DNA.

The notion of engineering characterizes an operation that aims at a product with pre-determined functions, and all routes and tools are bent and focused for obtaining that function – for example, bacteria to make hydrogen, or a drug against insomnia. Let us see some examples: *Engineering a synthetic dual organism system for hydrogen production* (Waks and Silver, 2009); *Metabolic engineering for advanced biofuels production from* Escherichia coli (Atsumi and Liao, 2008); *Engineering microbes with synthetic biology frameworks* (Leonard *et al.*, 2008); *Metabolic engineering of microorganisms for biofuels production: from bugs to synthetic biology to fuels* (Lee *et al.*, 2008); *Light-energy conversion in engineered microorganisms* (Johnson and Schmidt-Dannert, 2008); *Towards systems metabolic engineering of microorganisms for amino acid production* (Park and Lee, 2008).

Those are also examples of novel forms of life, in the sense that these types of bacteria are new; that is, they did not exist before on Earth. The notion of alternative forms of life is emphasized directly in some papers, for example, see the work of Craig Venter's group titled *Genome transplantation in bacteria: Changing one species into another* (Lartigue *et al.*, 2007; Gibson *et al.*, 2009), or *Engineering microbes with synthetic biology frameworks* (Leonard *et al.*, 2008).

Also the terminology emphasizes this engineering aim, for example the terms *chassis* and *circuit* are used in the field of SB, like the naked chassis of an auto on which you put all parts – wheels and motor and brakes; or like a circuit where you can take away some parts and add some others. And indeed some of the titles reflect this assembly endeavor. See, for example: *An integrated cell-free metabolic platform for protein production and synthetic biology* (Jewett *et al.*, 2008); *Principles of cell-free genetic circuit assembly* (Noireaux *et al.*, 2003); *Engineering prokaryotic gene circuits* (Michalodimitrakis and Isalan, 2009); *Reconstruction of genetic circuits* (Sprinzak and Elowitz, 2005); *Toward scalable parts families for predictable design of biological circuits* (Lucks *et al.*, 2008); and *Scaling up digital circuit computation with DNA strand displacement cascades* (Qian and Winfree, 2011), a work based on DNA-based logic gates.

The work of Yaakov Benenson in Zurich focuses also on circuits and computational design (Hansen *et al.*, 2014, *Transplantation of prokaryotic two-component signaling pathways into mammalian cells*; Lapique and Benenson, 2014, *Digital switching in a biosensor circuit via programmable timing of gene availability*). The contribution of Richard Kitney of Imperial College in London should also be emphasized (Kitney and Freemont, 2012). At the London 2014 meeting, he introduced the notion of *"from oil-based feedstock – which gives all possible products to bio-based feedstock,"* looking namely at the importance of SB in plants. Also, the work in Jaramillo's group insists on the notion of computational design and circuits, see Carrera *et al.* (2012), where the authors infer the whole transcriptional network of *E. coli* and completely rewire it to create a new form of *E. coli* with a completely alternative genome but the same phenotype as the wild type. The automated design of circuits is also expressed in the works of Rodrigo's group: *De novo automated design of small RNA circuits for engineering synthetic riboregulation in living cells*, Rodrigo *et al.* (2012); and *Full design automation of multi-state RNA devices to program gene expression using energy-based optimization*, Rodrigo *et al.* (2013). In these two works, logic gates are realized using RNA only. This also appears in the example of the *iGEM* work (see infra) by a young team of researchers at Peking University (2007), who modified the bacterium *E. coli* so that it changes in color when light is switched on or off (see http://2009.igem.org/Main_Page). A novel operational unit is integrated in the original circuit – so that the final product (bacterium) is a new form of life.

To be cited, to give a further idea of the broad framework of SB work, the fact that in the last SB congress in London, as already mentioned, a considerable part of the presentations were covering various aspects of applied plant biochemistry (Luisi *et al.*, 2014).

However, bioengineering applications of SB also cover the field of medicine and pharmacology (see, for example, the work of S. Benner *et al.* (2005), titled *Synthetic biology for improved personalized medicine*; or the work by Sang Y. Lee and colleagues (2008) on the *Metabolic engineering of microorganisms for biofuels production: from bugs to synthetic biology to fuels*. At the last SB congress in London (2014), we had in this regard the interesting contribution of Jon Cesnut, with his work on genome engineering for mammalian cells, inserting new

DNA into double-stranded DNA, a technique based on DNA break, editing by cutting – for example, for multiple system atrophy (more details in Luisi *et al.*, 2014). Of course, this is an approach that works only for a disease caused by only one gene – and these cases are very few.

Also, see the work by Fussenegger's group (Weber *et al.*, 2009) on biotin-triggered genetic switch, which enabled dose-dependent vitamin H control in certain cell lines; the work by Chang and Keasling (2006) on *Production of isoprenoid pharmaceuticals by engineered microbes*; and that by G. Stephanopoulos' group (Ajikumar *et al.*, 2010) on terpenoid synthesis from microorganisms.

For the construction of these large systems, you need parts. In fact, in the field of SB the notion of "bio brick" has become very important: those are compo-nents, which are often commercially available. You can find nowadays several companies selling customized DNA coding sequences, customized genomes and plasmids (including large parts of the human genome), all kinds of RNA constructs, polymerase-based and ligase-based assembly kits, all kinds of qPCR, reverse transcription kits, Y-chromosme detection kits, and so on.

Let us attempt to summarize in terms of a simple illustration some of the main operations seen in the cited papers. To this aim, consider Figure 14.1. This may represent the genome of an organism containing, for the sake of simplicity, only four genetic elements **A, B, C** and **D**.

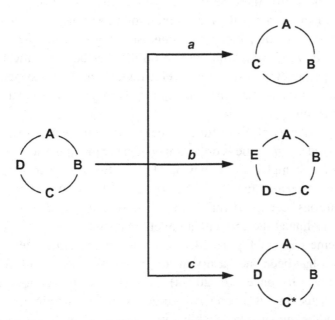

Figure 14.1 Schematization of the modularity operations of engineering SB.

One first possible procedure of SB is illustrated in Figure 14.1 (a), whereby one or more elements are eliminated, in what is generally called the "knockout procedure." The result is a genetically simpler novel organism, and the goal of the investigators is to maintain the viability and "life" of the original organism albeit simplifying the genome. A well-known example is the work of Venter's group with *Mycoplasma genitalium* (Fraser *et al.*, 1995), where the original 475 genes have been reduced to 380 with a series of knockout experiments.

In contrast to the mood of operation described before, Figure 14.1 (b) illustrates the case in which one gene (or one entire complex bio-brick) is added to a pre-existing organism. This is a more complex case, as the organism is now supposed to perform all his previous functions, plus the new ones due to the added **E** element.

The simplest case of this second manipulation is well known in classic bioengineering: if you want *E. coli* to fabricate insulin, you add to its genome the pig insulin gene. However, SB is able to give rise to more interesting cases. For example, in the work by Weber *et al.* (2009) – simplifying somehow – the *E. coli* repressor of the biotin biosynthesis operon was fused to the *Herpes simplex* transactivation domain to generate a biotin-dependent trans-activator: biotin-inducible transgene expression was functional in a variety of rodent, monkey, and human cell lines.

Let us consider now the case illustrated in Figure 14.1 (c). We have here – indicated with an asterisk – the replacement of a genomic part. We may obtain a more powerful functionality in the organism. A good example of this is the work by Lee *et al.* (2008), who operate on the metabolic engineering for the production of drugs at the whole cell level (a notion germane to systems biology) for enforcing – or removing – the existing metabolic pathways toward enhanced product formation.

Figure 14.1 (c) may also illustrate the case in which the new genetic element has been created *in situ* in the A organism by genetic manipulation of the original organism. For example, see the paper by D. Endy and co-workers titled, "*Refactoring bacteriophage T7*" (Chan *et al.*, 2005), in which more than one single module has been genetically changed. Also in this case, as in the previous one, we are dealing at the end with a novel organism.

The extreme case of Figure 14.1 (c) is when the entire genome is substituted. This has been the case with the work of Venter with *Micoplasma genitalium* (see Gibson *et al.*, 2008a). In this case, the entire genome has been synthesized in vitro, and exchanged with the original, natural one. The modified organism was viable, which was taken by the mass media (but not only) as an indication that life was synthesized from scratch. This is, of course, not true, as we are seeing in this experiment a kind of organ transplant. The synthesis from scratch of a cell would imply that all cellular components are also made in the lab, and this is far from being at hand. We will see in the next chapter some simple approximation of this enterprise to make a semi-synthetic living cell.

A case *per sé* is the synthesis of entire genomes. This particular field of activity is expanding very rapidly, with staggering results. Among the latest studies, one should mention the work by Boeke and Chandrisigaran (Annaluru *et al.*, 2014), who report the synthesis of a functional 272,871-base pair designer eukaryotic chromosome, synIII, which is based on the 316,617-base pair native *Saccharomyces cerevisiae* chromosome III. The synthetic work is attended by a series of genetic changes. Those in synIII include TAG/TAA stop-codon replacements, deletion of sub-telomeric regions, introns, transfer RNAs, transposons, and silent mating loci, as well as insertion of particular sites to enable genome scrambling. The mentioned deletions have the aim to see if these features are really needed, also with the idea to get to a possible minimal genome. And the more general question is: what can we do and what can't we do to the structure of a eukaryotic genome and still maintain a living cell?

We see here that this SB project brings an engineering perspective into yeast chromosome biology, and at the same time is capable of tackling fundamental questions in biology. It is expected that the genomic synthetic work will develop increasingly in this direction. In this regard, one should mention that there is now an international consortium (United Kingdom, Germany, United States, and France) for the work on yeast, the so-called *Saccharomyces cerevisiae 2.0* (Sc2.0) project (Dymond *et al.*, 2011; Annaluru *et al.*, 2014). The Sc2.0 project aims specifically at designing synthetic chromosomes and to generate eukaryotic cells driven by these fully synthetic chromosomes. A crucial engineered component of the Sc2.0 synthetic chromosomes is the system for Synthetic Chromosome Rearrangement and Modification (SCRaMbLE). The SCRaMbLE system generates a combinatorial diversity of genome complexity, at the same time with the capability of genome minimization. For the details of this work, which I am not able to discuss in a proper way, the reader is referred to the original literature. On finishing this section, it is proper to mention that on the subject of SB, and in particular, on engineering SB, there are many valid review articles. Let me cite here only a few, in an arbitrary way: Forster and Church, 2007; Khalil and Collins, 2010; Chiarabelli *et al.*, 2009, 2012; Luisi *et al.* (eds.), 2014; Cameron *et al.*, 2014; Church *et al.*, 2014.

Finally, I would like to add that iGEM students also carried some of this difficult work, as mentioned above about the yeast project (and in the biography of Paul Freemont).

14.3 A teaching phenomenon: iGEM

I would like to end these introductory pages on SB by mentioning the very good, successful teaching program iGEM. iGEM stands for *International Genetically Engineered Machine*, and it is a worldwide SB competition, initially aimed at undergraduate students, and then expanded to include high school students, entrepreneurs, and community laboratories, as well as postdocs ("overgraduates"). Many teams of students are being organized all over the world, each with one or more SB tutors, who provides them, in addition to professional advice, with the bio-bricks from the Registry of Standard Biological Parts, including plasmid backbones. Each team defines a particular SB project, and working in their own school over the summer, or whenever, they create different, and quite ambitious, SB constructs. Each year the international competition ends with the proclamation of a winner team, or more than one, since now awards have been diversified.

About the history: iGEM developed in 2003–2004 out of student projects at MIT, with a competition with five teams. Then other teams from outside the United States joined, so much so that in 2010, there were 130 competing teams in Europe, the Americas, and Asia. In January 2012, the iGEM Foundation became an independent non-profit organization located in Cambridge, Massachusetts, United States.

In 2014, iGEM celebrated the "10-year anniversary" (from Wikipedia) with one massive event, the Giant Jamboree. Over 2,300 participants gathered at the Hynes Convention Center in Boston, Massachusetts, to celebrate Synthetic Biology's brightest young minds. From all over the world 245 multidisciplinary teams worked all summer long to tackle today's toughest problems through synthetic biology based on standard parts. Figure 14.2 shows two young teams from that great last meeting.

As examples of this work, I have already mentioned the team at Beijing University (winner of 2007), with the construction of *E. coli* bacteria. This was modified so that it changes in color when light is switched on/off (see http://2009.igem.org/Main_Page) and also the building of synthetic mitochondria by a team at the University of San Francisco, made up of high school students, in 2008.

Some relevant results are shown in Side Box 14.1, which gives a simplified and much shortened list of the winning teams in 2013 and 2014 iGEM. To know more about the competition, see the article by Vilanova and Porcar, 2014.

Figure 14.2 Two young teams from the last iGEM anniversary meeting in Boston, 2014.

The great success of iGEM, and the great enthusiasm that it generates in young students, gives good hopes for our future research in life sciences and SB in particular, and the organizers of iGEM are to be complimented for their vision and success. My only small gripe is about the notion of "engineered machine" in the title, as this links to a mechanistic view of life – life as a machine that can be engineered as a watch or a small car. Life is not a machine, as we have argued in Part I of this book, and it is to be hoped that our young enthusiastic students will be given some philosophical hints to look at life in a proper, non-mechanistic sense.

Side Box 14.1

Recent iGEM activities
Some winner teams of iGEM 2013

iGEM Teams Sponsored by GenScript	Team Project	Team Awards
Aachen (Germany)	Team Aachen developed a novel biosensor system called Cellock Holmes that detects pathogenic bacteria on solid surfaces using a low-cost, rapid and portable technique.	Gold Medal Best Supporting Software, overgrad. Safety Commendation, overgrad. Best Measurement Project
UC Berkeley and UCSF (USA)	Team UCSF & UCB used cellular communities to model collective behaviors emergent from individual autonomous rules. Building upon quorum sensing mechanisms in the bioluminescent bacteria V. fischeri, they constructed a community averaging circuit in yeast.	Gold Medal Best Presentation, undergrad.
TU-Eindhoven (Netherlands)	Team TU-Eindhoven engineered a new way to make genetically-engineered bacteria more resilient for use in environmental or biomedical applications, by designing two Clickable Outer Membrane Proteins that can allow bacteria to be easily encapsulated by a protective, biocompatible PEG shell.	Gold Medal Best New Application Project, undergrad.
ETH-Zurich (Switzerland)	Team Zurich developed "Mosaicoli" cells that contain genetic circuits combining quorum sensing molecules with logic gates in order to model how complex patterns found in nature can emerge from simple rules, and to create new tools for creating biological computers.	Gold Medal Best Model, overgrad.
Cornell (USA)	Team Cornell designed a novel lead remediation / heavy metal filtration system including bacterial strains engineered to simultaneously express heavy metal transport proteins and metallothioneins.	Gold Medal

(*cont.*)

iGEM Teams Sponsored by GenScript	Team Project	Team Awards
Groningen (Netherlands)	Team Groningen developed a new bandage, called LactoAid, to prevent common burn wound infections without the use of antibiotics. *Lactococcus lactis* was genetically engineered to secrete three multimodal infection-preventing molecules upon detection of *Staphylococcus aureus* and *Pseudomonas aeruginosa* at the wound site.	Gold Medal
NTNU-Trondheim (Norway)	Team NTNU-Trondheim BioBricks that allow the insertion of transgenes into the photosynthetic organism *Synechocystis* sp. PCC 6803. to lay the groundwork for environmental innovations.	Gold Medal
Tufts (USA)	Team Tufts expressed a merRNA from *Bdellovibrio bacteriovorus* in *E. coli* to sequester cyclic di-GMP and inhibit biofilm formation.	Gold Medal
Ga Tech (USA)	Team Georgia Tech addressed the problem of dissolved methane entering water sources during fracking. They engineered E. coli to produce soluble methane monooxygenase (sMMO) in order to convert methane into the less volatile methanol, which can then be treated by further remediation pathways or extracted for commercial use.	Silver Medal
University of York (UK)	Team York developed EcoCADMUS (*E. coli* CAdmium DecontaMination Universal System), a new system to provide a safe and effective way to remove Cadmium and Sulfate from wastewater.	Silver Medal

Some winner teams of iGEM 2014

iGEM Teams Sponsored by GenScript	Team Project	Team Awards
Albert Ludwigs University Freiburg (Germany)	Team Freiburg developed the uniCAS universal toolkit that enables customizable gene regulation in mammalian cells utilizing the CRISPR/Cas9 system for specific targeting of DNA sequences, which can be matched with different effector domains for efficient gene activation or repression.	Regional Gold Medal Advance to World Championship Regional Finalist, Europe Best New BioBrick Part or Device, Engineered. World Championship Best Foundational Advance Project
Instituto Tecnológico y de Estudios Superiores de Monterrey (Mexico)	Team TecMonterrey engineered bacterial cells to fight cancer, by targeting anaerobic bacteria to the hypoxic tumor microenvironment where they can secrete therapeutic proteins that induce apoptosis selectively in tumor cells.	Regional Gold Medal Advance to World Championship Best Presentation Best Human Practices Best New BioBrick Part or Device, Engineered.
Shanghai Jiao Tong University (China)	Team SJTU-Bio-X created a Metabolic Gear Box combining light sensors with CRISPR interference technology to allow *in vivo* metabolic flux regulation. As a proof-of-principle, they demonstrated light-inducible repression of fatty acid synthesis.	Regional Gold Medal Advance to World Championship
Nanjing University (China)	Team Nanjing-China designed the Atrazine Elf, a bacterium that can sense, uptake, and degrade the ubiquitous herbicide atrazine from soil and water.	Regional Gold Medal Advance to World Championship
Hokkaido University (Japan)	Team Hokkaido created a Random Operon Shuffling Kit to mix and match transcriptional promoters and ribosomal binding sites in order to optimize gene expression.	Regional Silver Medal Advance to World Championship

iGEM Teams Sponsored by GenScript	Team Project	Team Awards
Turgut Ozal University (Turkey)	Team ATOMS-Turkiye sought to fight cancer by engineering gut *E. coli* to detect the EpCAM antigen present on cancer cell membranes and to express cancer-specific apoptotic proteins under the control of bacterial quorum-sensing mechanisms.	Regional Gold Medal

(*cont.*)

14.4 More on epistemology

All projects of iGEM are inspired by the genetic engineering work that we have outlined before. In terms of classic epistemic terminology, we can say that we are dealing here with *teleological* enterprises, where teleology means that the purpose (in particular, the function) is set at the very beginning – as engineers do when they have to construct a bridge, or an airplane.

In fact, one of the basic principles of engineering SB is the modularity, whereby life is an assembly of parts relatively independent from each other. These parts are interchangeable, and one may be able to assemble them in a modular fashion (see the already cited Chan *et al.*, 2005; and Chopra and Kamma, 2006). This viewpoint seems to correspond to a strong reductionist attitude.

The question is, how far can one go in SB with reductionism. As we have discussed in Part I of this book, life is an integrated system of parts, which are strongly interacting with each other – like the various organs in our body, or like the elements of the metabolic chart of any bacterium. Thus, if I decompose *E. coli* in, say, 500 bio bricks, I lose life, in the sense that none of these bio bricks has life. And it is also true that I cannot mix the 500 bio bricks in order to obtain the entire, fully working *E. coli*. The construction of a biological cell, even the simplest, is not a process under thermodynamic control. When a cell is being born in biology, the addition of the various parts in the ontogeny process must follow a very precise sequential pattern, one after the other in a programmed order.

It is then fair to say that even having at our disposal all possible bio bricks of this world, by their mechanical combination we can never reach the assembly of a living cell. It is also true, however, that the aim of SB is not to make life from scratch, but to modify the extant genome, in order to make "new" organisms provided with novel functions, as illustrated in Figure 14.1 (a) and (c).

In the analogy with human life, these two operations correspond to organ transplant or removal. It is perhaps different in the case of Figure 14.1 (b), in which we add some extra function. Here, the analogy with human life would be the addition of, say, two arms to the human body. This analogy is interesting, because it goes again to the relation between reductionism and the integrated systems view of life. In fact, the two new arms, in order to work properly, must be coordinated with the other organs –the brain, the heart, the lungs, and so on; they should constitute a unity with the entire body. Likewise, looking at Figure 14.1 (b), the new genomic structure should work as an entire integrated system. We see then an interesting characteristic of the epistemic operation of engineering SB: although the modularity operations are reductionist, the aim is to arrive at an integrated living system.

The idea of making new functions by adding genes takes us to another important epistemic characteristic of engineering SB. As mentioned already at the beginning of this section, we are generally witnessing a procedure under the profile of teleology: the structure/function is set at the very start and guides all the work.

It is worthwhile to mention at this point that teleology is not the way by which nature and biological evolution in particular are supposed to proceed. The function (for example, the eyesight, the spinning flagellum, the wings to fly. . .) in nature's evolution is never set a priori, but is the result and consequence of the contingent structure's development (see also Capra and Luisi, 2014). The primacy of the structure is the main operational procedure of nature: first the structure, and then, if this invented new structure happens to have a function that is advantageous for the organism, this will remain codified. This point has been discussed in detail in Part II of this book: there is no programmer or intelligent designer in the plans of nature: the amoeba, the bees, or the ants move about and do what they do adhering more or less blindly to a genetic program, determined by the laws of natural evolution. Yes, it may look like finality, but this is *teleonomy* instead of teleology, namely the more or less blind working of an implemented genetic program (see the section on finality in this book). In the case of microorganisms, insects, or other small animals deprived of mind, we can say that is teleonomy is at work.

Teleonomy is then not the procedure of engineering SB, which, by construct-ing its alternative forms of life, operates completely based on teleology – at least the bio-engineers do so. Let us move now to that part of SB where the function is generally not set a priori.

14.5 Chemical SB

You see things as they are and you say, "Why?"
But I dream things that never were, and I say, "Why not?"
George Bernard Shaw

With this kind of epistemic framework in mind, let us turn now to the other mentioned aspect of SB. It is, in fact, proper to say that SB has a double soul. One corresponds to the bioengineering approach outlined above, with the aim of changing the extant bacterial life to bend it to our needs. The other "soul" underlying question is "why did nature do things in a certain way, and not in another one?" Why 20 amino acids, and not 15, or 55? Why do nucleic acids contain ribose instead of glucose? Must mammalian hemoglobin be constituted by four chains, why not six or twelve? Why didn't nature make much simpler cells?

SB possesses the tools that may permit us to tackle this kind of philosophical question: let us synthesize the alternative form, and see whether there are some reasons why nature did so and not otherwise.

The experimental approach to this kind of questioning is by means of chemistry, or chemical procedures. We will see some examples. They are generally examples of "clean" SB, in the sense that they are being carried out without genetic manipulations of the living. The term *chemical synthetic biology* has been coined to represent this field (Luisi, 2007), and a book (Luisi and Chiarabelli, 2011), a paper (Chiarabelli *et al.*, 2012), and reviews (Chiarabelli *et al.*, 2013; Chiarabelli and Luisi, 2014) have appeared following this.

In this case, the construction of the new structure is set a priori. However, the function cannot be set a priori; actually, the functionality corresponding to this novel structure is not predictable – it is what we want to study. The question is then: which emergent properties are going to arise from such a novel SB structure?

From one general, philosophical point of view, this kind of questioning (why this and not that?) links to the dichotomy between *determinism and contingency*: are the things of nature the way they are, simply because there were no other ways to make them ("absolute determinism")? Alternatively, are they the way they are, due to contingency – something that some time ago, less properly and less fashionably, we used to call "chance"?

Take the example of the work by Albert Eschenmoser and collaborators at the Swiss Federal Institute of Technology in Zurich (ETH-Z). The question here was "why has nature chosen ribose as the sugar to associate to nucleic acids? Why not glucose, which is more stable, and more diffuse in nature?"

This is a typical "why this and not that" question, and can be tackled with SB, "simply" making in the lab the synthesis of nucleic acids containing pyranose instead of ribose (Eschenmoser, 2003; see also Bolli *et al.*, 1997a). The authors arrived at some important conclusions. For example, the pyranose-DNA was able to form a double helix, but this was much more stable than that formed by normal DNA, in the sense that the two strands could not be so easily separate from each other. Then obviously, with pyranose-DNA replication would have been much more difficult or impossible. In other words, these analog structures would finally impair replication. Therefore, as Eschenmoser puts it (2003):

... these systems could not have acted as functional competitors of RNA in Nature's choice of a genetic system, even though these six-carbon alternatives of RNA must have had a comparable chance of formation under conditions where RNA was formed.

In continuing their studies, Eschenmoser and his coworkers also found other structural analogs that turned out to be highly efficient informational base-pairing systems. As a way of conclusion, Eschenmoser arrives at this (2003):

While our experimental observations indicate that Nature, in selecting the molecular structure of her genetic system, had also other options besides RNA, the notion we naturally would be inclined to consider, namely, that RNA might be the biologically fittest of them all, remains a conjecture.[11]

This, in terms of the question, "why this. . . and not that." is consistent with the view that the structures chosen by nature are not necessarily the obligatory ones. This is, of course, in keeping with the idea of contingency.

Still in the vast camp of nucleic acid, there is the question of the four bases, A, T, G, and C. Are they the product of determinism, in the sense that the genetic code and DNA replication can only work with these particular chemical structures? Or then, again, are they the product of contingency, and the chemical structure might have been different?

This important question has been tackled by Steve Benner and his group, who synthesized nucleic acids built up with bases different from the canonical ones (Benner and Sismour, 2005). The Benner laboratory introduced the first expanded DNA alphabets in 1989, and developed these into an Artificially Expanded Genetic Information System (AEGIS), which enables the synthesis of proteins with more than 20 encoded amino acids, and provides insight into how nucleic acids form duplex structures (Young *et al.*, 2006; see also Kool and Morales, 2005).

But this is part of the wide field of unnatural base pairs (UBP), with very many groups and various research directions; see, for example (a very small sample), Malyshev *et al.*, 2012; Callaway, 2014; Fikes, 2014; and Pollack, 2014.

---

[11] In some of his work, although in a different context, Eschenmoser proposes the question "why this. . . and not that." In his 1999 review, he cites Einstein as saying: "We not only want to know *how* nature is (and *how* her transactions are carried through), but we also want to reach, if possible, a goal which may seem utopian and presumptuous, namely, to know why nature *is such and not otherwise.*"

And jumping from nucleic acids to proteins, take the already mentioned work in Yanagawa's group (Doi *et al.*, 2005) on proteins with a reduced alphabet of amino acids. Here again, the question, to which chemical SB is called to give an answer, is: is it really necessary that proteins are composed by 20 different amino acids? Why not 15, or perhaps only 10?

See, in this particular field, the work by Baker, 1997; Murphy *et al.*, 2000; and Fan and Wang, 2003. And we have to mention the work in Blaber's group (Longo *et al.*, 2013). Basically, most of this work could show – among other things – that enzymes "re-made" with only 10–12 amino acids instead of 20 may work rather well.

To the world of proteins belongs an even more fundamental question, once we look at the proteins that constitute our life: why these proteins and not others? This question, and the attempts to answer it, have led to the synthesis of proteins which do not exist in nature – the so-called "never born proteins" (Chiarabelli *et al.*, 2006a, 2006b). To this, we will devote a specific section below. The work of Chiarabelli *et al.*, although not involving genetic manipulation, is based on the present knowledge of molecular biology. We have seen in Part I a procedure to produce *de novo* proteins that uses only classic chemistry. This is the work of Chessari *et al.* (2006), a work that is a proof of concept that the biogenesis of ordered, folded proteins can be carried out by prebiotic means. You may remember from Figure 5.2 that the procedure is based on fragment condensation, according to which short co-oligopeptides, possibly produced by the NCA condensation, can be joined together. And although this condensation method gives in principle a combinatorial number of products, the introduction of a criterion of "natural" selection (elimination of all insoluble products) yields – at the end – a water-soluble, folded, thermodynamically stable, single protein.

To the camp of chemical SB belongs also the chemical synthesis of entire genomes. Well known is the case of Cello and co-workers, with the *de novo* chemical synthesis of a poliovirus genome, which had the same infectious properties as the natural one (Cello *et al.*, 2002). Here comes again the work of Craig Venter's group: they were able to synthesize the genome of the $\phi174$ phage starting from commercially available nucleotides (Smith *et al.*, 2003). This group has performed exceptional progress in the efficiency of synthesis of very large genomes (Gibson *et al.*, 2008a and 2008b). In 2010, they were able to carry out design, synthesis, and assembly of the 1.08-mega-base pair *Mycoplasma mycoides* JCVI-syn1.0 genome starting from digitized genome sequence information and its transplantation into a *M. capricolum* recipient cell to create new *M. mycoides* cells. The new cells, as the authors describe, have expected phenotypic properties and are capable of continuous self-replication (Gibson *et al.*, 2010). We have argued earlier that – although this should not be seen as the synthesis of new life – it is indeed a work of paramount importance. They were also able to synthesize a yeast chromosome – which is 272,871 base pairs long, representing about 2.5 percent of the 12-million-base-pair *S. cerevisiae* genome. Literature reports on genome synthesis are expanding rapidly; it is difficult to keep up. We have mentioned in the previous section the work of Boeke and Chandrisigaran's team on the genomic synthesis of part of the yeast chromosome (Dymond *et al.*, 2011; Annaluru *et al.*, 2014), and we have also mentioned the creation of an international consortium for the construction of novel synthetic yeast genomes. Adding an important point: in this case, engineering SB is also capable of addressing fundamental questions of basic science.

Jumping back from the synthesis of large genomes to simple molecules, we should mention the recent work in Martin Burke's lab. Although developed for organic chemistry at large, namely not specifically directed towards SB, the method is potentially relevant for the synthesis of small compounds of biological and therapeutic interest. They have used a single automated process to synthesize 14 distinct classes of small molecules from a common set of building blocks (Li *et al.*, 2015). This thanks to a machine, that they term "3D printer," which according to the authors should enable the synthesis of thousands of potentially useful molecules, including of course compounds of biological and therapeutic interest.

Going back to the main focus of this section, there is another important question that comes to mind, and that can be tackled with the tools of chemical SB. Again, this is, "why this and not that." In particular, "why are cells, even bacterial ones, so complex, being constituted by thousands of genes?" And: "isn't it possible to make simplified cells constituted, for example, by a few dozen of genes?"

These questions, which lead the way to the field of the "minimal cell/minimal life," take us back in a way to the origin of life. Chapter 15 is devoted entirely to this issue.

14.6 The never born proteins

Considering the general question within chemical SB of "why this and not that?" we came to mention the case of "our" proteins, where "our" means the proteins of our life. In fact, the question was why these proteins, and not others?

Let us dwell a little more on this, as by doing so we will arrive at another chapter of SB – and one which may give a possible answer to a philosophical question regarding contingency and determinism.

In order to do so, let us start with a question for high school students: in how many ways can one, in principle, realize a polypeptide chain with 100 residues?

Considering that each of the 100 boxes (the amino acid residues) can be one of the 20 different amino acids, the staggering answer is 20^{100}. And how many are "our" proteins on Earth? Considering the number of species we have (most probably around 10 million considering those not yet analyzed), and assuming an average of 1,000 genes/proteins for each species, we may roughly arrive at something like 10^{10} proteins on our Earth. Today's proteins are the product of billions of years of evolution, but we all accept the view that the present day's proteins have the imprints and basically similar structures and function of the early days of life. Indeed an impressive number – at the same time ridiculously small compared to the theoretical number of possible proteins calculated before. Probably for an extremely large fraction of these calculated, hypothetical structures are energetically impossible, but even if one in 10 billion would be energetically allowed, the resulting number would be staggering, say 20^{90} and clearly only a very limited number of these found their birth on Earth.

Just to present a simple-minded illustration about all this, let us consider our two extreme figures, 20^{90} and 10^{10}. The ratio between these two figures would correspond (within a few orders of magnitude) to the ratio between the whole sand of the Sahara desert and one single grain of sand. We humans thrive on a grain of sand (see Figure 14.3).

The Sahara desert, so vast and several kilometers deep, represents in our example an immense reservoir of proteins, which in principle wait to be "called to life." Let us call them – as in fact we did in our work-group – "never born

Figure 14.3 We humans (and, in general, the whole life on this planet) thrive on a single grain of sand! The ratio between one single grain of sand and the whole sand of the Sahara desert is very close to the ratio between the number of all the natural proteins existing on Earth and the number of all the theoretically possible proteins.

proteins," or NBP (Chiarabelli *et al.*, 2006a and 2006b). The name NBP is also implying that they are not in our data bank; and that these novel constructs are not designed by evolution.

There are occasionally claims that the number of proteins actually tested in the course of evolution is much more than the estimated 10^{10} figure. This consideration is, however, irrelevant for the sake of the present argument: even if an immensely larger number of proteins would have been tested by evolution, the fact is that our extant proteins are very few: and then, how and why have those few been selected, and how and why have all others been discarded?

Determinism or contingency? A strict deterministic view would require that our proteins had to be what they are due to the original selection of a set of features – for example, water solubility, stable folding, capable of binding, hydrodynamic properties, etc. If this were so, then our proteins would be characterized by a series of particular, specific properties that all other grains of sand would not have, or would have only sparingly.

The contingency view would say instead that our proteins happened to be what they are not because of an obligatory selection based on physical laws, but because of the vagaries of contingent factors, which governed their biogenesis (local concentration, pH, occasional catalysts, etc.) In this case, "our" proteins do not necessarily have to display peculiar thermodynamic or structural properties.

Talking about particular properties, let us focus first on folding, as this is the property that more than any other epitomizes the functionality of globular proteins:

Is folding one specific property of our globular proteins, or something that NBP proteins can share?

Here comes SB and the possibility of testing this experimentally: let us make a different grain of sand and see how the corresponding proteins look. In particular, let us measure their frequency of folding within a library of "never born proteins" (NBP)[12].

Given a 150-nucleotides-long DNA macromolecule, which in principle may code for a 50-residues-long polypeptide, and using the same kind of mathematical computation as before, we have in principle 4^{150} ways to realize it. If you ask your computer to make or pick up one of these by chance, the probability that by doing so you hit an extant gene (and by inference an extant 50-residues-long

---

[12] *Proteins not existing in nature* and *never born proteins* (NBP) are not the same thing. Proteins not existing in nature have been prepared before, and we have seen some examples in previous sections. However, they were generally based on partial randomization of existing scaffolds, namely of things already selected by evolution. Here we are talking about NBP, structures that have no evolutionary bias at all.

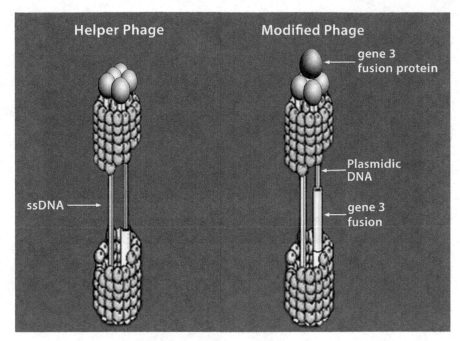

Figure 14.4 Phagemid vector system in phage display. On the right side, the modified phage showing the inserted DNA sequences (for more details refer to the original papers by Chiarabelli *et al.*, 2006a and 2006b).

protein) is as small as that ratio between a grain of sand and the whole Sahara desert – practically zero.

So, let us make such random DNA sequences, and let us proceed with them with the so-called phage display technique. This consists in inserting these DNA sequences in the genome of a phage; then, we will let the phage infect the *E. coli* (see Figure 14.4).

This has been done (Chiarabelli *et al.*, 2006a and 2006b), with the provision that in each of these totally random DNA sequences there is a constant obligatory sequence, a tripeptide (PRG), which is the substrate for thrombin, a well-known protease. The reason for the insertion of this peptide is to demonstrate the presence of a rigid folding: it is known that a rigidly folded protein is going to be resistant to the action of proteases, or much more so, than a chain, which is not stably folded. Thus, digestion of the NBP with thrombin provides a first criterion for stable folding.

Out of the obtained large library (*c.* 10^9 clones), a limited but statistically significant number of clones (*c.* 80) has been randomly selected, which permits us to derive the structure of the corresponding proteins (see Figure 14.5).

None of the corresponding 50-residues-long chains was in the data bank, not even under the most stringent criteria of similarity. The limited length (50

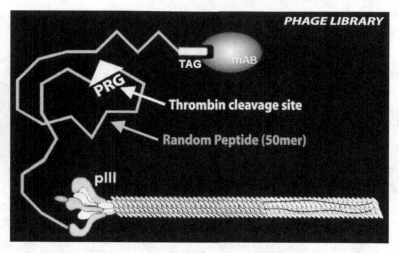

Figure 14.5 The phage display scheme for the production of never-born proteins. Proline-arginine-glycine (PRG) is a substrate for thrombin, and TAG is the antibody target. (Modified from Chiarabelli *et al.*, 2006a and 2006b.)

residues) was chosen as the critical length for folding – certainly not the most favorable one – and nevertheless a surprising high frequency of folding (according to the above first order criterion) was obtained. Actually, most of the investigated clones fell into two categories: the ones which were easily digested (more than 60 percent) and those which were not, even after prolonged thrombin digestion. The general conclusion was that about 20 percent of the structures were stably folded. The three-dimensional structure of these totally random *de novo* "never born proteins" (NBP) has been studied by theoretically utilizing the Rosetta method mostly, which reportedly is the best on the market (see Rohl *et al.*, 2004). Some results of these theoretical studies are shown in Figure 14.6.

At first sight, these NBP look remarkably similar to those, which we already know: note the familiar helical stretches, alternated by β-structures, gamma turns, and aperiodic loops.

Of course, the final definitive answer has to come from NMR data more than from theoretical predictions, but in the meantime a couple of these NBP have been amplified by PCR and studied by circular dichroism. The CD spectra are consistent with a folding having a significant helical structure, and in both cases the folding is thermo-reversible, as judged by the classic guanidinium chloride denaturation-renaturation test. This is shown in Figure 14.7.

At this point, we may remember the 44-residues-long protein, described in Part I of this book (see Figures 5.2 and 5.3), which was synthesized by a direct chemical method (Merrifield method and fragment condensation), again without

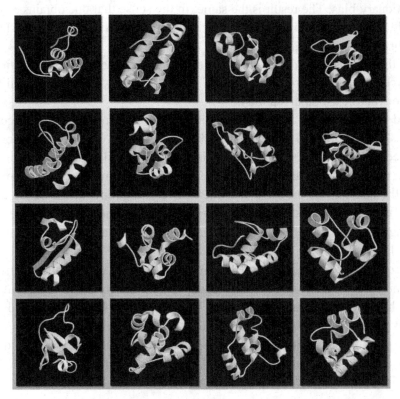

Figure 14.6 Some three-dimensional structures of totally random *de novo* proteins – namely, "never born proteins" (NBP) – as reported in theoretical studies (utilizing the so-called Rosetta method). (Adapted from Chiarabelli *et al.*, 2006a and 2006b.)

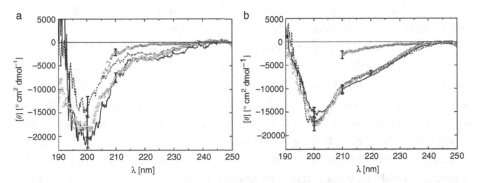

Figure 14.7 Spectra of proteins *NBP1* (**a**) and *NBP127* (**b**). Continued lines: native proteins; dashed lines: proteins at 60 °C; empty triangles: proteins in 6M urea; empty circles: proteins refolded after 12 hours. (From Chiarabelli *et al.*, 2006a and 2006b.)

evolutionary bias. The result is the same, in the sense that one obtains a water-soluble NBP with a thermodynamically stable folding with a significant helical content (Chessari *et al.*, 2006).

All these lines of evidence appear to suggest that "our" globular proteins are not exceptional at all from the point of view of folding, or water solubility, or thermodynamic globular stability.

And now let us go back to the question of determinism versus contingency. It is tempting to use the above data to say that "our" proteins were not selected because of structural prerequisites such as folding, water solubility, or thermodynamic stability. Then, the most likely interpretation stemming from the analysis of the NBP is a plea for contingency, in the sense that it suggests that our proteins are the way they are due to the accidents of contingency. And then, being fabricated in this way, they happened to foster life.

Of course, my colleagues in the ID party would say that yes, our proteins may have nothing worthwhile from the physicochemical point of view, but they might have been "chosen" just because they can foster life. Of course, there is no way to disprove it, and this is why ID is not a scientific theory.

The point may remain controversial. However, I would like to conclude this section in a somewhat positive way, in the sense that we have seen the relevance of SB for tackling questions of philosophical relevance.

14.7 The never born RNA

We have seen in the previous section that *de novo*, randomly produced proteins have a significant probability to assume a stable folding. This is very relevant for the origin of life field, as folding is the prerequisite for catalytic activity. The question is whether the same can be said for RNA.

To tackle this question, Anella *et al.* (2011) investigated the structural properties of totally *de novo* random RNAs, 97-nucleotides-long, of which 60 nucleotides were completely random. From this library of random sequences, 18 of them were randomly selected and analyzed. As a first criterion of stable folding, the newly developed "Foster assay" (de Lucrezia *et al.*, 2006a and 2006b) was used. This is based on the action of S1 ribonuclease digestion, an enzyme that is active only on the single RNA filaments, at various temperatures. Surprisingly, most of the RNAs were resistant to S1 nuclease digestion at up to 50 °C, and two sequences were resistant even at 70 °C, suggesting that these random RNA molecules could possess a stable secondary structure over a broad temperature range.

In this case, the library of RNA was obtained with classic molecular biology techniques. An important question is whether we can arrive at RNA macromolecules based on prebiotic methods, possibly using the concept of fragment condensation (Chessari *et al.*, 2006). We have considered this question already in Part I.

Concluding remarks

I have tried to present a *spaccato* of the SB field, and although not exhaustive, what we have seen gives an impressive view of the amplitude and richness of this novel scientific enterprise. The major characteristic of engineering SB is the modification of the extant bacterial life forms in order to bend them to our needs. It may represent an artificial, almost forceful possession of the products of nature and some may see in this a farther twist of Francis Bacon's view that nature must be at the complete service of man. And from this edge one can understand the origin of the objections and all bioethical criticism to SB, a controversy that includes the GMO and even staminal cells, and is probably destined to accompany us for a very long time.

Among the positive aspects of SB, I mention two at the level of social sciences, which are very apparent from the data presented. One is the character of international cooperation, as exemplified by the yeast chromosomes research consortium, which encompasses several nations in Europa and Asia and the Americas. An example indeed of how science should be and should operate – going beyond borders and passports.

The other aspect is the involvement of so many enthusiastic young people, as evidenced by the iGEM activities. The two things together give SB a social relevance, with an intensity that is new in science.

We have mentioned the "two souls" of SB, with the chemical SB souls that relate more directly to the question "why this and not that" in nature; and with SB as the tool to synthesize and explore "that." As such, we are witnessing here a new way to approach the mysteries and complexity of our chemical evolution, as we have seen with the example of the never born proteins. Determinism or contingency?

SB's main tool is the construction of modified life forms, playing with the genome of extant forms. What about the synthesis of life? The aim of SB is not the bottom up approach to life, and the clamor around the 2010 papers of Venter's group, with the mass media (and not only) talking about the synthesis of life, was rather quickly forgotten, and this was good. However, Venter's group experiment deserves a farther comment. The original bacterium, once deprived of its own genome, was not alive; the synthetic genome obtained in Venter's lab was also, obviously, not alive *per sé*. By putting these two non-alive parts together, a living organism came out. The same is true in a cloning experiment. Is this a direct demonstration that the emergent property of "life" may arise from the combination of inanimate matter? Not everybody would agree on this, but certainly, we all agree that is an important point of discussion. The bioethical confrontation still goes on, also alimented by the difficult issue of "patenting life," as we have seen in the conversation with Sarah Lau. Synthesizing new life? Is it possible in a softer way, without genetic manipulations of the extant life?

These questions link to the next and final chapter, which focuses directly on the question of the SB construction of minimal living cells.

Chapter 15

The minimal cell

Introduction

We are landing on the last section of this book, which concerns the synthetic biology approach to the minimal cell. It is in a way an approach that is in keeping with the arguments developed in Parts I and II of the book, in which we reached the conclusion that a proper way to tackle, experimentally and conceptually, the origin of life, is to focus on the biological cell.

About this, a particularly important contribution was the 1992 book by Harold Morowitz, *Beginning of Cellular Life*, where he expressly formulates that the first step toward the origin of life was the spontaneous condensation of amphiphilic molecules to form vesicles (or protocells). For this reason, I would like to insert at the end of this introductive section the conversation with Morowitz, which has the advantage of reconnecting us with the first two parts of this book.

In order to better understand this approach, let us consider Figure 15.1, which shows schematically two possible pathways: the first is the bottom-up approach, discussed at length in Part I. The main problem to make this pathway in the lab lies in the essence of contingency, which prevents us from knowing the detailed biogenesis structure of "our" proteins and nucleic acids.

The alternative route to the protocell, indicated in Figure 15.1, is a pathway of descending complexity – going from the modern cell to its most elementary precursor. This operation corresponds to the abstraction of eliminating conceptually all of what we may consider redundant with respect to the most basic, original functions of a primitive cell.

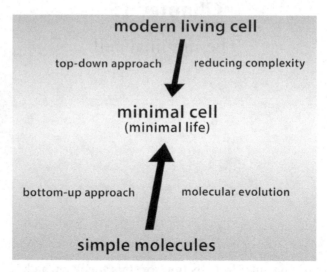

Figure 15.1 The two possible pathways to the minimal cell.

A note on terminology may be useful at this point, also keeping in mind
Figure 15.1 with the two alternative pathways to the minimal cell. The term
bottom-up is rather clear for the most common approach to the origin of
life, but the terminology of the alternative route to the minimal cell is less
clear. The term *top-down* was utilized in the early times, also by my group,
to indicate that we are coming from the "top" of a full-fledged cell to arrive
at a protocell. However, also in this case we are constructing a complex
system starting from a simple shell, and in this sense the terminology "top-
down" is not really correct. Really, an intermediate case permits us to
arrive at a protocell.

In this chapter, we will focus on this second working direction, which is an
exercise within chemical SB. In Part IV, we have seen already the first step of
this "road map" to the minimal cell, describing how complex biochemical
reactions can be hosted inside vesicles. We will see now whether and to what
extent the biochemical complexity of the hosted reactions can become the most
important cell operations, such as the expression of proteins, metabolism, and
possibly self-reproduction.

Conversation with Harold Morowitz

Harold Morowitz is a Robinson Professor at George Mason University, having previously been Professor of Molecular Biophysics and Biochemistry at Yale.

He has written extensively on the thermodynamics of living systems, and has been called the "present-day grandfather of the application of thermodynamics to biology." He is also well known for his popular books flavored with his sense of humor (*Mayonnaise and the Origins of Life, Cosmic Joy and Local Pain, The Thermodynamics of Pizza, Entropy and the Magic Flute*, and *The Kindly Dr. Guillotin...*).

Dr. Morowitz was principle investigator on the multi-institutional grant "From Geochemistry to the Origin of Life," which was centered at the Santa Fe Institute. In 1992, Oxford University Press published his book *The Emergence of Everything, How the World Became Complex*.

Most of us, working in the field of origin of life, have profited from his famous 1993 book, *Beginnings of Cellular Life: Metabolism Recapitulated Biogenesis*, Yale University Press. About himself, he likes to say, "When I was given the chance to choose my professorship title, I chose "Biology and Natural Philosophy." I guess that is how I think of myself as a scientist. I knew I was a biological generalist when I had published review papers on Mycoplasma as the smallest living organism and the humpback whale, *Megaptera novaeangliae*.

PLL: You have been for a long time one of the most active researchers on cellular life. The origin of cellular life and life on our Earth in general is still an open question. Do you think that we are closer now to understanding the origin of life than we were, say, 20–30 years ago?

HAROLD MOROWITZ: I think that we are closer but still have a long way to go. Let me add that I am reluctant to make my answers too personal. My job as a scientist is to ask what we are, where have we come from, and whither are we going. I also believe in trying to walk humbly with my fellow scientists. As you may have heard, Eric Smith and I, after some fifteen years of working together are finishing a book, *Biogenesis, Emergence of the Fourth Geosphere*. I think that our sympathy with your view of system thinking will show up. (I have not yet had an opportunity to see your book with Capra). As to progress, I think that filling in the metabolic chart, assembling the KEGG and Metacyc databases, the study of true autotrophs, and the geologists understanding of deep sea vents have given us the kind of background information that we need.

PLL: And what is, according to you, the most important experiment, or series of experiments, that have operated a quality jump in the field (field of the origin of life)?

HAROLD MOROWITZ: Studies demonstrating the existence of reductive chemoautotrophs and elaborating their intermediary metabolism with its universal features have provided a top down view of much greater detail and universality. I first learned of the reductive TCA cycle from Larry Hochstein around 1990. Vent microbiology may still point the way to important processes. Small molecule catalysis, particularly transition metal-carboxylic acid containing ligands, seem specially promising.

PLL: Most of the active research on the origin of life today is still focused on the RNA-world, with a hypothetical RNA that is self-replicating and evolving. Do you see any strong rationale in this focus, and possibility of progress?

HAROLD MOROWITZ: I think that the RNA worldview is important but is likely overdone. Emergence is almost certainly a series of phase transitions. I think of these as proceeding in time – and probably requiring subdivision for understanding:

A. Metabolism – organic chemistry – leading from geochemistry to intermediary metabolism.
B. RNA world – also the stage of polymers – (ribosomes are sort of mysterious) finally leading to DNA the possibility of individuality.
C. Vesicle world.
D. DNA + vesicle is cellular (vesicle could come between **A** and **B**).
E. Uncertain about polyphosphate.

PLL: Am I right to say that your second answer seems to go in the direction of metabolism without/prior to enzymes? If so, have you seen any significant spreading and acceptance of this idea in the field of origin of life?

HAROLD MOROWITZ: As to metabolism prior to enzymes, I don't know about acceptance, but I think I detect less hostility. Ligands between transition metals and carboxylic acids deserve a lot more study. My colleague V. Srinivasan is now working on this.

PLL: Talking about hydrothermal vents, etc.: they are only concerned with the production of low molecular weight compounds. We all agree, however, that life is based on ordered sequences of biopolymers – you can have all low molecular weight compounds of this world, and you are still far away from making life. Do you see some light on how to make ordered sequences of proteins and/or DNA/RNA by prebiotic means?

HAROLD MOROWITZ: The big gap in understanding is between metabolism and polymers. I really find ribosomes enigmatic. There may be a significant missing bit of theory in the physical chemistry of macromolecules. Where and how vesicles emerged is also a problem – but I wait for you to tell me about that one.

PLL: It is said about you that you negate now your earlier separation of religion and science, arguing that one can know God through the study of emergence. Is this true, and can you please say something more about that?

HAROLD MOROWITZ: Some years ago, I was being deposed as an expert witness in a Louisiana trial on teaching of creation and evolution. The attorney asked: "What are your religious beliefs?" Under oath, I replied: "I am a pantheist in the tradition of Spinoza".

15.1 The notion of the minimal cell

Although we will focus on a minimalist approach to the cell, it is proper to start by looking at the complexity of a real biological cell just to know what we are talking about. Figure 15.2 gives a "spaccato" of the metabolism of *E. coli*: hundreds and hundreds of reactions, which are mutually linked and mutually dependent on each other, building a three-dimensional system continuity. Where is life localized in this complex maze? The life of the bacterium is not localized in any single spot, or any metabolic cycle: it is the entire system – the system's view (see also Capra and Luisi, 2014).

With a simple organism like *E. coli*, we are dealing with a system encompassing more than 1,000 genes, and corresponding numbers of proteins and nucleic acids. Even if one considers much simpler modern cells, we have to deal with hundreds of genes and thousands of components inside the small compartment of a few micron dimensions. A staggering complexity. However, it is precisely this complexity, which elicits the question: is all this complexity necessary to make a living cell, can't we conceive to "make" a much simpler one?

This kind of question acquires a meaning particularly when we consider that at the origin of life, cells cannot have begun with such a complexity – not with hundreds of genes at one time. In fact, it is more reasonable to assume that the first cells must have begun with a much simpler structure.

This argument brings the notion of a minimal cell: defined as the cell that contains the minimal and sufficient number of components to be viable. The operational procedure to make something like that is schematized in Figure 15.3.

The notion of "viable" carries us to the notion of "living," and here we should re-open the discussion seen in Parts I and II of this book: "What is life?" However, we will not do so now. The more general idea is to make a proto-cellular system that displays some of the basic functions of a living cell, for example, some form of metabolism, and/or protein expression.

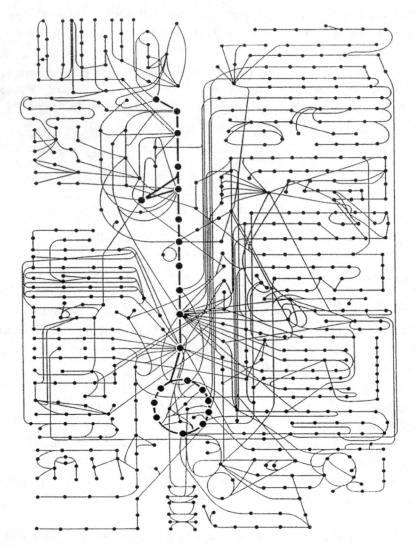

Figure 15.2 A section of the metabolic network of a bacterium (*E. coli*). Note that each point (each chemical compound) is linked to any other point via the complexity of the network. Each line of the network represents an enzymatic reaction.

We should also note that the phrase *basic functions of a cell* is rather vague. This is good, however, as it permits to circumscribe better the notion of minimal cell: in the sense that the term *minimal cell* does not define one single entity, but a large family of structures, which will differ from each other because of the different kind of primitive cellular functions. For example, we can have a protocell (this term is better than "cell" at this point) displaying a simple form of metabolism, another one displaying some form of homeostasis, or only

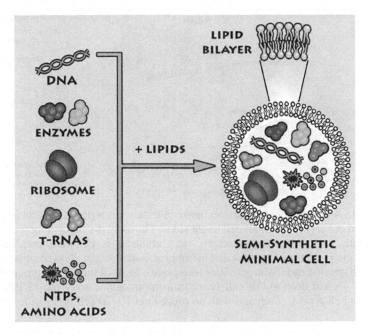

Figure 15.3 Experimental approach for constructing semi-synthetic minimal cell. (Reproduced with some modifications from Chiarabelli *et al.*, 2009, with the permission of Springer.)

part of protein expression, and so on. Each kind of these "limping cells" may actually represent an intermediate to the biogenesis of a primitive cell. In terms of nomenclature, one should consider the more general term *protocell*, often used in the literature. The term comes from the Greek *protos*, which means "first," or prior to. To me the term *protocell* conveys an historical connotation – the structure(s) that in the prebiotic molecular evolution came prior to the real cell, for example, a compartmented structure prior to the genome and/or reproduction. Protocell obviously does not denote a unique structure, but rather, as said before, a family of all possible preliminary cell constructs.

Considering that we are dealing in Figure 15.3 with a semi-synthetic procedure, and that we are utilizing macromolecules that already exist, the term *semi-synthetic cell* appears appropriate. The general term *artificial cells* has been used (Pohorille and Deamer, 2002). However, as I have explained, talking in general about SB, I prefer to reserve the term *artificial* to chemical systems as opposed to the biochemical ones.

Before going into the operational procedure of SB of the minimal cells, let us ask the question, how does the semi-synthetic structure of Figure 15.3 compare with real biological cells?

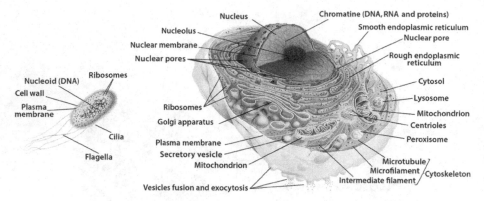

Figure 15.4 Schematic comparison between a bacterium (left side) and a small, idealized eukaryotic (animal) cell (right side). The main distinguishing feature of eukaryotes as compared to bacteria (and archaea) is compartmentalization, namely the presence of nucleus and membrane-bound organelles (compartments in which specific metabolic activities take place). Most of the animal groups have evolved a great diversity of cell types in a multicellular body (up to 100–150 different cell types), compared with no more than 10–20 cell types in fungi and plants.

To clarify this question, let us consider Figure 15.4, which schematizes the structures of the two best studied classes of cells, namely a bacterial cell, which do not have a compartmentalized nucleus and an idealized eukaryotic cell.

Bacteria and archaea are the smallest of all organisms, ranging from 0.5 to 2.0 µm in diameter, while most of the eukaryotic cells are generally one order of magnitude larger, ranging from 10 to 100 µm in diameter. In terms of the volume, eukaryotic cells can be as much as 1,000–1,500 times greater in volume than bacteria and archaea cells – so that the two cells represented in the Figure 15.4 are approximately in scale.

Our semi-synthetic cell, based on a simple liposome, may look like one of the proposed eukaryote ancestors,[1] being a structure without a cell wall,[2] still having a DNA free in the "cytoplasm" water pool of the liposome, and before the symbiotic "addition" of mitochondria.

---

[1] As reported in Part I, the *Lokiarchaeota*, a newly discovered archaea group, were recently proposed by Spang *et al.* (2015) as the closest known prokaryotic relatives of eukaryotes and plausible eukaryotes' ancestors. Spang and colleagues' findings suggest that the genomes of these archaea contain more eukaryotic-like genes than other known archaea. Further studies are needed to close the evolutionary gap between archaea and eukaryotes, increasing the precision with which we can identify when key cellular innovations such as the nucleus, mitochondrion, and endoplasmic reticulum first evolved.

[2] It is interesting to remark that for some authors the origin of eukaryotes, which arrived much later in evolution, has to be searched in ancestors which had lost the cell wall (Lane, 2009). Among eukaryotes, cell wall is not present in animals. Cell walls evolved independently in other eukaryotes groups, as fungi and plants. In these lineages, the cell wall is involved in the evolution of the multicellularity, terrestrialization, and vascularization.

Let us briefly consider now the operation for the SB of a minimal cell. We need first a compartment, and we already know from Figure 15.3 and more generally from Part IV that vesicles – and liposomes in particular – are the ideal structure for that. Then, we need to incorporate extant genes and/or enzymes, in the minimal and sufficient number, to arrive at the biological functions.

We will find later on in this chapter the most relevant literature on specific aspects of the minimal cell. Here, it is proper to mention a book of general character (Luisi and Stano, 2011), with a series of contributions from different groups and schools of thought. Thus, Boal and Forde, and, independently, Gardner and Davis, discuss the evolution of the cell's mechanical design. Peter More asks the question, "how small is small?" – a question posed also by Lancaster and Adams in a different context. In the same book, Monnard and Deamer consider the steps towards the first cellular life focusing on the membrane self-assembly process. Petra Schwille considers the pathway from minimal membrane systems to minimal cells in the case of giant vesicles, and Mavelli reconsiders the case of the minimal RNA cell (the ribocell), which we have mentioned in Part I of the book (see also Mavelli, 2012). Aside from this book, other authors consider the problematics of the minimal cells in general terms; see for example, the later work by Mansy and Szostack (2009). Quite interesting is also the approach of Blacka *et al.* (2013), in collaboration with David Deamer, who report on how nucleobases bind to and stabilize aggregates of a prebiotic amphiphile, which they consider as a possible mechanism for the emergence of protocells.

Finally, still on the subject of terminology, the notion of minimal cell should not be confused with the notion of minimal genome. To clarify this point, let us consider more in detail the meaning of the term *minimal genome*, as it is usually presented in the literature.

15.2 The minimal genome

The smallest sizes of a genome, as already mentioned, are those of *Mycoplasma genitalium* (580 kb – see Fraser *et al.*, 1995) and *Buchnera spp* (450 kb), with a value that agrees well with the calculations of Shimkets (1998), according to which the minimum genome size for a complete living organism should be approximately 600 kb. It is argued that these two organisms have undergone massive gene losses and that their limited encoding capacities are due to their adaptation to the highly permissive intracellular environments provided by the hosts (Islas *et al.*, 2004). They are, however, parasites and the next step of complexity concerns microbes with thousands of expressed proteins. Actually, one of the earliest attempts to describe the DNA/proteins minimal cell was by

Morowitz (1967). Based on the enzymatic components of primary metabolism, Morowitz estimated that the size of a minimal cell should be about one-tenth smaller than mycoplasma.

The values of DNA content of free-living prokaryotes can vary over a tenfold range, from 1,450 kb for *Halomonas halmophila* to the 9,700 kb genome of *Azospirillium lipoferum* sp59b. By way of comparison, *Escherichia coli* K-12 has a genome size of *c*. 4,640 kb and *Bacillus subtilis* 4,200 kb.

More in general, the question of the minimal genome has been considered by Mushegian and Koonin (1996); Shimkets (1998); Mushegian (1999); Koonin (2000); Kolisnychenko *et al.* (2002); Luisi *et al.* (2002); Gil *et al.* (2004); Islas *et al.* (2004).

Mushegian and Koonin (1996) calculated an inventory of 256 genes that represents the amount of DNA required to sustain a modern type of minimal cell under permissible conditions. This number, as indicated later by Koonin (2000), is quite similar to the values of viable minimal genome inferred by site-directed gene disruptions in *B. subtilis* (Itaya, 1995) and transposon-mediated mutagenesis knock-outs in *M. genitalium and M. pneumonia* (Hutchinson *et al.*, 1999).

Based on these and other data, some authors also provide indications for the existence of a more primitive, less regulated version of protein synthesis (Gavrilova *et al.*, 1976; Spirin, 1986).

Andres Moya and his group in Valencia arrived at the smaller number of 206 genes based on their work with *Buchnera sp.* and other organisms (Gil *et al.*, 2004).

How is the minimal genome, seen in these terms, linked to the question of this chapter, the minimal cell, which in turn is linked to the question of the origin of life? There is no direct connection: the figure of *c*. 200 genes for a full-fledged cell still corresponds to a formidable complexity, which is not consonant with the beginning of the origin of life. With the notion of minimal cell, we are focusing on protocells, the first rudimentary, "limping," not yet full-fledged cells – the question namely of the very origin of the cell itself.

Thus, let us turn our attention to the SB approach previously illustrated in Figure 15.3.

15.3 The road map to the minimal cell: protein expression in vesicles

The first approach found in literature to deal with the minimal cells is to insert into liposomes commercial kits used for protein expression. As mentioned in the previous chapter, the pioneering work was published by Oberholzer *et al.* (1999), but limited to poly (Phe). The term *minimal cell*, based on vesicles, had been already coined by the same authors a few years earlier (Oberholzer

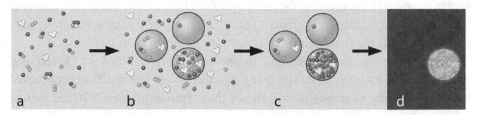

Figure 15.5 The cell-free protein synthesis machinery, containing the different macromolecules (*c.* 80) required for transcription, translation, tRNA charging, and energy recycling, in solution (**a**). After addition of vesicle-forming lipids, a variable aliquot of macromolecules was entrapped in newly formed vesicles (**b**). After removal or inhibition of external components (**c**), the protein of interest (in this case, EGFP) was synthesized inside vesicles containing all protein synthesis machinery and easily detected by the green fluorescence (**d**).

et al., 1995a, 1995b). Following that, a series of workers have successfully attempted the synthesis of entire proteins, by introducing the whole transcription-translation (T&T) machinery inside the liposome. Generally, the preferred protein was the green fluorescence protein (GFP), for obvious analytical reasons.

From Stano *et al.* (2011), we take, with some modifications, a scheme, Figure 15.5, which summarizes the operational procedure for arriving at the situation described in Figure 15.3. Ribonuclease was used to inhibit the protein synthesis externally to the liposomes. The analytical procedure in this field is important, perhaps not difficult *per sé*, but rich in important small details, and for this the reader is referred to the specialized reviews: Pohorille and Deamer (2002), Forster and Church (2006, 2007), Luisi *et al.* (2006), Kuruma *et al.* (2009), Stano *et al.* (2011, 2014), D'Aguanno *et al.* (2014).

In this chapter, we give more emphasis to what has been produced in the literature concerning protein expression inside liposomes, and a list of the most relevant papers is given in Table 15.1.

We have mentioned the pioneer work of Oberholzer et al. in the 1990s. The first report on the synthesis of a folded protein is signed by Yomo, Urabe, and coworkers (Yu *et al.*, 2001; Nakashima *et al.*, 2007), of a mutant GFP (actually, the pET-21-GFPmutl-His6 mutant) in lecithin liposomes. Large GFP-expressing vesicles, prepared by the film hydration method, were analyzed using flow cytometry as well as confocal laser microscopy. Yomo's group has continued this kind of work, see Sunami *et al.*, 2010.

In the later procedure by Oberholzer and Luisi (2002), all ingredients were added to a solution in which the vesicles were then being formed by the ethanol injection method, and EGFP (enhanced GFP) production was then evidenced inside the compartments. In this case, the sample was analyzed spectroscopically, monitoring the increase of the fluorescent signal of the EGFP. The disadvantage of this procedure is that entrapping efficiency is generally low, due to the small internal volume of liposomes obtained with this method.

This problem is partly avoided in the procedure utilized by Yomo and Nomura (Nomura *et al.*, 2003) by using giant vesicles. The reaction is observed by laser-scanning microscopy, and shows that expression of rsGFP (red shifted GFP) takes place with a very high efficiency (the concentration of rsGFP inside vesicles was greater than that in the external environment). The authors also showed that vesicles could protect gene products from external proteinase K. Nomura's group had started the work direction towards protocells already in 2001, by encapsulating giant DNA and its histone complexes (Nomura *et al.*, 2001; see also Tsumoto *et al.*, 2002 and Nomura *et al.*, 2002).

Based on the initial report on the expression of functional protein into liposomes by Yomo and coworkers (Yu *et al.*, 2001), the work by Ishikawa *et al.* (2004) represents another stage of the work on GFP expression. In fact, a two-stage genetic network is described, where the first stage is the production of T7 RNA polymerase, required to drive the GFP synthesis as the second stage.

Pietrini and Luisi (2004) described the synthesis of GFP by mixing two or more initial reagents, utilizing water in oil emulsions. This system is less interesting from the biological point of view (organic solvent instead of water); however, as we have discussed in Part IV of this book, it is very interesting from the point of view of compartment chemistry, as this kind of compartments can fuse with each other and exhibit no leakage. Fiordemondo and Stano (2007) have further developed this emulsion work. They could show that lecithin can also be used as surfactant for water-in-oil (w/o) droplets; and that protein synthesis can be achieved in such droplets either by incorporating simultaneously all components for transcription and translation reactions in a single emulsion, or by mixing four different emulsions, each containing only parts of the molecular machinery. They could also show that in the presence of excess lecithin, preformed w/o compartments can undergo spontaneous division, reducing the average droplet size, increasing the number of droplets, and continuing the protein expression after the division.

Going back to vesicles, of particular interest is the work by Noireaux and Libchaber (2004), with GV. Again, a plasmid encoding for two proteins was used; in particular, the authors introduced EGFP and α-hemolysin genes. At variance with the cascading network described above, the second protein (α-hemolysin) does not have a direct role in protein expression, but is involved in a different task. In fact, it is able to self-assemble as a heptamer in the bilayer, and so it was possible to feed the inner aqueous core of the vesicles, realizing

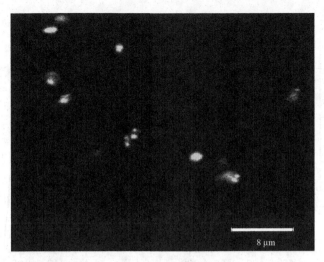

Figure 15.6 Confocal microscopy of GFP expression in liposomes. (From Murtas *et al.*, 2007.)

a long-lived bioreactor, where the expression of the reported EGFP was prolonged up to four days. Interesting is also the previous work by the same group, in which a cell-free genetic circuit assembly (without vesicles) is described (Noireaux *et al.*, 2003). In particular, the authors engineered transcriptional activation and repression cascades, in which the protein product of each stage is the input required to drive or block the following stage.

The development of the PureSystem by Ueda and coworkers (see *infra*) has demonstrated the synthesis of GFP by confocal microscopy (Murtas *et al.*, 2007; see Figure 15.6), as well as the expression of membrane proteins within liposomes (Kuruma *et al.*, 2009).

An interesting question, in this field, is about the minimal size of a liposome in which protein synthesis can take place. This can give an indication of the initial minimal size of viable protocells. To this aim, the expression of GFP was attempted, and successfully, in liposomes having a diameter of 200 nm (de Souza *et al.*, 2009). In this case, confocal microscopy could not be used (because of the low resolution inherent in this method), and the reaction was monitored by following the fluorescence intensity signal, as shown in Figure 15.7. Surprisingly, the synthesis yield was at least six times higher in liposomes than in bulk water.

To this rich list, one should add the work by Mansy *et al.* (2008) on a template-directed synthesis of a genetic polymer in a model protocell, and by Shohda and Sugawara (2006), on the DNA polymerization on the inner surface of a giant liposome for synthesizing an artificial cell model. (Several other investigators should be mentioned, like Saito *et al.*, 2007; Kita *et al.*, 2008; and Hosoda *et al.*, 2008.)

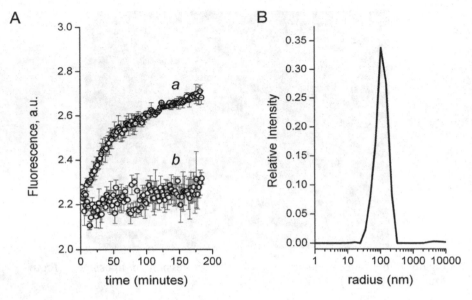

Figure 15.7 *Panel A* shows the time course of the fluorescence intensity due to the synthesis of the green fluorescence protein inside vesicles (a) over a negative control (b). *Panel B* illustrates the average size of the vesicles where the synthesis takes place. (From de Souza *et al.*, 2009.)

See other names and details in Table 15.1. The table is limited to protein synthesis, but one should add that there is an important collateral work. For example, RNA oligomers have been synthesized inside liposomes, starting from mononucleotides monophosphates, poly(adenylic acid), and poly(uridylic acid) (Rajamani *et al.*, 2007). This work was done without chemical activation of the nucleotides, with the aim of describing a laboratory model for a primordial evolution of the RNA world.

Note also from Table 15.1 that the first studies on protein expression in liposomes were performed with commercial cell extracts, for example, the Promega, which, as we have noted earlier, are unknown in terms of composition. Things changed with the report from the group of Ueda in Tokyo, with the development of the so-called PureSystem. This is a minimal TX-TL system, containing only 37 enzymes and *c.* 80 macromolecules, two dozen small molecules – including ATP as an energy source, plus organic buffer and inorganic salts (Shimizu *et al.*, 2001, 2005). The main point is that all components are given and listed in terms of concentration, so that finally the chemists working in the area do not have to deal with an unknown mixture. This important development was immediately signaled in the early protocell work (Oberholzer and Luisi, 2002), but utilized expressly only a couple of years later. In fact, most of the work after *c.* 2004 has used the PureSystem (see Table 15.1). It is interesting to note that, according to the authors of this work (Shimizu *et al.*, 2001, 2005), the minimal number of 37 enzymes can be further reduced in principle. However, what would be really necessary and important is a minimal enzyme, with a somewhat enriched genetic set that is capable of self-reproducing. Talking about a simpler system, we have to recall the experiments reported in Chapter 8 about the minimal ribosome: it is possible to utilize a smaller and simpler version of ribosomes inside the vesicles, in principle. All this represents a work in progress, mostly in Ueda's group.

Work on the minimal cell is progressing in several other laboratories, both from the experimental and theoretical point of view; see for example, Miller and Gulbis (2015), Mavelli *et al.* (2014), Blain and Szostak (2014), as well as all entries of Table 15.1 for the last two years.

As a final point, let me add that the minimal cell project is primarily a theme of basic science. However, application potentiality is there, in consideration of the fact that liposomes are already used as a drug delivery system. Therefore, one can envisage a scenario where there exists an entire synthetic cell with an internal metabolism, which is delivered as a therapeutic unit. People are working already under this perspective; as such, this is something to be discussed in the near future.

15.4 A confederacy of protocells

All the work previously seen focuses on one single cell, or protocell, at a time. At this point, one citation by Harold Morowitz, from whom we have seen the conversation at the beginning of the chapter, appears to be timely (Morowitz, 1992, p. 54):

Sustained life is a property of an ecological system rather than a single organism or species. Traditional biology has tended to concentrate attention on individual organisms rather than on the biological continuum. The origin of life is thus looked for as a unique event in which an organism arises from the surrounding milieu. A more ecologically balanced point of view would examine the proto-ecological cycles and subsequent chemical systems that must have developed and flourished while objects resembling organisms appeared.

This is a message indicating two additional concepts: mutual cell interactions, and the relation with the environmental niche. In fact, generally cells do not live in isolation, but in colonies. This behavior gives them considerable advantages, and this in turn may suggest that in the study of the origin of life, we should perhaps also consider colonies from the very start. For example, would an assembly of minimal cells present some advantages with respect to a single one? In effect, in most of the literature on the origin of cells there is no explicit discussion about the relevance of interactions of primitive cells with each other, and not much mention on the relation with the environment.

Table 15.1 *Synthesis of proteins inside vesicles*

Description of the system	Target and Results	References
Liposomes of POPC containing ribosomes and other components for protein synthesis	Synthesis of poly (Phe). Detection: Radioactive Labelling	Oberholzer *et al.*, 1999
Liposomes of egg lecithin, cholesterol and DSPE-PEG 5000 containing cell extracts	Expression of mutant-GFP. Detection: Flow Cytometry, Fluorescence Microscopy	Yu *et al.*, 2001
Cell-sized liposomes of POPC containing cell extracts initially mixed in water	Expression of GFP. Detection: Batch Fluorescence	Oberholzer & Luisi, 2002
Giant Vesicles (DOPC/POPC 10:1) containing cell extracts	Expression of rsGFP (red shifted GFP). Detection: Fluorescence Microscopy	Nomura *et al.*, 2003
A genetic network in two steps encapsulated in liposomes composed of lipid mixture (a)	In the first step it produces (T7 RNA) polymerase which in turn induces proteic synthesis of the second step(GFP). Detection: Flow Cytometry	Ishikawa *et al.*, 2004
Cell extracts of *E. coli* encapsulated in liposomes, transferred in a stock solution containing amino acids and nucleotides	Expression of channel protein α-hemolysin from *S. aureus* inside of vesicles resolves the limitations of energy and the entry of substrates (the reactor can keep up to four days the expression). Detection: Fluorescence Microscopy	Noireaux & Libchaber, 2004
Lipid blend (b), containing a DNA templated	Expression of GFP through PURE SYSTEM. Detection: Flow Cytometry	Sunami *et al.*, 2006
Vesicles of POPC	Expression of GFP through PURE SYSTEM. Detection: Fluorescence Microscopy	Murtas *et al.*, 2007
Giant vesicles containing the PURE SYSTEM	Production of GFP. Detection: Fluorescence Microscopy	Saito *et al.*, 2007
Liposomes composed of lipid mixture (c), containing both the PURE SYSTEM and genes for membrane proteins	Expression of two membrane proteins: GPAT (glycerol-3-phosphate acyltransferase) and LPAAT (lysophosphatidic acid acyltransferase). Detection: Radioactive Labelling	Kuruma *et al.*, 2009

Table 15.1 (*cont.*)

Description of the system	Target and Results	References
Liposomes composed of lipid mixture (d), containing replicase and β-galactosidase (+) RNA encoding Qβ replicase	Synthesis of Qβ from RNA. Detection: Flow Cytometry	Kita *et al.*, 2008
Liposomes composed of lipid mixture (d), containing the PURE SYSTEM and plasmid for β-glucuronidase	Synthesis of β-glucuronidase. Detection: Flow Cytometry	Hosoda *et al.*, 2008
Liposomes of POPC, minimal size (100 nm radius)	Synthesis of GFP with PURE-SYSTEM. Detection: Batch Fluorescence	de Souza *et al.*, 2009
Giant vesicles containing the PURE SYSTEM	Synthesis of GFP from DNA and RNA. Detection: Fluorescence Microscopy	Saito *et al.*, 2009
Liposomes of DOPC containing cell extracts	Synthesis of GFP from RNA. Detection: Fluorescence Microscopy	Yamaji *et al.*, 2009
Liposomes composed of lipid mixture (e), containing the PURE SYSTEM	Synthesis of GFP from DNA. Detection: Flow Cytometry and Fluorescence Microscopy	Sunami *et al.*, 2010
Giant vesicles (GUV) formed from POPC/cholesterol 9:1 containing the PURE SYSTEM and cell extracts	Synthesis of GFP from RNA. The increased sizes of the compartment positively influence the reaction rate. Detection: Flow Cytometry	Nishimura *et al.*, 2012
Cell-sized unilamellar Liposomes containing both the PURE SYSTEM and genes for proteic synthesis	Starting from mutated genes, synthesis of Ebg (*E. coli* protein of unknown function, which evolved into mutant protein with β-galactosidase catalytic activity), and expression of two mutant-gfp (GFPuv2, GFPuv5) and GUS (β-glucuronidase). Detection: FACS (Fluorescence-Activated Cell Sorter)	Nishikawa *et al.*, 2012
Cell-sized unilamellar Liposomes containing the PURE SYSTEM	Synthesis of mutant α-hemolysin which has only two point mutations but expresses a 30-fold increase in its activity compared to wild-type. Detection: FACS (Fluorescence-Activated Cell Sorter)	Fujii *et al.*, 2013

Table 15.1 (*cont.*)

Description of the system	Target and Results	References
Giant vesicles (GUVs) of POPC containing the PURE SYSTEM and the plasmids encoding the target protein	Synthesis of EmrE (membrane protein). The integration of protein into the membrane increases with decreasing vesicle volume and is indipendent from concentration of DNA Detection: Fluorescence Microscopy and FACS (Fluorescence-Activated Cell Sorter)	Soga *et al.*, 2013
Liposomes containing PURE SYSTEM and other components for protein synthesis	1) Synthesis of the Sec YEG translocon from DNA. 2) From Sec YEG translocation of two proteins (pOMP (A) and Yid C) to the membrane. 3) Synthesis of protein Lep B, (able to digest the N-terminal signal sequence of p OMP (A). Detection: SDS PAGE	Matsubayashi *et al.*, 2014
Liposomes containing an encapsulated translation cell-free system	Synthesis of α-hemolysin and other membrane proteins from mutated DNA. Detection: FACS (Fluorescence-Activated Cell Sorter)	Fujii *et al.*, 2014

Table notes:
1 EggPC:cholesterol:DSPE-PEG5000 (1.5:1:0.08) (molar ratio)
2 POPC:PLPC:SOPC:SLPC:cholesterol:DSPE-PEG5000 (129:67:48:24:180:14)
3 POPC:POPE:POPG:cardiolipin (50.8:315.6:11.5:2.1)
4 POPC:cholesterol:DSPE-PEG5000 (58:39:3)
5 (different ratios of POPC/POPG/POPE):cholesterol:DSPE-PEG5000 (58:39:3)

Abbreviations:
DOPC 1,2-dioleoyl-sn-glycero-3-phosphatidylcholine
DOPG 1,2-dioleoyl-sn-glycero-3-phosphatidylglycerol
DSPE-PEG5000 1,2-distearoyl-sn-glycero-3-phosphatidylethanolamine-PEG5000
EggPC phosphatidylcholine extracted from egg yolk
GPAT glycerol-3-phosphate acyltransferase
LPAAT lysophosphatidic acid acyltransferase
PLPC 1-palmitoyl-2-lauryl-sn-glycero-3-phosphatidylcholine
POPC 1-palmitoyl-2-oleoyl-sn-glycero-3-phosphatidylcholine
POPE 1-palmitoyl-2-oleoyl-sn-glycero-3-phosphatidylethanolamine
POPG 1-palmitoyl-2-oleoyl-sn-glycero-3-phosphatidylglycerol
SOPC 1-stearoyl-2-oleoyl-sn-glycero-3-phosphatidylcholine

The simplest case is the interaction between two, or more, vesicles, giving rise to fusion and to the corresponding mixing of their content, with the arising of a greater biological complexity. We have mentioned the fusion among vesicles in Part IV of this book, and we can add here that experimental investigation on vesicle fusion has been carried out by mixing cationic and anionic vesicles (Thomas and Luisi, 2004, Caschera *et al.*, 2010; Caschera and Noireaux, 2014). More recently, this research was successfully expanded in the groups of Yomo and Hanczyc who reported, for the first time, the synthesis of a functional protein inside vesicles obtained by the fusion of two different vesicle populations (Caschera *et al.*, 2011).

This kind of view has been extended to giant vesicles (GV), where the term *colonies* has been introduced (Carrara *et al.*, 2012). In this work, GV colonies have been constructed by binding GVs to each other via electrostatic interactions. In particular, anionic GVs (POPC + sodium oleate), when treated with poly-L-arginine (PLA), form large aggregates that can be firmly attached to a solid hydrophobic support. These GV colonies can be seen in first approximation as models for cells bound to each other. GV colonies are stable against the flow, capture solutes in the colony interstices, and can recruit new vesicles. GVs in the colony can also fuse and mix their inner content. The membrane of PLA-treated GV colonies resulted in being permeable to small and large phosphate-bearing molecules (like ADP, tRNAs), providing the GV colonies with an advantage when compared to isolated GVs (Carrara *et al.*, 2012).

Those are very preliminary data, but it is enough to show that the study of colonies of protocells is a pathway for future research.

15.5 About the statistics of entrapment

Let us now resume our discussion about Table 15.1. Several research groups have been able to express proteins inside liposomes, and in particular the enzyme kit known as PureSystem has been instrumental for a considerable advancement. As already mentioned, this system consists of 37 enzymes, and a total of 80–82 macromolecules, in addition to a series of low molecular weight compounds. This is a minimal system, and this means that a liposome, to be viable, must contain all 82 macromolecular components, at least one copy of them. And this poses immediately a thorny question: how is it possible that 80-odd macromolecular components are going to be simultaneously inserted into a single compartment, which can be as small as 100 nm radius, as in the example of Figure 15.3? What is the probability of such a simultaneous entrapment?

At first sight, based on common sense, one would be tempted to answer that this probability is tendentiously zero. However, the experiments, as we have

seen, give a positive response. It is not zero, but something sizable. We have then
to dwell a little more on this point.

To this aim, let us recall first how we perform the entrapment – something we
have seen in detail in Part IV of this book. Two procedures are the most common.
We can start from the aqueous solution of the PureSystem, and liposomes are
formed *in situ* thanks to the swelling of phospholipid films. Alternatively,
a concentrated aliquot of the lipid in ethanol (or methanol) can be added to the
solution, whereby vesicles form spontaneously *in situ*. In both cases, when
vesicles form and close, they mechanically entrap the solutes.

What can we say about the statistics of the entrapment? The simplest view is to
assume a random sampling, so that each molecule is being incorporated as a single,
independent event – and this would give rise to a classic Poisson distribution.
In particular, as argued in detail elsewhere (de Souza *et al.*, 2009), the average
number of solute molecules found in a vesicle of volume V is simply $\mu = N_A \, C_{bulk} \, V$,
where N_A is the Avogadro's number and C_{bulk} the bulk solute concentration.
It follows that the probability $p(n)$ of finding n molecules in a vesicle is given by
the Poisson statistics, i.e., $p(n) = e^{-\mu} \, \mu^n / n!$, and that the co-entrapment probability for
k molecules is the product of the k individual entrapment probabilities (de Souza
et al., 2009; see also D'Aguanno *et al.*, 2015).

Results of these calculations are shown in Figure 15.8. As intuitively expected,
the co-entrapment of one (or several copies) of the entire PureSystem kit is highly
improbable inside small vesicles.

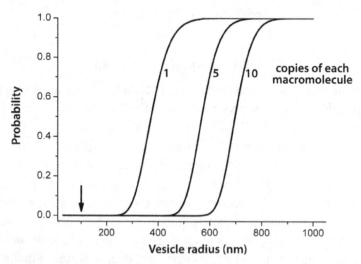

Figure 15.8 Calculation of the probability of simultaneous incorporation of 82
different macromolecular components inside the same compartment, according
to classic Poisson distribution. The arrow indicates the size of the smallest
liposomes used in our experiments. (Modified from de Souza *et al.*, 2009, with
permission, and from de Souza *et al.*, 2012.)

Note from this figure that with vesicles as small as 100 nm radius, the probability of incorporating one single copy of all macromolecular components is zero, and becomes sizable only with vesicle dimensions around 300–400 nm radius. However, considering that for protein synthesis one needs more copies of ribosomes, even with five copies, the probability can be set equal to zero until *c.* 500 nm radius. In this respect, one question we have not yet clarified is about the minimal number of ribosome copies which are necessary to give a measurable viability.

More in general, there is something interesting to note: nobody in the literature has noticed this incongruence, namely that according to classic statistics, the experiments should not have functioned.

Consider also that in order for the protein synthesis to occur, not only the macromolecules should be incorporated, but also all necessary low molecular weight compounds – amino acids, mono-nucleotides, ATP, metal ions, and each of them in a sizable μM concentration. Introduction of all these components in the calculation would be not so easy, but clearly, this effect would shift the curves of Figure 15.8 further towards the right-hand side, making the viability difficult also for micrometer size vesicles. These low molecular weight components are usually in large excess, and therefore should not give a problem in the statistics of incorporation inside the liposomes.

Therefore, the Poisson statistic tells us that reaction should not have occurred in 100–200 nm vesicles. Instead, the reaction worked well (. . . and we all believe that experiments are "stronger" than statistical theories). The weaker partner in the contradiction is the theoretical statistics, and thus, we can say at this point that the Poisson statistics do not work here. In fact, already the original entrapment data of ferritin (Berclaz *et al.*, 2001a, 2001b), which we have seen in the previous part of this book, gave some hint to the fact that the entrapment statistics of macromolecules in liposomes was anomalous (Luisi, 2006). So, how to reconcile theory with experiments?

15.6 A story of spontaneous overcrowding

Once ascertained that the classic Poisson distribution is not valid for our vesicular systems, we needed a way to discover the working statistics. In order to do that, we resumed the older work from the group, in which we had utilized electron microscopy and ferritin as marker for solute entrapment (Berclaz *et al.*, 2001a, 2001b). As you may remember, we have discussed the technique of the ferritin entrapment in Part III of this book, although for a different purpose. We repeated the entrapment experiments, starting from a solution of ferritin, adding lipid so as to produce vesicles *in situ*, and focusing on the distribution of ferritin inside the single vesicles. The main aim of this study was to confirm

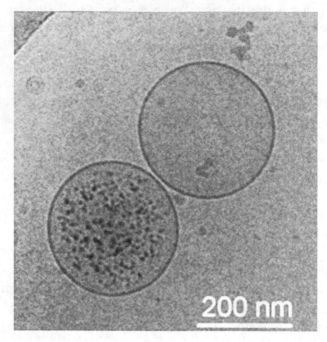

Figure 15.9 TEM micrograph showing the "all-or-nothing" incorporation of ferritin inside non-extruded POPC/oleate vesicles. (Adapted from Luisi *et al.*, 2010.)

experimentally that the Poisson distribution would hold; and then to figure out which one was the correct one.

Typical data are shown in Figure 15.9 (see also Figure 4.6 in Part I). Results were indeed quite surprising. In fact, the situation is characterized by a very large number of empty vesicles, and a very small percentage of overcrowded ones; at first approximation, idealizing the situation somehow, we could talk of the all-or-nothing mechanism of incorporation (Luisi *et al.*, 2010). The following simple-minded scheme (Figure 15.10) gives a qualitative idea of the very peculiar kind of entrapment. It is as if a few vesicles would act like attractors for solute molecules, up-taking ferritin out of solution and leaving the great majority of vesicles without a single solute molecule.

This effect has been investigated in detail under quite different concentrations of ferritin and/or lipid molecules, and the ensemble of results is depicted in Figure 15.11 (Luisi *et al.*, 2010).

The distribution of ferritin molecules inside vesicles appears to follow a kind of power law. Note in particular that vesicles filled with many ferritin molecules represent only a very small fraction of all vesicles (*c*. 0.1 percent). But consider that 0.1 percent of vesicles may correspond to a very large absolute number of

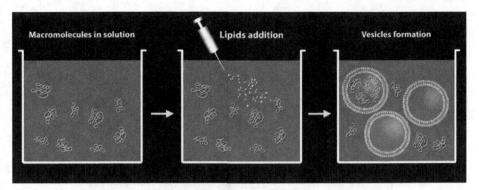

Figure 15.10 Idealized schematization of the all-or-nothing encapsulation mechanism according to the finding of the overcrowding of entrapped macro-molecular solutes following the formation and closure of the vesicles *in situ*. (Modified from de Souza et al., 2012.)

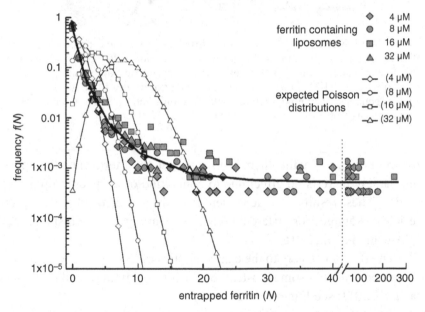

Figure 15.11 Actual distribution of ferritin inside liposomes, as determined by TEM. Notice the comparison between the originally expected theoretical Poisson distribution curves, and the observed behavior, corresponding to a kind of power law. (Adapted from Luisi *et al.*, 2010 and de Souza *et al.*, 2012.)

vesicles: by our operational concentration of about 30 mM lipid, and on the basis of an approximate aggregation number in the vesicles of 10^5 (lipids/vesicle), we may end up with several thousand overcrowded vesicles in one nano liter of solution.

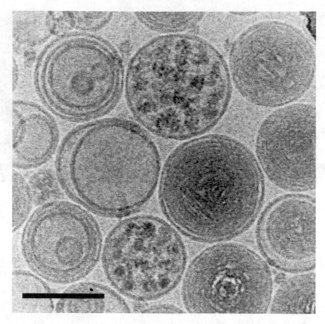

Figure 15.12 Micrographs of ribosome-filled lipid vesicles, prepared by the ethanol injection method in a 4.0 mm ribosome solution. After vesicle formation and free ribosomes removal by gel filtration chromatography, samples were imaged by cryo-TEM. Filled vesicles are present at a ratio of about 0.1–1% in a heterogeneous vesicle population containing uni- and multi-lamellar vesicles. Size bar represents 100 nm. (From de Souza *et al.*, 2011.)

Note from Figure 15.9 that there is no aggregation of ferritin on the liposome bilayer, namely there is no effect due to some kind of unspecific lipid-protein association. These results are independent from the procedure utilized to prepare the vesicles; examples for different preparations are given in Luisi *et al.*, 2010 and D'Aguanno *et al.*, 2015.

Is this an effect due solely to the characteristics of ferritin? Experiments have been repeated with ribosomes instead of ferritin, obtaining similar results (de Souza *et al.*, 2011; see Figure 15.12).

Actually, in this last work, it was possible to count the number of ribosomes entrapped in each vesicle. The number varies in a broad range, depending on the size of the ribosomes, (see Table S2 in de Souza, 2011), the highest values being around 100–200 individual ribosomes in each compartment. Many of the other liposomes have 20–30 internalized ribosomes – and, as already mentioned, many are empty or containing only one or two of them.

The higher number values corresponds to a concentration which is about 50 and 100 μm for smaller vesicles. This is remarkable, given that lipid vesicles were prepared in this work from 0.48 and 4.0 μm ribosome solutions. Therefore, ribosomes are concentrated by one or two orders of magnitude following lipo-some encapsulation. In that paper, the intriguing observation is mentioned that such a local ribosome concentration is close to that of ribosomes in *E. coli* (~ 20 μm). In the smallest filled liposomes, the internal space occupied by ribosomes varies between 20 and 60 percent (consider that the maximal theore-tical figure, in the approximation of packed spheres, should be 70 percent) (D'Aguanno *et al.*, 2015).

Also in the case of ribosomes, no aggregation with the lipid bilayer was observed, nor self-aggregation of ribosomes in the internal space of the lipo-somes (see also Figure 15.12).

In addition to proteins and ribosomes, other macromolecular complexes have been incorporated in liposomes with the phenomenon of spontaneous over-crowding (D'Aguanno *et al.*, 2015). This holds, for example, for a RNA-poly arginine complex, to a smaller degree for dextran (labelled with different fluorescent probes), and for enzymes. In particular, the case of carbonic anhy-drase and carboxy-fluorescein di-acetate has been studied as an example in which the substrate permeates inside the liposomes from the external medium; and the system of proteinase K and Bovine serum albumin has been studied as the case in which both enzyme and substrates are macromolecules (the effect here is larger). The case of enzymes is particularly interesting when the substrate is also a macromolecular compound, and as such, subjected to the phenomenon of overcrowding. In fact, an enzymatic reaction takes place in an efficient way only when the substrate concentration is of the order, or higher, of K_M. Thus, one can start from a situation in which there is no reaction in the bulk aqueous medium, because the substrate concentration is too low, and as soon as vesicles are formed *in situ*, reaction starts inside the compartment.

In all these last cases, the overconcentration effect was not as high as in the case of ferritin or ribosomes, and no systematic study has been carried out as yet in order to optimize the overcrowding effect. More recently, the case of simpler polymers, such as PEG of various polymerization degrees, has been studied (D'Aguanno *et al.*, in preparation). This last study was prompted also by the consideration that it is not clear yet whether and to what extent the phenomenon of overcrowding is restricted to macromolecules. In experiments with the low molecular weight solute calcein (0.67 kDa, used at a concentration of 1.25–10 μM), it was not possible to detect vesicles with fluorescence higher than the background (D'Aguanno *et al.*, 2015; D'Aguanno *et al.*, in preparation).

Studies are in progress with polymers of different length to answer the question of whether this kind of overcrowding effect has a molecular weight threshold.

This brings us directly to the other question regarding the mechanism of such a phenomenon. It is not clear now why and how the phenomenon of spontaneous overcrowding takes place, but a few groups are working on this. A first idea has been proposed in earlier papers (Luisi *et al.*, 2010, de Souza *et al.*, 2011), which must be further developed.

I believe that the mechanism should be based on the concomitance of kinetic and thermodynamic effects in the overall scenario of the dynamics of the closure of the liposomes. It is commonly accepted that liposomes arise from the closure of planar bilayer sheets. One first assumption is now that the rate of closure is slowed down by the presence of bound macromolecules. In other words, sheets with one or two bound proteins already would take longer to close, and this would give more time for the binding of more incoming protein molecules. Such a kinetic effect would not be enough for the observed intense overcrowding in some liposomes. A second assumption is necessary, based on the cooperativity (which is an autocatalytic effect) of intermolecular protein interactions. In particular, in those few liposomes that stochastically have incorporated a larger initial number of macromolecules, there will be a cooperative, non-linear addition of further macromolecules, driven by an increase of entropy (as for hydrophobic forces in water). In other words, the macromolecules attract each other and the more they do so, the higher is their initial number inside a closing compartment.

A simple-minded illustration is given in Figure 15.13, which gives only the case of two liposomes, one which remains empty and closes fast (*pathway a*); while the other, having reached a state of several bound macromolecules, cooperatively and auto-catalytically attracts more and more solute molecules, which results in an overcrowding situation (*pathway b*).

About this hypothesis, consider that, in water, hydrophobic forces determine the association of hydrophobic macromolecules, whereas hydrophilic

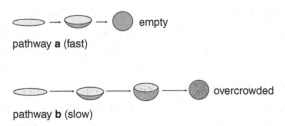

pathway **a** (fast)

pathway **b** (slow)

Figure 15.13 A possibile mechanism underlying the encapsulation of macromo-lecules, such as ferritin, inside lipid vesicles and the overcrowding effect. The **pathway a** represents the fast formation of empty vesicles. The closure process would be slower if there is one or more bound macromolecules; and in the extreme case we have the **pathway b**, where, by a stochastic event, some macromolecules are initially bound, and this elicits a corresponding autocatalytic mechanism of non-linear increase of the number of the solute macromolecules entrapped in the forming vesicle. For the sake of simplicity, the macromolecules surrounding vesicles were not shown here.

components tend to associate in solvents of low polarity. It is conceivable that in such an environment as a closing bilayer sheet, the dielectric constant is considerably lower than bulk water, so as to favor an association of hydrophilic macromolecules. This would sort out an additional autocatalytic effect, as the more proteins (or ribosomes) are bound, the more bulk water is excluded from the local environment, thereby further decreasing the dielectric constant.

Other works in the literature discuss, in a different context, the overcrowding in vesicles; see for example, Minton and Rivas, and Acerenza and Grana in the cited book edited by Luisi and Stano (2011), and also the work in the group of Christine Keating (Dominak *et al.*, 2010).

15.7 The origin of metabolism?

With the phenomenon of spontaneous overcrowding, our narrative moved away from the character of a review, to arrive into present-day research. It is a story in progress, where several aspects must still be clarified. There is no doubt that the phenomenon of the spontaneous overcrowding is a new finding that has relevance both in the field of vesicle research, and in the field of the origin of life. And it has an explanatory force in both these realms of inquiry.

In fact, let us consider first the following question: How can the phenomenon of the spontaneous overcrowding of single macromolecules explain the viability of liposomes for the synthesis of GFP, a case in which we have many different macromolecular components?

The answer is rather straightforward. Simply, the overcrowding effect has the power of increasing the local concentration by an order of magnitude with respect to the bulk. Reaction outside does not take place because of the excessive dilution, and this observation brings us to the other important conclusion of this finding, which concerns the origin of life, and in particular the stumbling block we have seen in the "compartmentalistic" point of view in Part I of this book. There, we have mentioned that it is very difficult to conceive of how one could reach the critical concentration threshold inside a protocell.

In fact, this permits us to go back to a hypothetical origin of life scenario. Consider a "warm pond" where the macromolecules of life – enzymes and nucleic acids – have been somehow synthesized, but are present as a very dilute solution. Under such diluted conditions, no reaction can take place in the bulk. However, if in this diluted solution some vesicles are formed *in situ*, the phenomenon of spontaneous overcrowding will be operative and will permit the reaction.

There is a nice, simple experiment which can be carried out in the laboratory to confirm the above: you take a cellular extract, with the plasmid of GFP, and dilute 1:100 with water. No protein synthesis takes place because of the excessive dilution. Then you add a small aliquot of phospholipids in methanol, to form spontaneous liposomes *in situ*. And then you will observe the green color appear in some of the liposomes. This is indeed an easy experiment to perform, and quite instructive.

You can ask where the prebiotic surfactant may come from in this scenario. We have talked about that in Part I, where we have described the ideas and the experiments of David Deamer with fatty acids as surfactants, possibly also found in the Murchison meteorite. And we have seen additional hypotheses for the biogenesis of surfactants.

15.8 And (why not?) the origin of life?

We have mentioned above that the idea to relate the spontaneous overcrowding to the origin of a pristine metabolism leads almost automatically to the idea of the very origin of life. Let us dwell on this point, making clear that we are in the realm of speculations, although these are initially based on solid micrographs.

To this point, more than considering the "simpler" PureSystem, let us go to the case of the cellular extracts used to express GFP in liposomes before the advent of Ueda's system.

There is no life in this too diluted, chaotic mixture. Now, as a further assumption, let us assume that vesicles are formed *in situ*, due to the biogenesis of prebiotic surfactants, such as fatty acids, as mentioned previously.

Then, the phenomenon of spontaneous overcrowding will take place. There will be then a combinatorial, stochastic, distribution of solutes inside these crowded protocells, which tendentiously will be all different from each other in terms of composition and concentration. We have mentioned this possible scenario already in Chapter 5, talking about the zero ground approach, and also about the notion of the confederacy of protocells, as discussed in Chapter 11, Section 11.5.

This scenario is what we can call the protocellular multiverse, borrowing metaphorically from the language of the cosmologists when they talk about a large number of universes, each with its own characteristics (Tegmark, 2003; Carr, 2007; Barrow, 2001; Susskind, 2005). They further assume that one of these universes may be the right one – having the characteristics of permitting life. Regarding the absolute numbers, consider that even if the overcrowded vesicles are only 0.1 percent of the total, a 30–50 mM solution of surfactant will contain thousands of billions of overcrowded vesicles – a real multiverse.

Our protocellular multiverse might be just that, a distribution of billions of closed entities, most of them sterile, and just a few – or only one – characterized

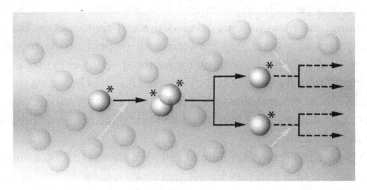

Figure 15.14 The protocell multiverse. The spontaneous overcrowding yields a vast variety of filled protocells with a combinatorial distribution of solutes, and only a few, or only one – as indicated with an asterisk in the figure – happen to be viable, with all ingredients and organization capable of life and self-reproduction, at the expense of non-viable protocells. After fusion and redistribution of the materials, two viable cells are formed from the first one, and the process can go further.

by the capability of displaying or starting life – in the form of a non-linear reproduction mechanism at the expenses of the non living protocells.

All this is illustrated in Figure 15.14. Here, the "rich" protocell which happens to have all ingredients in excess, binds and fuse with a "poor" one. These two viable cells activate then two more protocells, and the process goes on exponentially.

Concluding remarks

The notion of minimal cell was present already in the classic book by Harold Morowitz (1992), and the experimental work using vesicles started, independently, a few years later at the Swiss Federal Institute of Technology in Zurich, the ETHZ (Oberholzer *et al.*,1995a, 1995b, 1999), actually in a time where the laboratory construction of minimal cells still appeared a faraway vision. However, within a few years, several research groups have actively begun to work on it. Probably not so much because of the pioneer work at the ETHZ, but as a kind of convergent evolution, as it often happens: the time was ripe for it.

Where are we now, in terms of minimal semi-synthetic cell? And where are we going?

We have seen the work with the expression of GFP and other proteins. The synthesis of proteins is an important function of the cell, but probably we would all agree that this is not enough; we do not yet have in our hands a minimal living cell. The experimental systems described until now are not self-sustaining, the main missing level is the self-reproduction from within. The PureSystem is a great step forward with respect to the rough cellular extracts, but the attempts to construct a minimal set of enzymes that is self-reproducing is not yet at hand. And of course all is still based on the previous existence of DNA and enzymes, at least in some primitive form. We can only repeat here that this does not give us an answer to the origin of life, but to the origin of the first cells.

How to proceed to get to the minimal cell? We should link here with Part I of the book, with the idea of a very simple metabolism capable only of some form of homeostasis, with primitive forms of protocells even prior to genetic code and DNA. From this really prebiotic scenario, we can eventually reach the level of the minimal genetic cell.

More in general, the view of life and origin of life from the edge of the minimal cell corresponds to a system view of life, where we have and need to have, several components interacting with each other, and with the environment, as we have seen in Part II of the book. We have also mentioned, at the end of the last section, a scenario with a large variety of interacting cells or protocells, the idea namely of a "cellular multiverse". At this point in the research, however, this is little more than a speculation. But certainly a way to go further to design novel experiments.

Questions – and research proposals – for the reader

1. The PureSystem consists of 37 enzymes and a totality of *c*. 80 macro-molecules, and as such is the minimal kit to express proteins in vitro or inside liposomes. The system cannot self-replicate. Can you estimate how many more enzymes, and, in general which kind of the additional complexity you would need, in order to make self-reproduction possible in the PureSystem?

2. Let us go back to the green liposomes, obtained with the *E. coli* cellular extract (e.g., Pro omega). They produce the GFP. The question is: have they incorporated simply the translation-transcription machinery – or do they represent the entire reconstituted *E. coli*? Which kind of experiment (Petri disk or similar) would you devise, in order to clarify whether these viable liposomes have acquired the cellular functions of the entire microorganism?

3. Regardless of whether you found a positive answer to the second question, suppose now that you make your green liposomes by adding the plasmid for insulin. Do you expect that these liposomes also produce insulin? And would that be a system which has some potential economic advantage?

4. Suppose that you have found a partial positive answer to the second question – namely, the green liposomes contain much more than the protein expression ribosomal machinery. Then, you make a cellular extract as a mixture of two different cells. Would that be a way to obtain a hybrid cell, a new form of liposomal life? Or, possibly, in the phenomenon of overcrowding, would you get segregation in the two different liposomal lives?

As a way of conclusion

One of the issues addressed in this book regarded the ancient question of the definition of life. Ancient because it could have been asked at any time in the past; Renaissance people did not have to wait until Watson and Crick to deal intelligently with the question, "what is life?" Even so, we have seen a very broad spectrum of opinions. Not a surprise, since the quest for the origin of life provokes interest not only in life sciences (biology, chemistry, molecular biology, bioengineering, artificial life, and astrobiology) but also in philosophy and theology. Each one with its own language and traditions, and each group sticks to their own. No unifying view.

This may be so, but it does not mean that this is the way of good science. Good science should aim at a convergence towards a common denominator, and in the case of the question "what is life?," it should converge towards what Sir Stafford Beer sees as "the understanding of the organization of the living systems in relation to their unitary character" (Beer, in Maturana and Varela, 1980, p. 65).

In this book, I have emphasized that a way to do so is to see life as a systems view: to see life as a confederacy of components that in virtue of their specific mutual interaction form a collective, organized unity. The systems view also includes the interaction with the environment in the specific niche.

I offered the full expression of the systems view of life in Part II of this book, with the description and reappraisal of the theory of autopoiesis. I believe that this theory has identified the common denominator of life in all living organisms. When we say, "the living is an open, self-maintaining molecular system due to self-regeneration of the components within a boundary of its own making," we should recognize that there is no living system on our Earth that does not comply with such a statement. I also made the point that it is ontologically wrong to answer the question "what is life?" with the word *reproduction*. Reproduction is a property of life; it does not say anything about why and how an organism is alive. Reproduction is certainly the main mechanism and criterion of evolution and biodiversity, but not the criterion of life.

With Humberto Maturana's imprinting, we also stressed the concept of *operational closure* and the notion of *cognition* – the latter corresponding in my interpretation, to the specific interaction with the environment. At the level of humans, cognition goes into the notion of mind and consciousness, and this is in turn conducive to the problem of the "first person" in science. Not an easy matter, as apparent from the conversations with Humberto Maturana and Evan Thompson in this book. However, as complex as it is, this shows the possibility of an intense dialogue between biology and philosophy, for which the late Francisco Varela was pleading (after so many years, I still feel indebted to him for the long discussions on this subject). This is what the notion of life should be: something that links biology to the other domains of life science, including philosophy and cognitive science.

In the chapters about autopoiesis and cognition, but also in Part V, which is concerned with synthetic biology, the reader may have met with more philosophy than expected. This corresponds to my view, that philosophy is an integral part of science, and should not be considered a separate category. This point should actually be kept in mind in all our courses for undergraduate and graduate students.

Having said that, the definition of life is still a controversial subject, and we should add that the origin of life is still an unsolved problem – both conceptually and experimentally.

Why is this so? We all know the objective reasons that make the task so difficult. We want to construct – possibly even experimentally – a piece of history without any concrete signpost left to us. However, as pointed out in the book, there might be also epistemic and social reasons for the relatively slow pace of the progress in the field. One of these is the dominant paradigm of our generation. We are in the "century of the gene" – as Evelyn Fox-Keller (2000) aptly said – which brings about a sort of genetic determinism in most areas of life and medical science. To the point to arrive to the unfortunate equivalence between life and nucleic acids. This equation compels most of the present researchers to think only, or mostly, in terms of a pristine RNA, although they do not know where it may come from.

Also in the origin of life, the systems view may offer a helpful operational vision: to construct the very first vestiges of life, conceptually and experimentally, you have to design a system of components mutually organized in a first metabolic network, capable at least initially of some form of homeostasis. The idea that a single self-replicating molecule such as RNA can do the job alone was certainly a stroke of genius – but it was just a vision, which was then taken too seriously and without proper criticism by too many workers in the field. We have shown in this book – but it was already known in some part of literature – that the implementation of the idea of a self-replicating prebiotic RNA makes no sense from the chemical point of view. RNA should be a piece of the confederacy that eventually brings to the emergent property of the pristine cellular life.

The systems view leaves unsolved the biogenesis of the single components. And to that, I have emphasized in the book that to me the main stumbling block is our remarkable lack of information about the biogenesis of the macromolecular order. We need to understand how the ordered sequences of amino acids, and of the nucleotides, appeared in a prebiotic world.

The acceptance of this point, that the origin of life pivots on the biogenesis of sequential structural order, should compel the researcher to something clear and strong: that she/he cannot look for the origin of life starting from already ordered macromolecular sequences – like a self-replicating RNA, or a pristine ribosome, or a viroid. Of course, this kind of study may give and has given beautiful and interesting pieces of science: however, as far as the origin of life is concerned, metaphorically, it is similar to constructing a house by starting from the roof.

I have also shown in the book that the view of the compartmentalistic approach is not much better in this sense, as actually all this work on the minimal cell is based on extant enzymes and genes. Thus, one can say that all most important theories of the origin of life are lacking in the fundamental ingredient: the biogenesis of macromolecular order.

What to do in order to tackle this basic, fundamental stumbling block? I believe that one needs a reordering of the priority of inquiry on the origin of life, and for this reason, I have emphasized a return to "ground zero." In particular, it has been indicated in Part I of the book how ordered *de novo* protein molecules, and possibly in a similar way RNA macromolecules, can form prebiotically by way of fragment condensation of shorter sequences. It should then be possible to make a library of *de novo* proteins, possibly with catalytic properties, since such property is already present in simple dipeptides. This may bring to the idea of "proteases first" – a first family of proteins that can eventually evolve from a "ground zero" approach, and in principle produce chains with polymerase activity.

The possible elucidation of the biogenesis of orderly polypeptides and nucleotide sequences, however, would not solve the question of their mutual causal dependence – the old egg and chicken problem, the genetic code. This is something additional, generally bypassed by the RNA-world, and by me in this book – indeed a question that I was not able to clarify to myself.

I expressed in the book severe criticism to the view of life based on a pristine form of the prebiotic RNA-world, but I also see from literature and from "after-beer" discussions that there is the emerging notion of a pre-RNA world. If we all would agree that we need a prerequisite pre-RNA world, and, reasonably, one based on amino acids and peptides, the dichotomy between the "geneticist" and the "compartmentalistic" points of view would smooth over in favor of a unified search vision. This unified search vision should be, in my opinion, in the direction of a minimal cell, with minimal life seen as a systems view.

The last part of the book goes into synthetic biology, to arrive at the construction of the minimal cells. My original idea was to write only a couple of pages of introduction on synthetic biology – but then the richness of this subject took over, and the 20-plus pages on this subject practically wrote themselves. Indeed, synthetic biology, and the cousin fields of staminal cells and GMO, will revolutionize our technological life even more. If this will be for the better, I do not know.

And it is likely that one of these near future achievements will be the laboratory construction of minimal life from scratch, although this will likely involve using extant pieces of DNA and polypeptides. How will the assembly be actuated? To that, the finding of the spontaneous overcrowding may really be important for the first steps of a primitive metabolism. Here, the stochastic character of the macromolecular distribution may give rise to a consortium of many different protocells. It is at this point that I took the liberty of talking about the "cellular multiverse" scenario, a multiverse library of "limping" cells or protocells with a stochastic distribution of macromolecular components – with life arising only in the one(s) where the stochastic distribution had given the lucky combination.

This would still be synthetic biology. Would that lead to an understanding of how life started spontaneously on our planet? Probably not, for the reasons expressed above (show me first how the macromolecular sequential order is being made prebiotically, plus the genetic code. . .). The progress will be slow, partly because we need to reorganize our research strategy, and because real work is necessarily hard and slow. In fact, as for myself, I was not able to implement several research projects that came to my mind. I have collected and described those in this book, with the hope that the new generation of readers may do that, possibly contributing to the general movement forward.

Appendix

The open questions about the origin of life

We have discussed – throughout the book – some ideas for research projects that are not commonly present in the literature of today. This is, of course, only a small specimen of the questions we may still ask, and which have not been asked sufficiently, about the origin of life. Actually, the questioning on the questions which are not being asked, has been in my agenda for a long time, and at a certain point I decided to formalize this by instituting international meetings with the title *Open Questions about the Origin of Life* (OQOL). While the aim of the normal meetings on the origin of life was to shed light on the new findings in the field, such OQOL meetings would instead have been devoted to the shadow aspects, in the sense to bring forth the questions which, for one reason or another, are usually not asked. In turn, this "not being asked" is *per sé* an interesting matter, as it is connected to the Zeitgeist, to the dominating philosophy and way of thinking in this particular area of scientific inquiry. I have already mentioned, maybe too many times, that the DNA-centered reductionist view is the one which mostly pervades the area in this "century of the gene." In addition to that, there are in my opinion questions which are too difficult to tackle, both conceptually and experimentally – for example, the biogenesis of ordered macromolecular sequences, the origin of the kinetic control, or even the origin of the genetic code – and that therefore are almost never found in the titles of papers on the origin of life. Another point is that we scientists keep thinking along the same – our own – research lines for decades, we do not like to change. Several reasons, then, why one should make an effort to think at the OQOL.

The first of these meetings took place in Erice, Sicily, in 2006, where I had the collaboration of Pasquale Stano; the second one in San Sebastian, Spain, 2009, with Kepa Ruiz Mirazo; the third one in 2011 in Leicester, with the collaboration of Rainer Derek, and the last one in 2014 as a workshop of the *International*

Congress on the Origin of Life in Nara, Japan with the collaboration of Yutetsu Kuruma. The proceedings have been published until now by Alan Schwartz in a special issue of *The Origin of life and Biosphere*. Each meeting was between two and two-and-a-half days, and each question was discussed by three or four contributors.

These selected questions were chosen by correspondence with about 50 to 70 interested researchers in a kind of poll vote, asking them to formulate their "unanswered questions." From an initial list of about 20 to 30 questions, in a second correspondence round, the same persons were asked to choose the final questions to be discussed at the meeting.

The questions asked and discussed in these first four meetings are listed below. Notice that several of these questions are returning; indeed a good deal of repetitions, and this is unavoidable. Also notice, that the style of questioning is different from meeting to meeting. Moreover, as I observed before, some questions are really new, in the sense that you do not find them in the current literature (personally, I may not agree with some of the questions/arguments raised).

Selection of the open questions presented in the OQOL Workshop of Erice – 2006

1 On prebiotic low molecular weight compounds

Many low molecular weight compounds have been produced under alleged prebiotic conditions. Generally, they can be considered products under thermodynamic control (formed by "spontaneous" reactions because they are the most stable products under the given conditions). The "free ticket" of thermodynamic control is not sufficient, however: if a chemist is given all these compounds in any amount he wishes, he would be unable to make life. For making life, one needs a series of additional reactions and products under kinetic control – enzymes and nucleic acids are not with us because they are the most stable chains. Thus, the origin of life can be traced back to the origin of kinetic control. Do you agree with this statement; and how would you envisage the prebiotic evolutionary bridge between thermodynamic and kinetic control?

2 On Contingency vs. determinism

The proteins (or nucleic acids) existing on our Earth correspond to an infinitesimal part of the theoretically possible sequences; the ratio between possible and existing structures corresponds more or less to the ratio between the space of the universe and the space occupied by one hydrogen atom. The above ratio can be interpreted as an indication that our "few" proteins have not been selected primarily because of distinctive properties (such as thermodynamic or thermal stability, solubility, particular kinetic processes of formation, etc.) – but rather due to a most significant contribution of the vagaries of contingency. Do you agree with this statement, and with its corollary, that then life on our Earth, which is based on these "few" proteins, is not an obligatory pathway, but is largely based on contingency?

3 What do we learn from astrobiology?

The compounds found in meteorites and the cosmos correspond to thermodynamically stable products, synthesized under prebiotic conditions and are therefore of great significance to the origin of abiotic organic compounds on our Earth. The fact that, until now, no oligopeptides or nucleotides have been detected in cosmic material may signify that these oligomers do not tend to form spontaneously. Thus, what we learn from cosmic products is of limited interest regarding macromolecular prebiotic molecular evolution. Do you agree with this statement, and where would you see then the importance of astrobiology?

4 How to make prebiotically long hetero-peptides or hetero-nucleotides?

There are no or rather scanty reports in the literature on how to make, under prebiotic conditions, long – say, 30 residues – and specific sequences of co-oligopolypeptides (or polynucleotides) in many identical copies containing, say, five to six different amino acid residues or three to four bases (the Merrifield method cannot be considered a prebiotic method). Methods for homo-polypeptides (chains containing only one type of residue) have been described, but they are in principle not valid for mixtures of different amino acids – as all rules of copolymerization teach us. Random polymerization of mixtures of amino acids (which we also do not know how to make under prebiotic conditions) would produce a wild mixture of different chains, with circa zero probability to make two identical chains. Do you agree then that we do not know – neither conceptually nor experimentally – how to make macromolecular sequences in many identical copies under prebiotic conditions?

5 On chirality: no longer a problem?

The origin of homochirality in nature is usually debated in terms of two opposite views. According to a deterministic (*ex-lege*) approach, one of the two enantiomers has a lower intrinsic energy and therefore a greater probability of occurrence. The alternative is a stochastic process, according to which the selection of one enantiomer over the other out of a racemate was determined by contingency. From recent experiments, it appears that the breaking of symmetry may be achieved rather easily in the laboratory, possibly under prebiotic conditions. Do you agree then with the view, that the origin of homochirality in nature "is no longer a problem"?

6 On chirality: prior to the onset of macromolecules, or after?

Was homochirality in nature implemented originally at the level of bio-monomers; or only after, namely at the level of the separation of diastereomeric macromolecules originated from racemic bio-monomers? (Some authors assert that it is easier to physically separate diastereomeric macromolecules than enantiomeric monomers).

7 On the impact of the RNA-world on the origin of life

There is little doubt that the RNA-world has written some of the most significant and outstanding pages of modern molecular biology, and quite generally, shows the importance of macromolecular evolution. However, the importance and success of the RNA-world is restricted to the field of synthetic biology, whereas its impact in the field of the origin of life is negligible. The question "who/what made RNA?" is in fact still un-answered. Do you agree with these statements and in general with the point that we have learned very little or nothing about the origin of life from the RNA-world?

8 On the chemical reality of the RNA-world

There is a lot of emphasis in the RNA literature about a possible self-replicating RNA as the primary motor for the origin of life. However, when one puts chemical constraints to this view, one realizes that self-replication cannot be achieved by one single molecule (it needs at least two), and generally for any workable chemical system one needs RNA local concentrations of at least femtomoles, which still means billions of identical copies of this compound (and larger concentrations of the mono-nucleotides). Do you agree with this statement, and with the corollary that even in such a hypothetical scenario, such amounts of RNA can only come from an active previous cellular metabolism?

9 On the genetic code

We have not yet reached a generally accepted view on how the genetic code might have originated. Models that have been presented are generally theoretical scenarios without reliable experimental proof. It appears indeed to be a complex machinery (see, for example, B. Davies, 1999 and 2002; M. Di Giulio, 1998, 2001, and 2003, and Di Giulio and Medugno, 1999). On the other hand, some recent studies may indicate that the genetic code may be as old as 4 billion years. Is there anything solid that we can actually state at the moment, on the origin of the genetic code? Does it make sense to invoke a genetic code prior to the onset of cellular life?

10 On early cells

The simplest cells on our Earth contain at least 500–600 genes, and more generally, several thousand. This observation elicits the question of whether this high complexity is really necessary for the simplest form of cellular life, also in view of the fact that early cells in the origin of life and evolution could not have been as complex as modern cells. This would imply that the first early cells were alive (although perhaps in a kind of "limping" life form) with a much smaller number of genes. This, in turn, results in the possibility of constructing models of early cells in the laboratory, displaying a kind of primitive cellular life (self-maintenance + self-reproduction + evolvability) based on a number of genes, which is one order of magnitude smaller than the present-day simplest cells. Say, a living cell with 30–40 genes. Do you believe that this is indeed a possibility?

11 On theoretical models of the origin of life

There are many theoretical models of the origin of life which are based on notions of complexity. Particularly well known, and very often cited, is, for example, the model developed by Stuart Kauffman on the spontaneous origin of catalytic networks. One can say, however, that this and other theoretical models have had very little influence on the experimentalists in the field of the origin of life, mostly due to the fact that they have never been observed in the real world of organic chemistry. Do you agree?

12 On artificial life

Artificial life deals with life as it might have been, and researchers involved in this field aim to create forms of life that are different from "our" life based on

DNA and proteins. Up until now these efforts have not been very successful, and it almost appears that there are no forms of life simpler than "our" life. Do you have data to counteract such a (rather negative) statement; and do you accept the notion that alternative forms of life (still within the general category of metabolism + self-reproduction + evolvability) may be possible with different chemical systems?

Selection of the open questions presented in the OQOL Workshop of San Sebastian – 2009

1 Contingency versus determinism in the origin of life/origin of proteins

Premise. The origin of life is often seen in terms of two basic, opposite schemes, determinism and contingency. Generally, the two principles work hand in hand, as each "choice" made by contingency must then comply with the natural laws and, in turn, contingency arises from a given thermodynamic asset. However, when we ask the basic question of whether the origin of life follows an obligatory deterministic pathway (absolute determinism), or whether it is due to the vagaries of contingency, the two views become again drastically opposite to each other. More precisely, according to the deterministic view (as represented most notably by Christian de Duve), the origin of life is seen as an event of very high probability: actually, it had to come out inevitably from the starting and boundary conditions (the so-called "gospel of inevitability"). The opposite view (advocated, for example, by Jacques Monod), implies that the origin of life was due to the occurrence of several independent factors, each of them perhaps not un-deterministic, whose simultaneous and unpredictable interaction led to successive events, up to the origin of life.

The question. Do you agree that the choice between these two extreme points of view cannot be done on a rational, scientific basis, and is instead for each scientist a matter of philosophical or religious belief? And, if you do not agree, which scientific arguments would you offer in favor of one or the other lines of thought?

2 Is life an emergent property?

Premise. Although emergence is a notion with many complex sides, the general view is that emergent properties are those novel properties that arise when parts or components assemble together into a higher hierarchic order – novel in the sense that they are not present in the parts or components. Most modern scientists would consider cellular life an emergent property, as the single components are *per sé* not living.

The question. Do you think there are sufficient data now to say that life is indeed an emergent property, arising from the interactions and self-organization of non-living parts? Or do you still see a kind of "vitalistic" flavor in the statements that define life as an emergent quality?

3 Heterotrophic versus autotrophic scenarios

Premise. One of the important questions relating to the origin of life problem today is the heterotrophy/autotrophy dichotomy. In an (extreme) heterotrophic scenario, the organic material supposed to have accumulated in a prebiotic world by high-energy processes (such as those of the Miller type in a primordial atmosphere, or by impact delivery to the Earth from extraterrestrial sources) is assumed to generate the critical self-organization processes culminating in life's origin. In sharp contrast, in an (extreme) autotrophic scenario, this kind of organic material is considered irrelevant and it is, instead, postulated that the substrates and intermediates of the chemical processes that organized themselves toward life were generated through synthetic processes within self-organized structures (e.g., from free-energy reach C-1- or C-2-organics, combined with strong inorganic reductants).

The question. Do you see strong chemical arguments in favor of the one or the other scenario? And which experiments would you do/suggest, in order to possibly clarify this dichotomy?

4 On the origin of catalytic cycles

Premise. In a prebiotic scenario, like that assumed by Stanley Miller in his famous experiments, once given the initial conditions, prebiotic reactions flow towards the most stable compounds, being ruled by thermodynamic control. With the 'free ticket' of thermodynamic control, however, chemical prebiotic evolution would not have gone very far. In fact, the question of the origin of life can be abstracted as the question of the origin of enzyme-like controlled catalysis (eventually leading to genetically controlled catalysis), giving rise to sequential metabolic cycles, as opposed to chemically equilibrated reaction pathways.

The question. How do you envisage the origin of sequentially catalyzed reactions in a prebiotic scenario? And can you provide facts or scientific arguments, not simply beliefs, about this critical point?

5 *Plausibility of the RNA world*

Premise. The origin of life on the basis of a prebiotic family of RNAs is still a preferred scenario. This assumes, however, that RNA is formed prebiotically, while the question 'what made RNA?' is still unanswered. One might conclude that the prebiotic synthesis of RNA is still a chimera from the scientific point of view.

The question. Do you share these arguments and rather bleak view? Which experiments or arguments can you suggest to counteract these objections against the "prebiotic" RNA world?

6 *Minimal (proto-) cellular world?*

Premise. The simplest cells on Earth contain at least 500–600 genes, and more generally a few thousand. This elicits the question, whether this high complexity is really necessary for cellular life, also in view of the fact that early cells, conceivably, could not have been so complex. Until now, however, the construction of chemical synthetic cells has not been successful, and the attempts to make DNA/Protein "minimal cells" with extant genes and enzymes are still based on systems with approximately a hundred genes. In other words, we are still missing the view of the early protocells—the primitive structures from which modern cells may have arisen.

The question. Do you see a way around the conundrum, that a living cell has to contain several dozens of independent specific macromolecular species and that, nevertheless, this complexity is not reasonably possible in prebiotic times? And/or: how do you envisage the structure of the simplest, early cells?

7 *Molecular differences in the (proto-) cellular world?*

Premise. The main building blocks of membranes in present-day prokaryotes are rather different from one another: in bacteria (like in eukaryotes), phospholipids are made of fatty acids, linked to the glycerol group (G3P) by ester bonds, whereas the phospholipids of archeabacteria are isoprenoid derivatives linked to glycerol (the stereoisomer, G1P) through ether bonds. And consider the extremely important role of hopanoids and steroids in modern bacterial and eukaryotic membranes.

The question. Do you think that these radical molecular differences show that the issue of compartments was not relevant until late stages in the origin of life? Or do you consider that compartmentalization was still an early landmark, phospholipid diversity being easily explained as a later evolutionary adaptation to extreme environments, for instance?

8 *Life as unity or confederacy*

Premise. Many sciences have conventionally (if implicitly) referred to "life" as a unitary concept, and all too often, we speak of "the origin of life" as if it were essentially one kind of unified event: a transition from "no life" to "life" on Earth. An alternative premise would be that life is a collection of coupled but still distinguishable subsystems, each with its own recognizable dynamics and requirements for stability. In that case the origin of life could involve a sequence of transitions understandable in somewhat independent terms. For instance, one could take separately the appearance of self-reproducing systems and the formation of vesicles, biogenesis of proteins different from setting up metabolic cycles, origin of reductive power different from prebiotic chemistry, etc. The degree of both contingency and of what some have called "irreducible complexity" in life will depend strongly on how tightly or loosely its subsystems are coupled.

  The question. Do you agree with this possible alternative view of life origin? And if yes, what is the proper way to apply the notions of interdependency versus subsystem independence, in the understanding of both the modern function of life and of its origin? Can a different understanding of the organization and stability of life today lead to better sequences of investigations of life's origins? If "life" is not a totally unitary notion, but rather a confederacy of coupled processes, can this recognition help us define the nature and process of origins of life in ways that do not lead to contradiction and confusion?

9 *Defining the very origin of life*

Premise. Defining life in a universal way is notoriously a difficult or impossible task, but also the notion of "origin of life" appears to be rather confusing. Some authors talk about origin of life at the level of the origin of low molecular weight compounds, obtainable either through hypothermal vents; or the pyrite reaction; or by Miller's type of processes. However, you can have all the low molecular weight compounds of this world, and you will never be able to make life, as life only arises at the level of specific macromolecular sequences like enzymes, DNA, and RNA.

  The question. Do you agree that we should have a critical review of the terminology of "origin of life," and, for example, not use this term at the level of low molecular weight compounds (where we have "prebiotic chemistry," or origin of reductive power. . .), and restrict it instead to the level of the biogenesis of specific macromolecules and their interactions?

10 On the entire field of the origin of life (final discussion)

Premise. The picture given until now suggests that we have not yet clarified the prebiotic synthesis of RNA, nor the biogenesis of macromolecular sequences, nor the development of the genetic code, nor the structure of the early cells. And probably several other points of ignorance could be added.

   The question. Would you agree with the statement that from the conceptual point of view the field has not progressed much since the early experiments of Stanley Miller? And why do you think this is so/not so?

Selection of the open questions presented in the OQOL Workshop of Leicester – 2012

Please give your own comment to the premises listed below.

1 Universality – What properties of life are universal?

The Premise: At one extreme, there are researchers who contemplate life not based on carbon (e.g., silicon), who consider solvents for life alternative to water (e.g., formamide), and who entertain a possibility of life without boundary structures (e.g., a "living ocean" on Europa). At the other extreme, a number of scientists claim that the severe constraints on life imposed by physics and chemistry enforce DNA as the necessary genetic polymer and the universality of energy transduction systems and core metabolism. The question is then, whether DNA is a universal property of life. The purpose of considering this question is to flesh out the scientific arguments for and against the universality of different properties of life, and perhaps to reach a consensus on the likely commonality between all forms of life. This question is highly topical in view of the rapidly increasing number of recent discoveries of extra-solar planets and the quest for finding habitable ones among them.

2 Evidence – What would convince us that we have answered the question, "How did life begin?"

The Premise: The transition from non-living to living systems involves, as a key step, the formation of a self-bounded physical system that contains interacting molecules. In many proposed scenarios, the complexity of such an event is often underestimated, and it is typically taken for granted, starting from the separated components. However, compared to the large amount of work done for understanding the emergence and the evolution of functional molecules and networks (ribozymes, catalytic peptides, simple metabolic cycles, self-replicating molecules, hyper-cycles, and autocatalytic sets, etc.), much less has been done for understanding the physical mechanisms underlying the assembly of primitive cell-like structures. In particular, little attention has been given to go beyond the general and simplistic sentence, ". . . and later became encapsulated in a membrane-based compartment."

3 Emergence – How does Biology emerge from Chemistry?

The Premise: The transformation of inanimate matter to complex life is traditionally divided into two stages. The first, abiogenesis, involves the conversion of non-living material to simplest life, and the second, the biological phase, is the stage on which Darwinian evolution began to operate. A key issue with regard to the problem of the origin of life is to shed light on the physicochemical relationship between these two stages. Processes are normally characterized by driving forces and mechanisms so the question of the origin of life will be greatly clarified if both abiogenesis and biological evolution can be characterized in this way.

4 Evolution – Is prebiotic evolution Darwinian?

The Premise: There is no doubt that present-day life is highly open-ended. For example, a small bacterium with genome size of about 500,000 bases has essentially infinite number of possible states. This allowed for life to evolve finding solutions to a variety of problems, giving rise to the high diversity of present-day life, e.g., the number of viable strains, subspecies, and individual variants of the aforementioned bacterium. In contrast, closed systems, exhibiting only relatively few possible states, have a much more limited capacity to undergo evolution. What is the relationship between the openness of a system and its evolvability? Is there an optimal 'openness' in this respect, and how could one effectively utilize this insight in computational and experimental models?

5 Reproduction and metabolism – How did metabolism and genetic replication get married?

The Premise: Is it conceivable that metabolism preceded genetic replication, i.e., that there was a form of homeostatic cellular life prior to the invention of genes, and of gene replication? Or do we have to necessarily assume that the two came out more or less simultaneously?

6 Chemistry – Do we understand enough about prebiotic chemistry to formulate meaningful hypotheses about the origin of life?

The Premise: Since Miller's 1953 demonstration of the synthesis of amino acids from reduced gases using an electric discharge, there has been a continual focus in origins of life research to find novel 'prebiotic' ways of making the molecules of modern biochemistry, including lipids, amino acids, nucleotides, and their polymers. Despite almost 60 years of research, many obstacles remain. One possible reason is that the first replicating systems did not use similar compounds in their biochemistry. Indeed, amino acids are actually a rather small percentage of the total number of compounds formed in an electric discharge; more than 95 percent of the products of such experiments remain unidentified. In carbonaceous chondrites, which are often presented as evidence that the "molecules of life" are ubiquitous in the cosmos, compounds found in modern biochemistry again make up a vanishingly small fraction (<<1 percent) of the total small molecule organic inventory. Given that we know so little about the available compound types, is a direct reconstruction of a modern cell from such components the best way, or even a reasonable way to approach the problem of the origin of life? If not, what would we need to know to be able to approach this problem more productively?

7 Where? – Where did life begin?

The Premise: Evolution described by the Darwinian theory may be depicted as a phylogenetic tree and a postulated root representing the last common ancestor and its precursors. It is sometimes believed that solving the question of the origins of life will need to bring information about the lapse of time separating the origin from the last common ancestor in the postulated root. Scenarios have been proposed based on possible environments assumed on the early Earth (ocean, vents, etc.). Alternatively, physicochemical principles can be used to reach a conclusion about how self-organization proceeds and chemistry can be helpful in bringing to light the related processes.

Selection of the open questions presented in the OQOL Workshop of Nara – 2014

1 How can we make ordered sequences of amino acids, or mononucleotides by prebiotic means?

The Premise: All life is based on proteins and nucleic acids, which are ordered sequences of units. Not to be confused with a simple random polymerization or copolymerization. The question "how can we make by prebiotic means ordered sequences of amino acids, or mononucleotides?" is in fact never asked in the modern research on the origin of life.

Methods for homo-polypeptides (chains containing only one type of residue) have been described, but they are in principle not valid for mixtures of different amino acids – as all rules of copolymeration teach us. Random polymerisation of mixtures of amino acids would produce a wild mixture of different chains, with almost zero probability to make two identical chains. Do you agree then that we do not know how to make macromolecular sequences in many identical copies under prebiotic conditions? Do we have to wait for this orderly sequence until the genetic code has been developed?

2 Why is the origin of life still a mystery?

The Premise: Why is the origin of life still a mystery? Yes, we all in science accept Oparin's idea that life on Earth originated from inanimate matter via a series of chemical steps of increasing molecular complexity and functionality. However, the turning point of non-life to life has never been put into one experimental set up. There are, of course, several hypotheses, and this plethora of ideas means already that we do not have a convincing one. The most popular is with the RNA-world prebiotic scenario, which has the advantage of providing on paper a theoretical series of imaginary events, each with an unimaginably small probability. Take, for instance, the prebiotic production of a self-replicating RNA and its eventual transformation into a catalyst for DNA and independently for protein synthesis: why should this happen, and what about the genetic code? Aside from the problem of experimental implementation, don't you think we lack (until now) the capability of intellectually conceiving how the turning point really happened?

3 Is molecular crowding critical for the beginning of life?

The Premise: Quite a dense concentration of macromolecules in cells: Is it an essential condition for origin of life? In living cells, macromolecules concentrations reach 30 percent of cellular mass, and membrane proteins dominate a half of the membrane surface. Are these highly dense concentrations of macromolecules required for the beginning of life? And if so, how was the concentration acquired before the origin of life? Or, was it a result during the evolutionary process?

4 Can artificial life or synthetic biology contribute to the origin of life?

The Premise: Artificial life deals with life as it might have been, and researchers involved in this field aim to create forms of life which are different from "our" life based on DNA and proteins. Up until now these efforts have not been very successful, and it almost appears that there are no forms of life simpler than "our" life. On the other hand, the currently developing field, synthetic biology, has been expected to give a hint to the origin of life study by means of filling up the gap between chemical and biological evolutions. Especially, the attempt to create an artificial cell by combing the minimal, but necessary, biomolecules could figure out the border between "living" and "non-living" states. Do you have any data that imply alternative forms of life (still within the general category of metabolism + self-reproduction + evolvability) with molecules different from the biological ones? Or, do you think that synthetic biology research can provide a model or theory for the origin of life?

5 Prior to genetic code: Is the notion of prebiotic cells conceivable?

The Premise: The simplest cells on our Earth contain at least 500–600 genes, and more generally several thousand. This observation elicits the question, whether this high complexity is really necessary for the simplest form of cellular life, also in view of the fact that early cells in the origin of life and evolution could not have been as complex as modern cells. This would imply that the first early cells were alive (although perhaps in a kind of "limping" life form) with a much smaller number of genes. This, in turn, results in the possibility of constructing models of early cells in the laboratory, displaying a kind of primitive cellular life (self-maintenance + self-reproduction + evolvability) based on a number of genes, which is one order of magnitude smaller than the present-day simplest cells. Say, a living cell with 30–40 genes. Do you believe that this is indeed a possibility?

6 What are the physical mechanisms underlying the assembly of primitive cell-like structures?

The Premise: The transition from non-living to living systems involves, as a key step, the formation of a self-bounded physical system that contains interacting molecules. In many proposed scenarios, the complexity of such an event is often underestimated, and it is typically taken for granted, starting from the separated components. However, compared to the large amount of work done for understanding the emergence and the evolution of functional molecules and networks (ribozymes, catalytic peptides, simple metabolic cycles, self-replicating molecules, hyper-cycles, and autocatalytic sets, etc.), much less has been done for understanding the physical mechanisms underlying the assembly of primitive cell-like structures. In particular, little attention has been given to go beyond the general and simplistic sentence, "... and later became encapsulated in a membrane-based compartment".

References

Abel, D. L. (2002). Is life reducible to complexity? In G. Palyi, C. Zucchi, and L. Caglioti, eds., *Fundamentals of Life*. Elsevier, pp. 57–72.

Achilles, T. and von Kiedrowski, G. (1993). A self-replicating system from three starting materials. *Angew. Chem.*, 32, 1198–11201.

Adamala, K. and Szostak J. W. (2013). Nonenzymatic template-directed RNA synthesis inside model protocells. *Science*, 342, 1098.

Addiscott, T. (2011). Emergence or self-organization? Look to the soil population. *Commun. Integr Biol.*, 4(4): 469–470.

Aguilar, A. (2009). *What is Death? A scientific, philosophical and theological exploration of life's end*. Ateneo Pontificio Apostolorum.

Ajikumar P. K, Xiao W. H., Tyo K. E., Wang Y., Simeon F., Leonard E., Mucha O., Phon T. H., Pfeifer B., Stephanopoulos G.(2010). Isoprenoid pathway optimization for taxol precursor overproduction in Escherichia coli. *Science*, 330(6000): 70–74.

Akanuma, S., Kigawa, T., and Yokoyama, S. (2002). Combinatorial mutagenesis to restricted amino acid usage in an enzyme to a reduced set. *Proc. Natl. Acad. Sci. USA*, 99, 13549–13553.

Alberts, B., Bray, D., Lewis, J., *et al.* (1989). *Molecular Biology of the Cell*, 2nd edn. New York: Garland Publications.

Alberts, B., Johnson, A., Lewis, J., *et al.* (2002). *Molecular Biology of the Cell*, 4th edn. New York: Garland Publications.

(2007). *Molecular Biology of the Cell*, 5th edn. New York: Garland Publications.

Alexander, S. (1920). *Space, Time, and Deity*. London: Mamillan.

Allison, A. C. and Gregoriadis, G. (1974). Liposomes as immunological adjuvants. *Nature*, 252, 252–258.

Ambartsumian, T. G., Adamian, S. Y., Petrosia, L. S., and Simonian, A. L. (1992). Incorporation of water-soluble enzymes glucose-oxidase and urate oxidase into phosphatidylcholine liposomes. *Biol. Membr.*, 5, 1878–1887.

Anastasi, C., Buchet, F. F., Crowe, M. A., Parkes, A. L., Powner, M. W., Smith, J. M., and Sutherland, J. D. (2007). RNA: prebiotic product, or biotic invention? *Chem. Biodivers.*, 4(4): 721–739.

Anderson, G. and Luisi, P. L. (1979). Papain-induced oligomerization of alpha amino acid esters. *Helv. Chim. Acta*, 62, 488–494.

Anella, F. (2011). *Structural and functional exploration of the RNA sequence space. Implications for the origin of life and biotechnology*. Ph.D. thesis, University Roma Tre.

Anella, F., Chiarabelli, C., De Lucrezia, D., and Luisi, P. L. (2011). Stability studies on random folded RNAs ("never born RNAs"), implications for the RNA world. *Chemistry & Biodiversity*, 8, 1422–1432.

Anfinsen, C. B. and Haber, E. (1961). Studies on the reduction and re-formation of protein disulfide bonds. *J. Biol. Chem*, 236, 1361–1363.

Anfinsen, C. B., Haber, E., Sela, M., and White, F. H. Jr. (1961). The kinetics of formation of native ribonuclease during oxidation of the reduced polypeptide chain. *PNAS*, 47, 1309–1314.

Angelova, M. I. and Dimitrov, D. S. (1988). A mechanism of liposome electro-formation. *Progr. Colloid Polymer. Sci.*, 76, 59–67.

Annaluru, N., Muller, H., Mitchell, L. A., *et al.* (2014). Total synthesis of a functional designer eukaryotic chromosome. *Science*, 344(6179): 55–58.

Annesini, M. C., Di Giulio, A., Di Marzio, L., Finazzi-Agrò, A., and Mossa, G. (1992). *J. Liposome Res.*, 2, 455–467.

Annesini, M. C., Di Giorgio, L., Di Marzio, L., *et al.* (1993). *J. Liposome Res.*, 3, 639–48.

Annesini, M. C., Di Marzio, L., Finazzi-Agrò, A., Serafino, A. L., and Mossa, G. (1994). Interaction of cationic phospholipid-vesicles with carbonic anhydrase. *Biochem. Mol. Biol. Int.*, 32, 87–94.

Apte, P. (2002). Vedantic view of life. In G. Palyi, C. Zucchi, and L. Caglioti, eds., *Fundamentals of Life*. Elsevier, pp. 497–502.

Archibald, J. M. (2009). The puzzle of plastid evolution. *Current Biology*, 19(2): R81–R88.

Arinin, E. I (2002). Essence of organic life in Russian orthodox and modern philosophical tradition: beyond functionalism and elementarism. In G. Palyi, C. Zucchi, and L. Caglioti, eds., *Fundamentals of Life*. Elsevier, pp. 503–516.

Ashkenasy, G., Jagasia, R., Yadav, M., and Ghadiri, M. R. (2004). Design of a directed molecular network. *Proc. Natl. Acad. Sci.*, 101, 10872–10877.

Atmanspacher, H. and Bishop, R. (2002). *Between Chance and Choice, Interdisciplinary Perspectives on Determinism*. Imprint Academic.

Atsumi, S. and Liao, J. C. (2008). Metabolic engineering for advanced biofuels production from Escherichia coli. *Curr Opin Biotechnol.*, 19(5): 414–419.

Avetisov, V. V. and Goldanskii, V. I. (1991). Homochirality and stereospecific activity: evolutionary aspects. *Biosystems*, 25(3): 141–149.

Ayala, F. J. (1983). Beyond Darwinism? The challenge of macroevolution to the synthetic theory of evolution. In P. D. Asquith and T. Nickles, eds., *PSA 1982: Proceedings of the 1982 Biennial Meeting of the Philosophy of Science Association Symposia*, Vol. 2, pp. 275–292.

Baas, N. A. (1994). Emergence, hierarchies, and hyperstructures. In C. G. Langton, ed., *Artificial Life III, Santa Fe Studies in the Science of Complexity*, Vol. XVII. Addison-Wesley, pp. 515–537.

Bachmann, P. A. (1991). Self-replicating micelles: aqueous micelles and enzymatically driven reactions in reverse micelles. *J. Am. Chem. Soc.*, 113, 8204–8209.

Bachmann, P. A., Walde, P., Luisi, P. L., and Lang, J. (1990). Self-replicating reverse micelles and chemical autopoiesis. *J. Am. Chem. Soc.*, 112, 8200–8201.

Bachmann, P. A., Luisi, P. L., and Lang, J. (1992). Autocatalytic self-replication of micelles as models for prebiotic structures. *Nature*, 357, 57–59.

Bada, J. L. (1997). Meteoritics – extraterrestrial handedness? *Science*, 275, 942–943.

Bada, J. L. and Lazcano, A. (2002). Some like it hot, but not the first biomolecules. *Science*, 296, 1982–1983.

(2003). Prebiotic soup – revisiting the Miller experiment. *Science*, 300, 745–746.

Baeza, I., Ibáñez, M., Santiago, J. C., *et al.* (1990). Diffusion of Mn^{2+} ions into liposomes mediated by phosphatidate and monitored by the activation of an encapsulated enzymatic system. *J. Mol. Evol.*, 31, 453–461.

Baeza, I., Wong, C., Mondragón, R., *et al.* (1994). Transbilayer diffusion of divalent cations into liposomes mediated by lipidic particles of phosphatidate. *J. Mol. Evol.*, 39, 560–568.

Bain, A. (1870). *Logic*, Books II and III. Longmans, Green & Co.

Bak, P., Tang, C., and Wiesenfeld, K. (1987). Self-organized criticality: an explanation of 1/f noise. *Physical Review Letters*, 59(4): 381–384.

(1988). Self-organized criticality. *Physical Rev. A.*, 38, 364–374.

Ballard, D. G. H. and Bamford, C. H. (1956). Studies in polymerization. X. "The chain-effect." *R. Soc. Lond. A*, 236 (1206): 384–396.

Barabási, A.-L. and Réka, A. (1999). Emergence of scaling in random networks. *Science*, 286 (5439): 509–512.

Barbaric, S. and Luisi, P. L. (1981). Micellar solubilization of biopolymers in organic solvents. 5. Activity and conformation of α-chymorypsin in isooctane-AOT reverse micelles. *J. Am. Chem. Soc.*, 103, 4239–4244.

Barrow, J. D. (2001). Cosmology, life and the anthropic principle. *Ann. NY Acad. Sci.*, 950, 139–153.

Barrow, J. D. and Tipler, F. J. (1986). *The Anthropic Cosmological Principle*. Oxford University Press.

(1988). Action principles in nature. *Nature*, 331, 31–34.

Bartel, D. P. and Szostak, J. W. (1993). Isolation of new ribozymes from a large pool of random sequences. *Science*, 261, 1411–1418.

Bashan, A., Belousoff, M. J., Davidovich, C., and Yonath, A. (2010). Linking the RNA world to modern life: The proto-ribosome conception. *Orig. Life Evol. Biosph.*, 40, 425–429.

Bedau, M. A. (1997). Weak emergence. In J. Tomberlin, ed., *Philosophical Perspectives: Mind, Causation and World*, Vol. 11. Malden, MA: Blackwell, pp. 375–399.

Beer, S. (1980). Preface. In H. Maturana and F. J. Varela, *Autopoiesis and Cognition* (*see infra*).

Bell, E. A., Boehnke, P., Harrison, T. M., and Mao, W. L. (2015). Potentially biogenic carbon preserved in a 4.1 billion-year-old zircon. *Proc. Natl. Acad. Sci. USA*, 112(47): 14518–14521.

Belousoff, M. J., Davidovich, C., Bashan, A., and Yonath, A. (2010). On the development towards the modern world: a plausible role of uncoded peptides in the RNA world. In K. Ruiz-Mirazo and P. L. Luisi, eds., *Origins of life and evolution of biospheres*, 40 (Special Issue 4–5): pp. 415–419.

Ben Jacob, E., Becker, I., Shapira, Y., and Levine, H. (2004). Bacterial linguistic communication and social intelligence. *Trends Microbiol.*, 12, 366–72.

Benner, S. A. and Sismour, A. M. (2005). Synthetic biology. *Nature Rev. Gen.*, 6, 524–45.

Berclaz, N., Blöchliger, E., Müller, M., and Luisi, P. L. (2001a). Matrix effect of vesicle formation as investigated by cryotransmission electron microscopy. *J. Phys. Chem. B*, 105, 1065–1071.

Berclaz, N., Müller, M., Walde, P., and Luisi, P. L. (2001b). Growth and transformation of vesicles studied by ferritin labeling and cryotransmission electron microscopy. *J. Phys. Chem. B.*, 105, 1056–1064.

Bernal, J. D. (1951). *The Physical Basis of Life*. Routledge & Paul.

(1965). Molecular structure, biochemical function, and evolution. In T. H. Waterman and H. J. Morowitz, eds., *Theoretical and Mathematical Biology*. Blaisdell.

(1967). *The Origin of Life*. World Publishing Company.

(1971). *Der Ursprung des Lebens*. Editions Rencontre.

Bernard, C. (1865). *Introduction to the Study of Experimental Medicine*. Translated by H. C. Greene,1927. Henry Schuman.

Bernhardt, H. S. (2012). The RNA world hypothesis: the worst theory of the early evolution of life (except for all the others). *Biol Direct*, 7, 23.

Berti, D., Baglioni, P., Bonaccio, S., Barsacchi-Bo, G., and Luisi, P. L. (1998). Base complementarity and nucleoside recognition in phosphatidylnucleoside vesicles. *J. Phys. Chem. B*, 102, 303–338.

Berti, D., Luisi, P. L., and Baglioni, P. (2000). Molecular recognition in supramolecular structures formed by phosphatidylnucleosides-based amphiphiles. *Colloids Surf. A*, 167, 95–103.

Bianucci, M., Maestro, M., and Walde, P. (1990). Bell-shaped curves of the enzyme-activity in reverse micelles – a simplified model for hydrolytic reactions. *Chem. Phys.*, 141, 273–283.

Biebricher, K., Eigen, M., and Luce, R. (1981). Kinetic analysis of template, instructed and de novo RNA synthesis by Qbeta replicase. *J. Mol. Biol.*, 148, 391–410.

Birdi, K. S. (1999). *Self-Assembly Monolayer Structures of Lipids and Macromolecules at Interfaces*. Plenum Press.

Biron, J.-Ph. and Pascal, R. (2004). Amino acid N-Carboxyanhydrides: activated peptide monomers behaving as phosphate-activating agents in aqueous solution. *J. Am. Chem. Soc.*, Aug., 126(30): 9198–9199.

Bissel, R. A., Cordova, E., Kaifer, A. E., and Stoddart, J. F. (1994). A chemically and electrochemically switchable molecular shuttle. *Nature*, 369, 133.

Bitbol, M. and Luisi, P. L. (2004). Autopoiesis with or without cognition: defining life at its edge. *J. Royal. Soc. Interface*, 1, 99–107.

   (2011). Science and the self-referentiality of consciousness. In *Conciousness and the Universe*. Cambridge, MA: Cosmology Science.

Blacka, R. A., Blosser, M. C., Stottrup, B. L., *et al.* (2013). Nucleobases bind to and stabilize aggregates of a prebiotic amphiphile, providing a viable mechanism for the emergence of protocells. *PNAS*, 110 (3): 13272–13276.

Blain, J. C. and Szostak, J. W. (2014). Progress toward synthetic cells. *Annu. Rev. Biochem.*, 83, 11.1–11.26.

Blain, J. C., Ricardo, A., and Szostak, J. W. (2014). Synthesis and nonenzymatic template-directed polymerization of 2'-amino-2'-deoxythreose nucleotides. *J. Am. Chem Soc.*, 136, 2033–2039.

Blocher, M., Walde, P., and Dunn, I. J. (1999). Modeling of enzymatic reactions in vesicles: the case of alpha-chymotrypsin. *J. Biotechnol. Bioeng.*, 62, 36–43.

Blocher, M., Liu, D., and Luisi, P. L. (2000). Liposome-assisted selective polycondensation of α-amino acids and peptides: the case of charged liposomes. *Macromolecules*, 33, 5787–5796.

Blocher, M., Hitz, T., and Luisi, P. L. (2001). Stereoselectivity in the oligomerization of racemic Tryptophan N-Carboxyanhydride (NCA-Trp) as determined by isotopic labelling and mass spectrometry. *Helv. Chim. Acta*, 84, 842–848.

Blöchliger, E., Blocher, M., Walde, P., and Luisi, P. L. (1998). Matrix effect in the size distribution of fatty acid vesicles. *J. Phys. Chem.*, 102, 10383–10390.

Böhler, C., Bannwarth, W., and Luisi, P. L. (1993). Self-replication of oligonucleotides in reverse micelles. *Helv. Chim. Acta*, 76, 2313–2320.

Boicelli, C. A., Conti, F., Giomini, M., and Giuliani, A. M. (1982). Interactions of small molecules with phospholipids in inverted micelles. *Chem. Phys. Lett.*, 89, 490–496.

Boiteau, L., Plasson, R., Collet, H., *et al.* (2002). Molecular origin of life: when chemistry became cyclic. The primary pump, a model for prebiotic emergence and evolution of petides. In G. Palyi, C. Zucchi, and L. Caglioti, eds., *Fundamentals of Life*. Elsevier, pp. 211–218.

Bolli, M., Micura, R., and Eschenmoser, A. (1997a). Pyranosyl-RNA: chiroselective self-assembly of base sequences by ligative oligomerization of tetranucleotide-2',3'-cyclophosphates (with a commentary concerning the origin of biomolecular homochirality). *Chem. Biol.*, 4, 309–320.

Bolli, M., Micura, R., Pitsch, S., and Eschenmoser, A. (1997b). Pyranosyl-RNA: further observations on replication. *Helv. Chim. Acta*, 80, 1901–1951.

Bonaccio, S., Walde, P., and Luisi, P. L. (1994a). Liposomes containing purine and pyrimidine bases: stable unilamellar liposomes from phosphatidyl nucleosides. *J. Phys. Chem.*, 98, 6661–6663.

Bonaccio, S., Cescato, C., Walde, P., and Luisi, P. L. (1994b). Self-production of supra-molecular structures. In G. R. Fleischaker *et al.*, eds., *Liposomes from Lipidonucleotides and from Lipidopeptides*. Kluwer Academic, pp. 225–259.

Bonaccio, S., Wessicken, M., Berti, D., Walde, P., and Luisi, P. L. (1996). Relation between the molecular structure of phosphatidyl nucleosides and the morphology of their supramolecular and mesoscopic aggregates. *Langmuir*, 12, 4976–4978.

Bonaccio, S., Capitani, D., Segre, A. L., Walde, P., and Luisi, P. L. (1997). Liposomes from phosphatidyl nucleosides: an NMR investigation. *Langmuir*, 13, 1952–1956.

Böttcher, B., Lucken, U., and Graber, P. (1995). The structure of the H^+-ATPase from chloroplasts by electron cryomicroscopy. *Biochem. Soc. Trans.*, 23, 780–785.

Bourgine, P. and Stewart, J. (2004). Autopoiesis and cognition. *Artificial Life*, 10(3): 327–345.

Bozic, B. and Svetina, S. (2004). A relationship between membrane properties forms the basis of a selectivity mechanism for vesicle self-reproduction. *Eur. Bioph. J.*, 33, 565–571.

Brack, A. (ed.) (1998). *The Molecular Origin of Life*. Cambridge University Press.

Brasier, M. D., Green, O. R., Jephcoat, A. P., *et al.* (2002). Questioning the evidence for Earth's oldest fossils. *Nature*, 416, 76–77.

Briggs, T. and Rauscher, W. (1973). An oscillating iodine clock. *J. Chem. Educ.*, 50, 496.

Britt, R. R. (2000). Are we all aliens? The new case for panspermia. http://www.space .com.

Broad, C. D. (1925). *The Mind and Its Place in Nature*. Routledge and Kegan.

Brunner, J., Graham, D. E., Hauser, H., and Semenza, G. (1980). Ion and sugar perme-abilities of lecithin bilayers: Comparison of curved and planar bilayers. *J. Membr. Biol.*, 57, 133–141.

Buchet, F. F. and Sutherland, J. D. (2006). Synthesis of pyrimidic nucleotides under potentially prebiotic conditions. *Origins of Life and Evolution of the Biosphere*, 36, 259.

Bucknall, D. G. and Anderson, H. L. (2003). Polymers get organized. *Science*, 302, 1904–1905.

Buhse, T., Nagarajan, R., Lavabre, D., and Micheau, J. C. (1997). Phase-transfer model for the dynamics of "micellar autocatalysis." *J. Phys. Chem. A*, 101, 3910–3917.

Buhse, T., Lavabre, D., Nagarajan, R., and Micheau, J. C. (1998). Origin of autocatalysis in the biphasic alkaline hydrolysis of C-4 to C-8 ethyl alkanoates. *J. Phys. Chem. A.*, 102, 10552–10559.

Bujdak, J., Slosiarikova, H., Texler, N., Schwendinger, M., and Rode, B. M. (1994). On the possible role of montmorillonites in prebiotic peptide formation. *Monats. Chem.*, 125, 1033–1039.

Bujdak, J., Eder, A., Yongyai, Y., Faybikova, K., and Rode, B. M. (1995). Peptide chain elongation: a possible role of montmorillonite in prebiotic synthesis of protein precursors. *Orig. Life Evol. Biosph.*, 5, 431–441.

Burmeister, J. (1998). Self-replication and autocatalysis. In A. Brack, ed., *The Molecular Origin of Life*. Cambridge University Press, pp. 295–310.

Butler, P. J. (1999). Self-assembly of tobacco mosaic virus: the role of an intermediate aggregate in generating both specificity and speed. *Phil. Trans. R. Soc. Lond.*, 354 (1383): 537–550.

Caffrey, M. (2015). A comprehensive review of the lipid cubic phase or in meso method for crystallizing membrane and soluble proteins and complexes. *Acta Crystallogr. F. Struct. Biol. Commun.*, 71, 3–18.

Cairns-Smith, A. G. (1977). Takeover mechanisms and early biochemical evolution. *Biosystems*, 9, 105–109.

(1978). Precambrian solution photochemistry, inverse segregation, and banded iron formations. *Nature*, 276, 808–809.

(1982). Genetic Takeover and the Mineral Origins of Life. Cambridge University Press.

(1985). *Seven Clues to the Origin of Life*. Cambridge University Press.

(1990). *Seven Clues to the Origin of Life*, 2nd edn. Cambridge University Press.

(2008). Chemistry and the missing era of evolution. *Chemistry*, 14(13): 3830–3839.

Cairns-Smith, A. G. and Walker, G. L. (1974). Primitive metabolism. *Curr. Mod. Biol.*, 5 (4): 173–186.

Cairns-Smith, A. G., Hall, A. J., and Russell, M. J. (1992). Mineral theories of the origin of life and an iron sulphide example. *Orig. Life Evol. Biosph.*, 22, 161–180.

Calderone, C. T. and Liu, D. R. (2004). Nucleic acid-templated synthesis as a model system for ancient translation. *Curr. Opin. Chem. Biol.*, 8, 645–653.

Callaway, E. (2014). Scientists Create First Living Organism with "Artificial" DNA. *Nature News, Huffington Post*.

Cameron, D. E., Caleb, J., Bashor, J., and Collins, J. (2014). A brief history of synthetic biology. *Nature Reviews Microbiology*, 12, 381–390.

Capra, F. (2002). *The Hidden Connections*. Harper Collins.

Capra, F. and Luisi, P. L. (2014). *The Systems View of Life: A Unifying Vision*. Cambridge University Press.

Caretta, N. (2005). Ipotesi sull'origine della vita: la chimica proteica prebiotica basata su un insieme ridotto di amminoacidi. Thesis 2004/2005, Department of Biology, University Roma Tre.

Carey, M. V. and Small, D. M. (1972). Micelle formation by bile salts. Physical-chemical and thermodynamic considerations. *Arch. Intern. Med.*, 130, 506–527.

Carr, B. (2001). Life, the cosmos and everything. *Phys. World*, 14, 23–25.

Carr, B., ed. (2007). *Universe or Multiverse?* Cambridge University Press.

Carrara, P., Stano, P., and Luisi, P. L. (2012). Giant vesicle "colonies": a model for primitive cell communities. *Chembiochem*, 13, 1497–1502.

Carrera, J., Elena, S. F., and Jaramillo, A. (2012). Computational design of genomic transcriptional networks with adaptation to varying environments. *Proc. Natl. Acad. Sci. USA*, 109, 15277–15282.

Caschera, F. and Noireaux, V. (2014) Integration of biological parts toward the synthesis of a minimal cell. *Current Opinion in Chemical Biology*, 22, 85–91.

Caschera, F., Stano, P., and Luisi, P. L. (2010). Reactivity and fusion between cationic vesicles and fatty acid anionic vesicles. *J. Colloid Interface Sci*, 345, 561–565.

Caschera, F., Sunami, T., Matsuura, T., *et al.* (2011). Programmed vesicle fusion triggers gene expression. *Langmuir*, 27, 13082–13090.

Caselli, M., Maestro, M., and Morea, G. (1988). A simplified model for protein inclusion in reverse micelles. *SANS measurements as a control test. Biotech. Prog.*, 4, 102–106.

Cech, T. R. (2011). The RNA worlds in context. *Cold Spring Harb Perspect Biol.* doi:10.1101/cshperspect.a006742.

Cello, J., Paul, A. V., and Wimmer, E. (2002). Chemical synthesis of poliovirus cDNA: generation of infectious virus in the absence of natural template. *Science*, 297, 1016–1018.

Celovsky, V. and Bordusa, F. (2000). Protease-catalyzed fragment condensation via substrate mimetic strategy: a useful combination of solid-phase peptide synthesis with enzymatic methods. *J. Pept. Res.*, 55, 325–329.

Cevc, G. and Blume, G. (1992). Lipid vesicles penetrate into intact skin owing to the transdermal osmotic gradients and hydration force. *Biochim Biophys Acta.*, 1104(1): 226–232.

Cevc, G. and Marsh, D., eds. (1987). *Phospholipid Bilayers – Physical Principles and Models*, Vol. 5. New York: John Wiley & Sons.

Chakrabarti, A. C., Breaker, R. R., Joye, G. F., and Deamer, D. W. (1994). Production of RNA by a polymerase protein encapsulated within phospholipid vesicles. *J. Mol. Evol.*, 39, 555–559.

Chalmers, D. (1995). Facing up to the problem of consciousness. *Journal of Consciousness Studies*, 2(3): 200–219.

Chan, L. Y., Kosuri, S., and Endy, D. (September 13, 2005). Refactoring bacteriophage T7. *Mol. Syst. Biol.*, 1: 18.

Chang, M. C. Y. and Keasling, J. D. (2006). Production of isoprenoid pharmaceuticals by engineered microbes. *Nature Chemical Biology*, 2, 674–681.

Chapman, K. B. and Szostak, J. W. (1995). Isolation of a ribozyme with 5'-5' ligase activity. *Chem. Biol.*, 2, 325 433.

Chen, I. A. and Szostak, J. W. (2004). A kinetic study of the growth of fatty acid vesicles. *Bioph. J.*, 87, 988–998.

Chen, I. A., Roberts, R. W., and Szostak, J. W. (2004). The emergence of competition between model protocells. *Science*, 305, 1474–1476.

Chen, I. A., Salehi-Ashtiani, K., and Szostak, J. W. (2005). RNA catalysis in model protocell vesicles. *J. Am. Chem. Soc.*, 127(38): 13213–13219.

Chen, Y., Ma, P., and Gui, S. (2014). Cubic and hexagonal liquid crystals as drug delivery systems. *Biomed Res Int*. doi:10.1155/2014/815981.

Cheng, J. (2012). Synthesis of polypeptides by ring-opening polymerization of α-aminoacids. *Top Curr. Chem.*, 310: 1–26.

Cheng, Z. and Luisi, P. L. (2003). Coexistence and mutual competition of vesicles with different size distributions. *J. Phys. Chem. B*, 107(39): 10940–10945.

Chessari, S., Thomas, R., Polticelli, F., and Luisi P. L. (2006). The production of *de novo* folded proteins by a stepwise chain elongation: a model for prebiotic chemical evolution of macromolecular sequences. *Chemistry & Biodiversity*, 3(11): 1202–1210.

Chiarabelli, C. and Luisi, P. L. (2014). Chemical synthetic biology. *Science Progress*, 97, 48–61.

Chiarabelli, C., Vrijbloed, J. W., Thomas, R. M., and Luisi, P. L. (2006a). Investigation of *de novo* totally random biosequences, Part I: a general method for in vitro selection of folded domains from a random polypeptide library displayed on phage. *Chemistry and Biodiversity*, 3, 827–839.

Chiarabelli, C., Vrijbloed, J. W, De Lucrezia, D., et al. (2006b). Investigation of *de novo* totally random biosequences, Part II: on the folding frequency in a totally random library of *de novo* proteins obtained by phage display. *Chemistry and Biodiversity*, 3, 840–859.

Chiarabelli, C., Stano, P., and Luisi, P. L. (2009). Chemical approaches to synthetic biology. *Curr. Opin. Biotech.*, 20, 492–497.

Chiarabelli, C., Stano, P., Anella, F., Carrara, P., and Luisi, P. L. (2012). Approaches to chemical synthetic biology. *FEBS Letters*, 586, 2138–2145.

Chiarabelli, C., Stano P., and Luisi, P. L. (2013). Chemical synthetic biology: a mini-review. *Frontiers in Microbiotechnology, Ecotoxicology and Bioremediation*, 4, 285. doi:10.3389/fmicb.2013.00285.

Christidis, T. (2002). Probabilistic causality and irreversibility: Heraclitus and Prigogine. In H. Atmanspacher and R. Bishop, eds., *Between Chance and Choice*. Academic Imprint.

Chungcharoenwattana, S. and Ueno, M. (2004). Size control of mixed egg yolk phosphatidylcholine (EggPC)/oleate. *Chem. Pharm. Bull.*, 52, 1058–1062.

(2005a). New vesicle formation upon oleate addition to preformed vesicles. *Chem. Pharm. Bull.*, 53, 260–262.

(2005b). Effect of preformed egg phosphatidylcholine vesicles on spontaneous vesiculation of oleate micelles. *Colloid Pol. Sci.*, 283, 1180–1189.

Chopra, P. and Kamma, A. (2006). Engineering life through Synthetic Biology. *In Silico Biol.*, 6(5): 401–410.

Church, G. M., Elowitz, M. B., Smolke C. D., Voigt, C. A., and Weiss, R. (2014). Realizing the potential of synthetic biology. *Nature Reviews Molecular Cell Biology*, 15, 289–294.

Chyba, C. F. and Sagan, C. (1992). Endogenous production, exogenous delivery and impact-shock synthesis of organic molelcules: an inventory for the origin of life. *Nature*, 355, 125–132.

Chyba, F. and McDonald, G. D. (1995). The origin of life in the solar system: current issues. *Ann. Rev. Earth Planet. Sci.*, 23, 215–249.

Cisar, J. O., Xu, D. Q., Thompson, J., Swaim, W., Hu, L., Kopecko, D. J. (2000). An alternative interpretation of nanobacteria-induced biomineralization. *Proc Natl Acad Sci USA*, 97, 11511–11515.

Cistola, D. P., Hamilton, J. A., Jackson, D., and Small, D. M. (1988). Ionization and phase-behavior of fatty-acids in water. *Application of the Gibbs phase rule. Biochemistry*, 27, 1881–1888.

Cohlberg, J. A. and Nomura, M. (1976). Reconstitution of Bacillus stearothermophilus 50S ribosomal subunits from purified molecular components. *The Journal of Biological Chemistry*, 251, 209–221.

Coleman, P. (2007). Frontier at your fingertips. *Nature*, 446, 379–385.

Collet, H., Bied, C., Mion, L., and Commeyras, A. (2010). Chem Inform Abstract: A New Simple and Quantitative Synthesis of α-Amino Acid-N- carboxyanhydrides (Oxazolidine-2,5-diones). *Chem. Inform.*, 28(15).

Commeyras, A., Collet, H., Boiteau, L., *et al.* (2002). Prebiotic synthesis of sequential peptides on the Hadean Beach by a molecular engine working with nitrogen oxides as energy sources. *Polymer International*, 51, 661–665.

Commeyras, A., Boiteau, L., Vandenabeele-Trambouze, O., and Selsis, F. (2005). Peptide emergence, evolution and selection on the primitive Earth. In M. Gargaud, B. Barbier, H. Martin and J. Reisse, eds., Lectures in Astrobiology – Vol. I: From Prebiotic Chemistry to the Origins of Life on Earth. Springer-Verlag (Part II, Chap. 4), pp. 517–545.

Conway-Morris, S. (2003). *Life's Solution, Inevitable Humans in a Lonely Universe*. Cambridge University Press.

Cooper, G. W., Onwo, W. M., and Cronin, J. R. (1992). Alkyl phosphonic acids and sulfonic acids in the Murchison meteorite. *Geochim. cosmochim. Acta.*, 56, 4109–4115.

Cooper, G., Kimmich, N., Belisle, W., Sarinana, J., Brabham, K., and Garrel, L. (2001). Carbonaceous meteorites as a source of sugar-related organic compounds for the early Earth. *Nature*, 414, 879–883.

Cooper, S. J. (2008). From Claude Bernard to Walter Cannon. Emergence of the concept of homeostasis. *Appetite*, 51(3): 419–427.

Corliss, J. B., Baross, J. A., and Hoffman, S. E. (1981). An hypothesis concerning the relationship between submarine hot springs and the origin of life. *Oceanologica acta*, 4, *Suppl.*, 59–69.

Coveney, P. and Highfield, R. (1990). *The Arrow of Time*. W. H. Allen.

Crans, D. C. and Levinger, N. E. (2012). The Conundrum of pH in water nanodroplets: sensing pH in reverse micelle water pools. *Acc. Chem. Res.*, 45, 1637–1645.

Crick, F. (1966). *Of Molecules and Men*. University of Washington Press.

Crick, F. H. C. (1968). The origin of the genetic code. *J Mol Biol.*, 38, 367–379.

(1980). *The Astonishing Hypothesis. The Search of the Soul from a Chemical Perspective*. Scribner.

Cronin, J. R. and Pizzarello, S. (1997). Enantiomeric excesses in meteoritic amino acids. *Science*, 275, 951–955.

Crusats, J., Claret, J., Díez-Pérez, I., *et al.* (2003). Chiral shape and enantioselective growth of colloidal particles of self-assembled meso-tetra(phenyl and 4-sulfonato-phenyl) porphyrins. *Chem. Commun.*, 13, 1588–1589.

Cullis, P. R., Hope, M. J., Bally, M. B., *et al.* (1987). Liposomes as pharmaceuticals. In M. J. Ostro ed., *Liposomes. From Biophysics to Therapeutics*. Marcel Dekker, pp. 39–72.

D'Aguanno, E., Altamura, E., Mavelli, F., *et al.* (2015). Physical routes to primitive cells: An experimental model based on the spontaneous entrapment of enzymes inside micrometer-sized liposomes. *MDPI Life*, 5, 969–996.

Damasio, A. R. (1999). *The Feeling of What Happens*. Harcourt.

Damer, B. and Deamer, D. (2015). Coupled phases and combinatorial selection in fluctuating hydrothermal pools: a scenario to guide experimental approaches to the origin of cellular life. *Life*, 5, 872–887.

Damiano, L. (2006). *L'unità in Dialogo: Autoorganizzazione, Autopoiesi, Enazione e Relazione Cognitiva*. Doctoral Thesis, University of Bergamo Press.

Damiano, L. and Luisi, P. L. (2010). Verso una ridefinizione autopoietica della vita. *Orig. Vita Evol. Biosph.*, 40, 145–149.

Davidson, A. R. and Sauer, R. T. (1994). Folded proteins occur frequently in libraries of random amino acid sequence. *Proc. Natl. Acad. Sci. USA*, 91, 2146–2150.

Davidson, A. R., Lumb, K. J., and Sauer, R. T. (1995). Cooperatively folded proteins in random sequence libraries. *Nature Structural Biology*, 2, 856–864.

Davies, B. (1999). Evolution of the genetic code. *Progr. Biophys. Mol. Biol.*, 72, 157–243.
(2002) Molecular evolution before the origin of species, Progr. *Biophys. Mol. Biol.*, 79, 77–133.

Davies, P. (1999). *The Fifth Miracle: The Search for the Origin and Meaning of Life*. Simon & Schuster.
(2007). *Cosmic Jackpot*, Houghton Mifflin; also appeared as *The Goldilocks Enigma: Why is the Universe Just Right for Life?* Allen Lane, 2006.

Dawkins, R. (1990). *The Blind Watchmaker: Why the Evidence of Evolution Reveals a Universe without Design*. Penguin Books.
(2002). *How Life Began: The Genesis of Life on Earth*. Cambridge, MA: Foundation for New Directions.

Day, W. (2002). *How Life Began: the Genesis of Life on Earth*. Cambridge, MA: Foundation for New Directions.

Deamer, D. W. (1985). Boundary structures are formed by organic components of the Murchison carbonaceous chondrite. *Nature*, 317, 792–794.
(1998). Possible starts for primitive life. In A. Brack, ed., *The Molecular Origins of Life*. Cambridge University Press.

Deamer, D. W. and Pashley, R. M. (1989). Amphiphilic components of the Murchison carbonaceous chondrite: surface properties and membrane formation. *Orig. Life Evol. Biosph.*, 19, 21–38.

Deamer, D. W., Harang-Mahon, E., and Bosco, G. (1994). Self-assembly and function of primitive membrane structures. In S. Bengtson, ed., *Early Life on Earth. Nobel Symposium No. 84*. Columbia University Press, pp. 107–123.

Decher, G. (1997). Fuzzy nano-assemblies: toward layered polymeric multicomposites. *Science*, 277, 1232–1237.

Decker, P., Schweer, H., and Pohlmann, R. (1982). Identification of formose sugars, presumable prebiotic metabolites, using capillary gas chromatography/gas chromatography-mass spectrometry of n-butoxime trifluoroacetates on OV-225. *J. Chromatogr.*, 225, 281–291.

de Duve, C. (1991). *Blueprint for a Cell: The Nature and the Origin of Life*. Neil Patterson Publishers.

(2002). *Life Evolving: Molecules, Mind and Meaning*. Oxford University Press.

(2005). *Singularities*. Cambridge University Press.

de Duve, C. and Miller, S. (1991). Two-dimensional life? *Proc. Natl. Acad. Sci.*, 88, 10014–10017.

De Kruijff, B., Cullis, P. R., and Verkleij, A. J. (1980). Non-bilayer lipid structures in model and biological membranes. *Trends Bioch. Sci.*, 5, 79–81.

De Lucrezia, D., Franchi, M., Chiarabelli, C., Gallori, E., and Luisi, P. L. (2006a). Investigation of *de novo* totally random biosequences, Part III: RNA Foster: a novel assay to investigate RNA folding structural properties. *Chemistry and Biodiversity*, 3, 860–868.

De Lucrezia, D., Franchi, M., Chiarabelli, C., Gallori, E., and Luisi, P. L. (2006b). Investigation of *de novo* totally random biosequences, Part IV: folding properties of *de novo*, totally random RNAs. *Chemistry and Biodiversity*, 3, 869–877.

De Napoli, M., Nardis, S., and Paolesse, R. (2004). Hierarchical porphyrin self-assembly in aqueous solution. *J. Am. Chem. Soc.*, 126, 5934–5935.

de Souza, T. P., Stano, P., and Luisi, P. L. (2009). The minimal size of liposome-based model cells brings about a remarkably enhanced entrapment and protein synthesis. *Chembiochem Eur. J. Chem. Biol.*, 10, 1056–1063.

de Souza, T. P., Steiniger, F., Stano, P., Fahr, A., and Luisi, P. L. (2011). Spontaneous crowding of ribosomes and proteins inside vesicles: a possible mechanism for the origin of cell metabolism. *Chem. Biochem.*, 12, 2325–2330.

de Souza, T. P., Stano, P., Steiniger, F., *et al.* (2012). Encapsulation of ferritin, ribosomes, and ribo-peptidic complexes inside liposomes: insights into the origin of metabolism. *Orig. Life Evol. Biospheres*, 42, 421–428.

Diedrich, G., Spahn, C. M. T., Stelzl, U., *et al.* (2000). Ribosomal protein L2 is involved in the association of the ribosomal subunits, tRNA binding to A and P sites and peptidyl. *Embo Journal*, 19, 5241–5250.

Diener, T. O. (1971). Potato spindle tuber "virus." IV. A replicating, low molecular weight RNA. *Virology*, 45(2): 411–428.

Di Giulio, M. (1998). Reflections on the origin of the genetic code: a hypothesis. *J. Theor. Biol.*, 191, 191–196.

(2001). The non universality of the genetic code: the universal ancestor was a progenote. *J. Theor. Biol.*, 209, 345–349.

(2003). The early phases of the genetic code origin: conjecture on the evolution of coded catalysis. *Orig. Life Evol. Biosph.*, 33, 479–489.

Di Giulio, M. and Medugno, M. (1999). Physicochemical optimization in the genetic code origin as the number of codified amino acids increase. *J. Mol. Evol.*, 49, 1–10.

Dodevski, I., Nucci, N. V., Valentine, K. G., *et al.* (2014). Optimized reverse micelle surfactant system for high-resolution NMR spectroscopy of encapsulated proteins and nucleic acids dissolved in low viscosity fluids. *J. Am. Chem. Soc.*, 136, 3465–3474.

Doi, N., Kakukawa, K., and Yanagawa, H. (2005). High solubility of random-sequence proteins consisting of five kinds of primitive amino acids. *Protein Engineering, Design & Selection*, 18, 279–84.

Dolgin, E. (2015). Synthetic biology: Safety boost for GM organisms. *Nature*, 517 (7535): 423. doi: 10.1038 / 517423a.

Domazou, A. S. and Luisi, P. L. (2002). Size distribution of spontaneously formed liposomes by the alcohol injection method. *J. Liposome Res.*, 12(3): 205–220.

Dominak, L. M., Omiatek, D. M., Gundermann, E. L., Heien, M. L., and Keating, C. D. (2010). Polymeric crowding agents improve passive biomacromolecule encapsulation in lipid vesicles. *Langmuir*, 26 (16): 13195–131200.

Dubois, L. H. and Nuzzo, R. G. (1992). Synthesis, structure, and properties of model organic-surfaces. *Ann. Rev. Phys. Chem.*, 43, 437–463.

Dworkin, J. D., Deamer, D. W., Sandford, S., and Allmandola, L. (2001). Self-assembling amphiphilic molecules: synthesis in simulated interstellar/precometary ices. *Proc. Natl. Acad. Sci.*, 98, 815–819.

Dymond, J. S., Richardson, S. M., Coombes, C. E., *et al.* (2011). Synthetic chromosome arms function in yeast and generate phenotypic diversity by design. *Nature*, 477 (7365): 471–476.

Dyson, F. J. (1985). *Origins of Life*. Cambridge University Press.

Eddy, S. R. (2002). Non-coding RNA genes and the modern RNA world. *Cell*, 109, 137–140.

Eichhorn, U., Bommarius, A. S., Drauz, K., and Jakubke, H.-D. (1997). Synthesis of dipeptides by suspension-to-suspension conversion via thermolysin catalysis: from analytical to preparative scale. *J. Pept. Sci.*, 3, 245–251.

Eigen, M. (1971). Self-organization of matter and the evolution of biological macromolecules. *Naturwissenschaften*, 58, 465–523.

Eigen, M. and Schuster, P. (1977). Hypercycle – principle of natural self-organization. A. Emergence of hypercycle. *Naturwissenschaften*, 64, 541–565.

(1979). *The Hypercycle: A Principle of Natural Self-Organization*. Springer Verlag.

Eigen, M. and Winkler-Oswatitisch, R. (1992). *Steps Towards Life*. Oxford University Press.

Eigen, M., Gardiner, W., Schuster, P., and Winkler-Oswatitsch, R. (1981). The origin of genetic information. *Sci. Am.*, 244(4): 88–92.

El Seoud, O. A. (1984). In P. L. Luisi and B. Straub, eds., *Reverse Micelles*. Plenum Press.

Ellis, G. (2005). Physics ain't what it used to be. *Nature*, 438, 739–740.

Engels, F. (1877). *Anti-Duehring*. Translated by Austin Lewis. Chicago: Charles H. Kerr & Company, 1907.

(1883). *Dialectics of Nature. Notes and Fragments*. Translated by Clemens Dutt. Moscow: Progress Publishers, 1st edn., 1934. Included in Karl Marx and Frederick Engels, *Collected Works*, Volume 25 (Engels), published in 1987 by Lawrence & Wishart.

(1894). *Herrn Eugen Dühring's Umwalzung der Wissenschaft*. Dietz Verlag. English translation (*Herr Eugen Dühring's Revolution in Science*) was included in: Karl Marx and Friedrich Engels, *Collected Works*. Volume 25 (Engels), published in 1987 by Lawrence & Wishart.

Erickson, J. C. and Kennedy R. M. (1980). Effects of histidyl-histidine and polyribonucleotides on glycine condensation in fluctuating clay environments. *Abstracts Papers Am. Chem. Soc.*, 179, 43.

Ericsson, B., Larsson, K., and Fontell, K. (1983). A cubic protein-monoolein-water phase. *Biochim. Biophys. Acta.*, 729, 23–27.

Erwin, D. H. (2003). Life's solution – inevitable humans in a lonely universe. *Science*, 302, 1682–1683.

Eschenmoser, A. (1999). Chemical etiology of nucleic acid structure. *Science*, 284, 2118–2124.

(2003). Creating a perspective for comparing. In *Proceedings of the J. Templeton Foundation "Biochemistry and Fine-tuning."* Harvard University, October 10–12, 2003.

Eschenmoser, A. and Kisakürek, M. V. (1996). Chemistry and the origin of life. *Helv. Chim. Acta.*, 79, 1249–1259.

Fadnavis, N. W. and Luisi, P. L. (1989). Immobilized enzymes in reverse micelles: studies with gel-entrapped Trypsin and alpha-Chymotrypsin in AOT reverse micelles. *Biotechnol. Bioeng.*, 33, 1277–1282.

Falbe, J. (1987). *Surfactants in Consumer Products. Theory, Technology and Applications*. Springer Verlag.

Famiglietti, M., Hochköppler, A., Wehrli, E., and Luisi, P. L. (1992). Photosynthetic activity of cyanobacteria in water-in-oil microemulsions. *Biotechnol. Bioeng.*, 40, 173–178.

Famiglietti, M., Hochköppler, A., and Luisi, P. L. (1993). Surfactant-induced hydrogen production in cyanobacteria. *Biotechnol. Bioeng.*, 42, 1014–1018.

Fan, K. and Wang, W. (2003). What is the minimum number of letters required to fold a protein? *J. Mol. Biol.*, 328, 921–926.

Fanelli, D. and McKane, A. J. (2008). Thermodynamics of vesicle growth and instability. *Physical Review E*, 78, 051406.

Farre, L. and Oksala, T., eds. (1998). Emergency, complexity, hierarchy, organisation. Selected papers from the ECHO III Conference (ESPOO, Finland), *Acta Polytechnica Scandi.*, 91.

Fendler, J. H. and Fendler, E. J. (1975). *Catalysis in Micellar and Macromolecular Systems*. Academic Press.

Ferris, J. P. (1998). Catalyzed RNA synthesis for the RNA world. In A. Brack, ed., *The Molecular Origin of Life*. Cambridge University Press, pp. 255–256.

Ferris, J. P. and Ertem, G. (1992). Oligomerization reaction of ribonucleosides on montmorillonite: reaction of 5′-phosphorimidazolide of adenosine. *Science*, 257, 1387–1389.

   (1993). Montmorillonite catalysis of RNA oligomer formation in aqueous solution: a model for the prebiotic formation of RNA. *J. Am. Chem. Soc.*, 115, 12270–12275.

Ferris, J. P., Sanchez, R. A., and Orgel, L. E. (1968). Studies in prebiotic synthesis. III, Synthesis of pyrimidines from cyanoacetilene and cyanate. *J. Mol. Biol.*, 33, 693–704.

Ferris, J. P., Donner, D. B., and Lobo, A. P. (1973). Possible role of hydrogen cyanide in chemical evolution. The oligomerization and condensation of hydrogen cyanide. *J. Mol. Biol.*, 74, 511–518.

Ferris, J. P., Wos, J. D., Nooner, D. W., and Oró, J. (1974). Chemical evolution. 21. Amino-acids released on hydrolysis of HCN oligomers. *J. Mol. Evol.*, 3, 225–231.

Ferris, J. P., Joshi, P. C., Edelson, E. H., and Lawless, J. G. (1978). HCN: a plausible source of purines, pyrimidines and amino acids on the primitive earth. *J. Mol. Evol.*, 11, 293–311.

Field, R. J. (1972). A reaction periodic in time and space. *J. Chem. Educ.*, 49, 308–311.

Fikes, B. J. (2014). Life engineered with expanded genetic code. *The San Diego Union-Tribune*.

Fiordemondo, D. and Stano, P. (2007). Lecithin-based water-in-oil compartments as dividing bioreactors. *ChemBioChem*, 8, 1965–1973.

Fischer, A., Oberholzer, T., and Luisi, P. L. (2000). Giant vesicles as models to study the interactions between membranes and proteins. *Biochim. Biophys. Acta*, 1467, 177–188.

Fleischaker, G. (1988). Autopoiesis: the status of its system logic. *Biosystems*, 22, 37–49.

Fletcher, P. D. and Robinson, B. H. (1981). *Ber. Bunsenges. Phys. Chem.*, 85, 863.

Foldvari, M., Geszles, A., and Mezei, M. (1990). *J. Microencapsul.*, 7, 479–489.

Folk, R.L. (1993). Sem imaging of bacteria and nannobacteria in carbonate sediments and rocks. *Journal of sedimentary petrology*, 63, 990–999.

Folsome, C. E. (1979). *The Origin of Life: A Warm Little Pond*. W. H. Freeman & Co.

Fontell, K. (1990). Cubic phases in surfactant and surfactant-like lipid systems. *Colloid Polym. Sci.*, 268, 265–285.

Forster A. C., and Symons, R. H. (1987). Self-cleavage of plus and minus RNAs of a virusoid and a structural model for the active sites. *Cell*, 49(2): 211–220.

Forster, C. A., and Church, G. M. (2006). Towards synthesis of a minimal cell. *Mol. Syst. Biol.*, 2, 45. doi: 10.1038/msb4100090.

(2007). Synthetic biology projects in vitro. *Genome Research*, 17, 1–17.

Föster, S. and Plantenberg, T. (2002). From self-organizing polymers to nanohybrid and biomaterials. *Angew. Chem. Int. Ed. Engl.*, 41, 688–714.

Fox, G. E., Tran, Q., and Yonath, A. (2012). An exit cavity was crucial to the polymerase activity of the early ribosome. *Astrobiology*, 12, 57–60.

Fox, S. W. (1988). *The Emergence of Life*. Basic Books.

Fox, S. W. and Dose, K. (1972). *Molecular Evolution and the Origin of Life*. W. H. Freeman.

Fox-Keller, E. (2002). *The Century of the Gene*. Harvard University Press.

Fraenkel-Conrat, H. and Williams, R. C. (1955). Reconstitution of active tobacco mosaic virus from its inactive protein and nucleic acid components. *Proc. Nat. Acad. Sci. USA.*, 41, 690–698.

Franceschi, F. J. and Nierhaus, K. H. (1990). Ribosomal protein-l15 and protein-l16 are mere late assembly proteins of the large ribosomal-subunit – analysis of an Escherichia coli mutant lacking l15. *Journal of Biological Chemistry.* 265, 16676–16682.

Franz, M.-L. von (1988). *Psyche und Materie*. Daimon Verlag.

Fraser, C. M., Gocayne, J. D., White, O., *et al.* (1995). The minimal gene complement of Mycoplasma genitalium. *Science*, 270, 397–403.

Freitas, R. A., Jr., and Merkle, R. C. (2004). *Kinematic Self-Replicating Machines*. Landes Bioscience.

Fry, I. (1999). *The Emergence of Life on Earth: A Historical and Scientific Overview.* London: Free Association Books.

(2000). *Emergence of Life on Earth: A Historical and Scientific Overview.* New Brunswick, NJ: Rutgers University Press.

(2011). The role of natural selection in the origin of life. *Origins of Life and Evolution of Biospheres*, 41(1): 3–16.

Fujii, S., Matsuura, T., Sunami, T., Kazuta, Y., and Yomo, T. (2015). In vitro directed evolution of alpha-hemolysin by liposome display. *Biophysics*, 11: 67–72.

Funqua, C., Parsek, M. R., and Greenberg, E. P. (2001). Regulation of gene expression by cell-to-cell communication: acyl-homoserine lactone quorum sensing. *Ann. Rev., Genet.*, 35, 439–468.

Ganti, T. (1975). Organization of chemical reactions into dividing and metabolizing units: the chemotons. *BioSystems*, 7, 15–21.

(1984). *Chemoton elmélet 1. kötet. A fluid automaták elméleti alapjai.* Translated as *Chemoton Theory, Vol. 1., Theory of Fluid Automata.* OMIKK.

(2003). *The Principles of Life.* Oxford University Press.

Gao, X. and Huang, L. (1995). Cationic liposome-mediated gene transfer. *Gene Ther.*, 2 (10): 710–722.

Gardner, P. M. and Davis, B. G. (2011). Approaches to building chemical cells/chells: examples of relevant mechanistic "couples." In P. L. Luisi and P. Stano, eds., *The Minimal Cell.* Springer.

Gavrilova, L. P., Kostiashkina, O. E., Koteliansky, V. E., Rutkevitch, N. M., and Spirin, A. S. (1976). Factor-free (non-enzymic) and factor-dependent systems of translation of polyuridylic acid by *E. coli* ribosomes. *J. Mol. Biol.*, 101, 537–552.

Gennis, R. B. (1989). *Biomembranes, Molecular Structure and Function.* Springer Verlag.

Ghosh, I. and Chmielewski, J. (2004). Peptide self-assembly as a model of proteins in the pre-genomic world. *Curr. Opin. Chem. Biol.*, 8, 640–644.

Gibson, D. G., Benders, G. A., Andrews-Pfannkoch, C., *et al.* (2008a). Complete chemical synthesis, assembly, and cloning of a Mycoplasma genitalium genome. *Science*, 319 (5867): 1215–20.

Gibson, D. G., Benders, G. A., *et al.* (2008b). One-step assembly in yeast of 25 over-lapping DNA fragments to form a complete synthetic *Mycoplasma genitalium* genome. *Proc. Natl. Acad. Sci. USA.*, 105(51): 20404–204009. doi:10.1073/pnas.0811011106.

Gibson, D. G., Glass, J. I., Lartigue, C., *et al.* (2010). Creation of a bacterial cell controlled by a chemically synthesized genome. *Science*, 329(5987): 52–56. doi:10.1126/science.1190719.

Gil, R., Silva, F. J., Peretó, J., and Moya, A. (2004). Determination of the core of a minimal bacteria gene set. *Microb. Molec. Biol. Rev.*, 68, 518–537.

Gilbert, R. J. C., Fucini, P., Connell, S., *et al.* (2004). Three-dimensional structures of translating ribosomes by cryo-EM. *Molecular Cell.* 14, 57–66.

Gilbert, W. (1986). The RNA world. *Nature*, 319, 618.

Gold, T. (1979). Terrestrial sources of carbon and earthquake outgassing. *Journal of Petroleum Geology* 1(3): 3–19.

Glotzer, S. C. (2004). Materials science. Some assembly required. *Science*, 306, 419–420.

Goodenough, U. and Deacon, T. W. (2006). Emergence and religious naturalism. In P. Clayton, ed., *Oxford Handbook of Science and Religion*. Oxford University Press.

Gorlero, M., Wieczorek, R., Adamala, K., *et al.* (2009). Ser-His catalyses the formation of peptides and PNAs. *FEBS Letters*, 583, 153–156.

Gould, S. J. (1989). *Wonderful Life*. Penguin Books.

Graf, A., Winterhalter, M., and Meier, W. (2001). Nanoreactors from polymer-stabilized liposomes. *Langmuir*, 17, 919–923.

Green, R. and Noller, H.F. (1997). Ribosomes and translation. *Annual Review of Biochemistry*, 66, 679–716.

Gregoriadis, G. (1976a). The carrier potential of liposomes in biology and medicine (first of two parts). *New Engl. J. Med.*, 295, 704–710.

(1976b). The carrier potential of liposomes in biology and medicine (second of two parts). *New Engl. J. Med.*, 295, 765–770.

Gregoriadis, G., ed. (1988). *Liposomes and Carriers of Drugs: Recent Trends and Progress*. New York: John Wiley & Sons.

Gregoriadis, G. (1995). Engineering liposomes for drug delivery: progress and problems. *Trends Biotechnol.*, 13(12): 527–537.

Groen, J., Deamer, D. W., Kros, A., and Ehrenfreund, P. (2012). Polycyclic aromatic hydrocarbons as plausible prebiotic membrane components. *Orig. Life Evol. Biosph.*, 42(4): 295–306.

Habraken, G. J. M., Peeters, M., Dietz, C. H. J. T., Koninga, C. E., and Heise, A. (2010). How controlled and versatile is N-carboxy anhydride (NCA) polymerization at 0 °C? Effect of temperature on homo-, block- and graft (co)polymerization. *Polym. Chem.*, 1, 514–524.

Häckel, E. (1866). *Allgemeine Anatomie der Organismen*. Walter de Gruyer.

Haines, T. H. (1983). Anionic lipid headgroups as a proton-conducting pathway along the surface of membranes: a hypothesis. *Proc. Natl. Acad. Sci. USA*, 80, 160–164.

Haldane, J. B. S. (1929). The origin of life. *Rationalist Annual*, 148, 3–10.

(1954). The origin of life. *New Biol.*, 16, 12–27.

Halling, P. J., Eichhorn, U., Kuhl, P., and Jakubke, H.-D. (1995). Thermodynamics of solid-to-solid conversion and application to enzymic peptide synthesis. *Enzyme Microb. Technol.*, 17, 601–606.

Hampl, H., Schulze, H., and Nierhaus, K.H. (1981). Ribosomal components from escherichia-coli 50-s subunits involved in the reconstitution of peptidyltransferase activity. *Journal of Biological Chemistry*, 256, 2284–2288.

Han, D. and Rhee, J. S. (1986). *Biotechnol. Bioeng.*, 27, 1250–1255.

Hanczyc, M. M. and Szostak, J. W. (2004). Replicating vesicles as models of primitive cell growth and division. *Curr. Opin. Chem. Biol.*, 8(6): 660–664.

Hanczyc, M. M., Fujikawa, S. M., and Szostak, J. W. (2003). Experimental models of primitive cellular compartments: encapsulation, growth, and division. *Science*, 302, 618–622.

Hansen, J., Mailand, E., Swaminathan, K. K., Schreiber, J., Angelici, B., and Benenson, Y. (2014). Transplantation of prokaryotic two-component signaling pathways into mammalian cells. *Proc. Natl. Acad. Sci. USA*, 111(44): 15705–15710. doi:10.1073/pnas.1406482111.

Hansler, M. and Jakubke, H.-D. (1996). Nonconventional protease catalysis in frozen aqueous solutions. *J. Pept. Sci.*, 2, 279–289.

Harada, S. and Schelly, Z. A. (1982). Reversed micelle of dodecylpyridinium iodide in benzene. Pressure-jump relaxation kinetic and equilibrium study of the solubilization of 7,7,8,8-tetracyanoquinodimethane. *J. Phys. Chem.*, 86, 2098–2102.

Hargreaves, W. R. and Deamer, D. W. (1978a). Liposomes from ionic, single-chain amphiphiles. *Biochemistry*, 17, 3759–3768.

(1978b). In D. W. Deamer, ed., *Light Transducing Membranes: Structure, Function and Evolution*. Academic Press, pp. 23–59.

Hargreaves, W. R., Mulvhill S. J., and Deamer, D. W. (1977). Synthesis of phospholipids and membranes in prebiotic conditions. *Nature*, 266, 78–80.

Häring, G., Luisi, P. L., and Meussdoerffer, F. (1985). Solubilization of bacteria cells in organic solvents via reverse micelles. *Biochem. Biophys. Res. Commun.*, 127, 911–915.

Häring, G., Pessina, A., Meussdoerffer, F., Hochköppler, A., and Luisi, P. L. (1987). Solubilization of bacterial cells in organic solvents via reverse micelles and microemulsions. *Ann. Biochem. Eng.*, 506, 337–344.

Hawker, C. J. and Frechet, J. M. J. (1990). Preparation of polymers with controlled molecular architecture – a new convergent approach to dendritic macromolecules. *J. Am. Chem. Soc.*, 112, 7638–7647.

Hayatsu, R., Studier, M. H., Moore, L. P., and Anders, E. (1975). Purines and triazines in the Murchison meteorite. *Geochim. Cosmochim. Acta*, 39, 471–488.

Hecht, M. H., Das, A., Go, A., Bradley, L. H., and Wei, Y. (2004). De novo proteins from designed combinatorial libraries. *Protein Science*, 13, 1711–1723.

Heinen, W. and Lauwers, A. M. (1997). The iron-sulfur world and the origins of life: abiotic thiol synthesis from metallic iron, H_2S and CO_2; a comparison of the thiol generating $FeS/HCl(H_2S)/CO_2$-system and its $Fe^0/H_2S/CO_2$-counterpart. *Proc. Royal Netherlands Acad. Arts Sci.*, 100, 11–25.

Herz-Fischler, R. (1998). *A Mathematical History of the Golden Number*. New York: Dover.

Hilborn, R. C. (1994). *Chaos and Non Linear Dynamics*. Oxford University Press.

Hilhorst, R., Spruijt, R., Laane, C., and Veeger, C. (1984). Rules for the regulation of enzyme-activity in reversed micelles as illustrated by the conversion of apolar steroids by 20-beta-hydroxysteroid dehydrogenase. *Eur. J. Biochem.*, 144, 459–466.

Hirwschmann, H. and Hanson, K. R. (1971). *Top Stereochem.*, 36: 329–399.

Hochköppler, A. and Luisi, P. L. (1989). Solubilization of soybean mitochondria in AOT/isooctane water-in-oil microemulsions. *Biotechnol. Bioeng.*, 33, 1477–1481.

(1991). Photosynthetic activity of plant cells solubilized in water-in-oil microemulsions. *Biotechnol. Bioeng.*, 37, 918–921.

Hochköppler, A., Pfammatter, N., and Luisi, P. L. (1989). Activity of yeast cells solubilized in water-in-oil microemulsions. *Chimia*, 43, 348–350.

Holden, C. (2005). Vatican astronomer rebuts cardinals' attack on Darwinism. *Science*, 309, 996–997.

Holland, J. H. (1998). *Emergence: From Chaos to Order*. Oxford University Press.

Holm, N. G. and Andersson, E. M. (1998). Hydrothermal systems. In A. Brack, ed., *The Molecular Origin of Life*. Cambridge University Press.

Horowitz, N. and Miller, S. (1962). Origins of life: the primal self-organization. In L. Zechmeister, ed., *Progress in the Chemistry of Natural Products*, Vol. 20. Springer Verlag, pp. 423–459.

Horowitz, P. and Sagan, C. (1993). Five years of Project META: an all-sky narrow-band radio search for extraterrestrial signals. *Astrophys. J.*, 415, 218–233.

Hosoda, K., Sunami, T., Kazuta, Y., Matsuura, T., Suzuki, H., and Yomo, T. (2008). Quantitative study of the structure of multilamellar giant liposomes as a container of protein synthesis reaction. *Langmuir*, 24, 13540–13548.

Hoyle, F. and Wickramasinghe, C. (1999). Astronomical origins of life – steps towards panspermia. *Astrophys. Space Sci.*, 268, Preface, VII–VIII.

(2000). *Astronomical Origins of Life – Steps Towards Panspermia*. Dordrecht, NL: Kluwer Academic.

Huang, S. S. (1959). Occurrence of life in the universe. *Amer. Sci.*, 47, 397–402.

Huber, C. and Wächtershäuser, G. (1997). Activated acetic acid by carbon fixation on (Fe, Ni)S under primordial condition. *Science*, 276, 245–247.

Huntley, H. E. (1970). *The Divine Proportion: A Study in Mathematical Beauty*. New York: Dover.

Hutchinson, C. A., Peterson, S. N., Gill, S. R., *et al.* (1999). Global transposon mutagenesis and a minimal mycoplasma genome. *Science*, 286, 2165–2169.

Ikehara, K. (2002). Origins of gene, genetic code, protein and life: comprehensive view of life systems from a GNC-SNS primitive genetic code hypothesis; *J. Biosci.* 27, 165–186.

(2005). Possible steps to the emergence of life: the (GADV)-protein world hypothesis. *The Chemical Record*, 5, 107–118.

(2009). Pseudo-replication of [GADV]-proteins and origin of life. *Int. J. Mol. Sci.*, 10, 1525–1537.

Ikehara, K., Omori, Y., Arai, R., and Hirose, A. (2002). A novel theory on the origin of the genetic code: a GNC-SNS hypothesis. *J. Mol. Evol.*, 54, 530–538.

Imre, V. E. and Luisi, P. L. (1982). Solubilization and condensed packaging of nucleic acids in reversed micelles. *Biochem. Biophys. Res. Commun.*, 107, 538–545.

Ishikawa, K., Sato, K., Shima, Y., Urabe, I., and Yomo, T. (2004). Expression of cascading genetic network within liposomes. *FEBS Lett.*, 576, 387–390.

Islas, S., Becerra, A., Luisi, P. L., and Lazcano, A. (2004). Comparative genomics and the gene complement of a minimal cell. *Orig. Life Evol. Biosph.*, 34 (1–2): 243–256.

Israelachvili, J. N. (1992). *Intermolecular and Surface Forces*, 2nd edn. Academic Press.

Israelachvili, J. N, Mitchell, D. J., and Ninham, B. W. (1977). Theory of self-assembly of lipid bilayers and vesicles. *Biochim. Biophys. Acta*, 470, 185–201.

Issac, R. and Chmielewski, J. (2002). Approaching exponential growth with a self-replicating peptide. *J. Am. Chem. Soc.*, 124, 6808–6809.

Itaya, M. (1995). An estimation of the minimal genome size required for life. *FEBS Lett.*, 362, 257–260.

Itojima, Y., Ogawa, Y., Tsuno, K., Handa, N., and Yanagawa, H. (1992). Spontaneous formation of helical structures from phospholipid-nucleoside conjugates. *Biochemistry*, 31, 4757–4765.

Jacob, F. (1982). *The Possible and the Actual*. University of Washington Press.

Jaeger, L., Wright, M. C., and Joyce, G. F. (1999). A complex ligase ribozyme evolved in vitro from a group I ribozymes domain. *Proc. Natl. Acad. Sci.*, 96, 14712–14717.

Jakubke, H.-D. (1987). Peptides: design, synthesis, and biological activity. In S. Udenfried and J. Meienhofer, eds., *The Peptides: Analysis, Synthesis, Biology*, Vol. 9. Academic Press.

(1995). Hydrolysis and formation of peptides. In K. Drauz and H. Waldmann, eds., *Enzyme Catalysis in Organic Synthesis*, Vol. 1. Wiley-VCH, pp. 431–458.

Jakubke, H.-D., Kuhl, P., and Könnecke, A. (1985). Basic principles of protease-catalyzed peptide bond formation. *Angew. Chem. Int. Ed. Engl.*, 24, 85–93.

Jakubke, H.-D., Eichhorn, U., Hansler, M., and Ullmann, D. (1996). Non-conventional enzyme catalysis: application of proteases and zymogens in biotransformations. *Biol. Chem.*, 377, 455–464.

Janiak, M. J., Small, D. M., and Shipley, G. G. (1976). Nature of the thermal pretransition of synthetic phospholipids: dimyristoyl- and dipalmitoyllecithin. *Biochemistry*, 15, 4575–4580.

Jeon, K. W., Lorch, I. J., and Danielli, J. F. (1970). Reassembly of living cells from dissociated components. *Science*, 167(3925): 1626–1627.

Jewett, M. C., Calhoun, K. A., Voloshin, A., Wuu, J. J., and Swartz, J. R. (2008). An integrated cell-free metabolic platform for protein production and synthetic biology. *Mol. Syst. Biol.*, 4, 220.

Jimenez-Prieto, R., Silva, M., and Perez-Bendito, D. (1998). Approaching the use of oscillating reactions for analytical monitoring. *Analyst*, 123, 1R–8R.

Jiménez, J. I., Xulvi-Brunet, R., Campbell, G. W., Turk-MacLeod, R., and Chen, I. A. (2013). Comprehensive experimental fitness landscape and evolutionary network for small RNA. *Proc. Natl. Acad. Sci. USA*, 110, 14984–14991.

Johnson, E. T., and Schmidt-Dannert, C. (2008). Light-energy conversion in engineered microorganisms. *Trends Biotechnol.*, 26(12): 682–689.

Johnston, W. K., Unrau, P. J., Lawrence, M. S., Glasner, M. E., and Bartel, D. P. (2001). RNA-catalyzed RNA polymerization: accurate and general RNA-templated primer extension. *Science*, 292(5520): 1319–1325.

Joyce, G. F. and Orgel, L. E. (1986). Nonenzymatic template-directed synthesis on RNA random copolymers – poly(C, G) templates. *J. Mol. Biol.*, 188, 433–441.

Juarrero, A. and Rubino, C. A. (2010). *Emergence, Complexity, and Self-Organization: Precursors and Prototypes (Exploring Complexity)*. Paperback. ISCE Publishing.

Kajander, E. O. and Çiftçioglu, N. (1998). Nanobacteria: An alternative mechanism for pathogenic intra- and extracellular calcification and stone formation. *Proc. Natl. Acad. Sci. USA*, 95(14): 8274–8279.

Kaler, E. W., Murthy, A. K., Rodriguez, B. E., and Zasadzinski, J. A. N. (1989). Spontaneous vesicle formation in aqueous mixtures of single-tailed surfactants. *Science*, 245, 1371–1374.

Kamtekar, S., Shiffer, J. M., Xiong, H. Y., Babik, J. M., and Hecht, M. H. (1993). Protein design by binary patterning of polar and nonpolar amino acids. *Science*, 262, 1680–1685.

Kaszuba, M. and Jones, M. N. (1999). Hydrogen peroxide production from reactive liposomes encapsulating enzymes. *Biochim. Biophys. Acta*, 1419, 221–228.

Kato, A., Yanagisawa, M., Sato, Y. T., Fujiwara, K., and Yoshikawa, K. (2012). Cell-sized confinement in microspheres accelerates the reaction of gene expression. *Scientific Report*, 2, 283.

Kauffman, S. A. (1986). Autocatalytic set of proteins. *J. Theor. Biol.*, 119, 1–24.

(1993). *The Origins of Order: Self Organization and Selection in Evolution*. Oxford University Press.

Kawamura, K. (2002). The origin of life from the life of subjectivity. In G. Palyi, C. Zucchi, and L. Caglioti, eds., *Fundamentals of life*. Elsevier, pp. 56–76.

Kawamura, K. and Kamoto, F. (2000). Condensation reaction of hexanucleotides containing guanine and cytosine with water soluble carbodiimide. *Nucleic Acid Symp. Ser.*, 44, 217–218.

Kent, S. (1999). Chemical protein synthesis by solid phase ligation of unprotected peptide segments. *J. Am. Chem. Soc.*, 121, 8720–8727.

Kenyon, D. and Mills, G. (1996). The RNA world: a critique. *Review Article Origins & Design*, 17: 1.

Khalil, A. S. and Collins, J. J. (2010). Synthetic biology: applications come of age. *Nature Reviews Genetics*, 11, 367–379. doi:10.1038/nrg2775.

Kiedrowski, G. von (1986). A self-replicating hexadeoxynucleotide. *Angew. Chem. Int. Ed. Engl.*, 25, 932–925.

(1993). Minimal replicator theory I: parabolic versus exponential growth. In D. H. Berlin, ed., *Bioorganic Chemistry*, Vol. 3. Springer Verlag, pp. 115–146.

Kikuchi, A., Aoki, Y., Sugaya, S., *et al.* (1999). Development of novel cationic liposomes for efficient gene transfer into peritoneal disseminated tumor. *Human Gene Therapy*, 10(6): 947–955.

Kim, J. (1984). Concepts of supervenience. *Phil. Phen. Res.*, 45, 153–176.

Kimura, M. (1983). *The Neutral Theory of Molecular Evolution*. Cambridge University Press.

Kita, H., Matsuura, T., Sunami, T., Hosoda, K., Ichihashi, N., *et al.* (2008). Replication of genetic information with self-encoded replicase in liposomes. *Chem. Bio. Chem.*, 9, 2403–2410.

Kitano, H. (2007). Towards a theory of biological robustness. *Molecular Systems Biology*, 3(1).

Kitney, R. and Freemont, P. (2012). Synthetic biology – the state of play, *FEBS Letters*, 586, 2029–2036.

Klee, R. (1984). Micro-determinism and concepts of emergence. *Phil. Sci.*, 51, 44–63.

Knenvolden, K., Lawless, J. G., Pering, K., *et al.* (1970). Evidence for extraterrestrial amino acids and hydrocarbons in the Murchison meteorite. *Nature*, 228, 923–926.

Kobayashi, K. and Kanaizuka, Y. (1977). Reassembly of living cells from dissociated components in Bryopsis. *Plant & Cell Physiol.*, 18, 1373–1377.

Kobayashi, K., Tsuchiya, M., Oshima, T., and Yanagawa, H. (1990). Abiotic synthesis of amino acids and imidazole by proton irradiation of simulated primitive earth atmospheres. *Origins of Life and Evolution of the Biosphere*, 20, 99–109.

Kobayashi, K., Kaneko, T., Saito, T., and Oshima, T. (1998). Amino acid formation in gas mixtures by high-energy particle irradiation. *Origins of Life and Evolution of the Biosphere*, 28, 155–165.

Kolisnychenko, V., Plunkett, G. III, Herring, C. D., *et al.* (2002). Engineering a reduced Escherichia coli genome. *Genome Res.*, 12, 640–647.

Kondepudi, D. K. and Prigogine, I. (1981). Sensitivity of non-equilibrium systems. *Physica A*, 107, 1–24.

Kondepudi, D. K., Prigogine, I., and Nelson, G. (1985). Sensitivity of branch selection in nonequilibrium systems. *Phys. Lett. A*, 111, 29–32.

Kondepudi, D. K., Kaufman, R., and Singh, N. (1990). Chiral symmetry breaking in sodium chlorate crystallization. *Science*, 250, 975.

Kondo, Y., Uchiyama, H., Yoshino, N., Nishiyama, K., and Abe, M. (1995). Spontaneous vesicle formation from aqueous-solutions of didodecyldimethylammonium bromide and sodium dodecyl-sulfate mixtures. *Langmuir*, 11, 2380–2384.

Kool, E. T. and Morales, J. (2005). Efficient replication between non-hydrogen-bonded nucleoside shape analogs. In M. Simon, ed., *Emergent Computation: Emphasizing Bioinformatics*. Springer, pp. 88–98.

Koonin, E. V. (2000). How many genes can make a cell: the minimal-gene-set concept. *Annu. Rev. Genomics Human Genet.*, 1, 99–116.

Koshland, D. E., Jr. (2002). The seven pillars of life. *Science*, 295, 2215–2216.

Krupkin, M., Bashan, A., and Yonath, A. (2014). Glimpse into the origin of life: what was first, the genetic code or its products, the proteins? In G. Trueba, ed., *Why Does Evolution Matter? The Importance of Understanding Evolution*. Cambridge Scholars Publishing, pp. 87–100.

Kuiper, T. B. H. and Morris, M. (1977). Searching for extraterrestial civilizations. *Science*, 196, 616–621.

Kullmann, W. (1987). *Enzymatic Peptide Synthesis*. CRC Press.

Kunin, V. (2000). A system of two polymerases – a model for the origin of life. *Origins of Life and Evolution of the Biosphere*, 30(5): 459–468.

Kuruma, Y., Stano, P., Ueda, T., and Luisi, P. L. (2009). A synthetic biology approach to the construction of membrane proteins in semi-synthetic minimal cells. *Biochim. Biophys. Acta*, 1788, 567–574.

Lahav, M. and Leiserowitz, L. (1999). Spontaneous resolution: from three-dimensional crystals to two-dimensional magic nanoclusters. *Angew. Chem. Int. Ed. Engl.*, 38, 2533–2536.

Lancaster, W. A. and Adams, M. W. W. (2011). The influence of environment and metabolic capacity on the size of a microorganism. In P. L. Luisi and P. Stano, eds., *The Minimal Cell*. Netherlands: Springer, pp. 93–103.

Landau, E. M. and Luisi, P. L. (1993). Lipid cubic phases as transparent, rigid matrices for the direct spectroscopic study of immobilized membrane proteins. *J. Am. Chem. Soc.*, 115, 2102–2106.

Landweber, L. F. and Pokrovskaya, I. D. (1999). Emergence of a dual-catalytic RNA with metal-specific cleavage and ligase activities: the spandrels of RNA evolution. *Proc. Natl. Acad. Sci.*, 96, 173–178.

Lane, N. (2009). *Power, Sex, Suicide*. Oxford University Press.

Langton, C. G., ed. (1988). *Artificial Life: Proceedings of an Interdisciplinary Workshop on the Synthesis and Simulation of Living Systems*. Addison-Wesley.

Langton, C. G. (1990). Computation at the edge of chaos: phase transitions and emergent computation. *Physica*, D42, 12–37.

Langton, C. G., ed. (1993). *Artificial Life III: Proceedings of the Third Interdisciplinary Workshop on the Synthesis and Simulation of Living Systems*. Addison-Wesley.

(1995). *Artificial Life: An Overview*. MIT Press.

Lapique, N. and Benenson, Y. (2014). Digital switching in a biosensor circuit via programmable timing of gene availability. *Nature Chemical Biology*. doi:10.1038/nchembio.1680.

Larsson, K. (1989). Cubic lipid-water phases: structures and biomembrane aspects. *J. Phys. Chem.*, 93, 7304–7314.

Lartigue, C., Glass, J. I., Alperovich, N., *et al.* (2007). Genome transplantation in bacteria: changing one species to another. *Science*, 317(5838): 632–638.

Lasch, J., Laub, R., and Wohlrab, W. (1991). How deep do intact liposomes penetrate into human skin? *J. Controll. Release*, 18, 55–58.

Lasic, D. D. (1995). In R. Lipowsky and E. Sackmann, eds., *Handbook of Biological Physics*, Vol. 1. Elsevier, pp. 491–519.

Lawless, J. G. and Yuen, G. U. (1979). Quantification of monocarboxylic acids in the Murchison carbonaceous meteorite. *Nature*, 282, 396–398.

Lawrence, D. S., Jiang, T., and Levett, M. (1995). Self-assembling supramolecular complexes. *Chem. Rev.*, 95, 2229–2260.

Lazcano, A. (2003). Just how pregnant is the universe? *Science*, 299, 347–348.

(2004). An answer in search of a question. A review of *How Life Began: The Genesis of Life on Earth*, by William Day. *Astrobiology*, 4(4): 469–471.

Lazcano, A., and Bada, J. L. (2003). The 1953 Stanley L. Miller experiment: fifty years of prebiotic organic chemistry. *Orig. Life Evol. Biosph.*, 33, 235–242.

Lazzara, S. (2001). *Vedi Alla Voce Scienza*. Manifesto Libri.

le Doux, J. (2002). *Synaptic Self: How Our Brains Become Who We Are*. Viking Books.

Lee, D. H., Granja, J. R., Martinez, J. A., Severin, K., and Ghadiri, M. R. (1996). A self-replicating peptide. *Nature*, 382, 525–528.

Lee, D. H., Severin, K., Yokobayashi, Y., and Ghadiri, M. R. (1997). Emergence of symbiosis in peptide self-replication through a hypercyclic network. *Nature*, 390, 591–594.

Lee, S. K., Chou, H., Ham, T. S., Lee, T. S., and Keasling, J. D. (2008). Metabolic engineering of microorganisms for biofuels production: from bugs to synthetic biology to fuels. *Curr Opin Biotechnol.*, 19(6):556–563.

Lehn, J.-M. (2002). Toward self-organization and complex matter. *Science*, 295(5564): 2400–2403. doi:10.1126/science.1071063.

Leonard, E., Nielsen, D., Solomon, K., and Prather, K. J. (2008). Engineering microbes with synthetic biology frameworks. *Trends Biotechnol.*, 12, 674–681.

Leser, M. E. and Luisi, P. L. (1989). Liquid 3-phase micellar extraction of peptides. *Biotech. Techniques*, 3, 149–154.

(1990). Application of reverse micelles for the extraction of amino acids and proteins. *Chimia*, 44, 270–282.

Levashov, A. V., Klyachko, N. L., Psbezhetski, A. V., et al. (1989). Biochim. *Biophys. Acta*, 988, 221–256.

Levy, M. and Ellington, A. D. (2003). Peptide-template nucleic acid ligation. *J. Mol. Evol.*, 56, 607–615.

Lewontin, R. C. (1970). The units of selection. *Annu. Rev. Ecol. Syst.*, 1, 1–18.

(1993). *The Doctrine of DNA – Biology as an Ideology*. Penguin Books.

Li, J., Ballmer, S. G., Gillis, E. P., et al. (2015). Synthesis of many different types of organic small molecules using one automated process, *Science*, 347(6227): 1221–1226.

Li, T. and Nicolaou, K. C. (1994). Chemical self-replication of palindromic duplex DNA. *Nature*, 369, 218–221.

Li, Y., Zhao, Y., Hatfield, S., et al. (2000). Dipeptide seryl-histidine and related oligopeptides cleave DNA, protein, and a carboxyl ester. *Bioorg. Med. Chem.*, 12, 2675–2680.

Li, X. and Chmielewski, J. (2003). Peptide self-replication enhanced by a proline kink. *J. Am. Chem. Soc.*, 125, 11820–11821.

Lifson, S. (1997). On the crucial stages in the origin of animate matter. *J. Mol. Evol.*, 44, 1–8.

Lindahl, T. (1993). Instability and decay of the primary structure of DNA. *Nature*, 362 (6422): 709–715.

Lindblom, G. and Rilfors, L. (1989). Cubic phases and isotropic structures formed by membrane lipids – possible biological relevance. *Biochem. Biophys. Acta*, 988, 221–256.

Lindsey, J. S. (1991). Self-assembly in synthetic routes to molecular devices – biological principles and chemical perspectives – a review. *New J. Chem.*, 15, 153–180.

Liu, A. P. and Fletcher, D. A. (2009). Biology under construction: in vitro reconstitution of cellular function. *Nature Reviews*, 10, 644–670.

Livio, M. (2002). *La Sezione Aurea. Storia di un numero e di un mistero che dura da tremila anni*, Rizzoli.

Lohrmann, R. and Orgel, L. E. (1973). Prebiotic activation processes. *Nature*, 244, 418.

Lonchin, S., Luisi, P. L., Walde, P., and Robinson, B. H. (1999). A matrix effect in mixed phospholipid/fatty acid vesicle formation. *J. Phys. Chem. B*, 103, 10910–10916.

Longo, L. M., Lee, J., and Blaber, M. (2013). Simplified protein design biased for prebiotic amino acids yields a foldable, halophilic protein. *PNAS*, 110(6): 2135–2139.

Love, S. G. and Brownlee, D. E. (1993). A direct measurement of the terrestrial mass accretion rate of cosmic dust. *Science*, 262, 550–553.

Lovelock, J. E. (1979). *Gaia: A New Look at Life on Earth*. Oxford University Press.

Luci, P. (2003). *Gene cloning expression and purification of membrane proteins*. ETH-Z Dissertation Nr. 15108, Zurich.

Lucks, J. B., Qi, L., Whitaker, W. R., and Arkin, A. P. (2008). Toward scalable parts families for predictable design of biological circuits. *Curr. Opin. Microbiol.*, 11(6): 567–573.

Luhmann, K. (1984). *Soziale Systeme*. Suhrkamp.

Luisi, P. L. (1979). Why are enzymes macromolecules? *Naturwissenschaften*, 66, 498–504.

 (1985). Enzyme hosted in reverse micelles in hydrocarbon solution. *Angew. Chem.*, 24, 439–450.

 (1993). Defining the transition to life: self-replicating bounded structures and chemical autopoiesis. In W. Stein and F. J. Varela, ed., *Thinking About Biology*, SFI Studies in the Sciences of Complexity. Addison-Wesley-Longman.

 (1996). Self-reproduction of micelles and vesicles: models for the mechanisms of life from the perspective of compartmented chemistry. *Adv. Chem. Phys.*, 92, 425–438.

 (1997). Self-reproduction of chemical structures and the question of the transition to life. In C. B. Cosmovici, S. Bowyer, and D. Werthimer, eds., *Astronomical and Biochemical Origins and the Search for Life in the Universe*. Editricc Compositori, pp. 461–468.

 (2001). Are micelles and vesicles chemical equilibrium systems? *J. Chem. Educ.*, 78, 380–384.

 (2003a). Contingency and determinism. *Phil. Trans. R. Soc. Lond., A*, 361, 1141–1147.

 (2006). *The Emergence of Life. From Chemical Origins to Synthetic Biology*. Cambridge University Press.

 (2007). Chemical aspects of synthetic biology. *Chemistry and Biodiversity*, 4, 603–621.

Luisi, P. L. and Chiarabelli, C. (2011). *Chemical Synthetic Biology*. Hoboken, NJ: John Wiley & Sons.

Luisi, P. L. and Magid, L. (1986). Solubilization of enzymes and nucleic acids in hydrocarbon micelar solutions. *Crit. Rev. Biochem.*, 20, 409–474.

Luisi, P. L. and Stano, P., eds. (2011). *The Minimal Cell. The Biophysics of Cell Compartment and the Origin of Cell Functionality*. Dordrecht: Springer.

Luisi, P. L., and Straub, B., eds. (1984). *Reverse Micelles*. Plenum Press.

Luisi, P. L. and Varela, F. J. (1990). Self-replicating micelles – a chemical version of minimal autopoietic systems. *Orig. Life Evol. Biosph.*, 19, 633–643.

Luisi, P. L. and Walde, P., eds. (2000). *Giant Vesicles, Perspectives in Supramolecular Chemistry*. New York: John Wiley & Sons.

Luisi, P. L., Henninger, F., Joppich, M., Dossena, A., and Casnati, G. (1977a). Solubilization and spectroscopic properties of α-chymotrypsin in cyclohexane. *Biochem. Biophys. Res. Commun.*, 74, 1384–1389.

Luisi, P. L., Pellegrini, A., and Walsoe, C. (1977b). Pepsin-catalyzed coupling between aromatic amino acid residues. *Experientia*, 33, 796.

Luisi, P. L., Giomini, M., Pileni, M. P., and Robinson, B. H. (1988). Reverse micelles as hosts for proteins and small molecules. *Biochim. Biophys. Acta*, 947, 209–246.

Luisi, P. L., Scartazzini, R., Haering, G., and Schurtenberger, P. (1990). Organogels from water-in-oil microemulsions. *Colloid Polymer Sci.*, 268, 356–374.

Luisi, P. L., Lazcano, A., and Varela, F. (1996). In M. Rizzotti, ed., *Defining Life: the Central Problem in Theoretical Biology*. University of Padova, pp. 149–165.

Luisi, P. L., Oberholzer, T., and Lazcano, A. (2002). The notion of a DNA minimal cell: a general discourse and some guidelines for an experimental approach. *Helv. Chim. Acta*, 85(6): 1759–1777.

Luisi, P. L., Stano, P., Rasi, S., and Mavelli, F. (2004). A possibile route to prebiotic vesicle reproduction. *Artificial Life*, 10, 297–308.

Luisi, P. L., Ferri, F., and Stano, P. (2006). Approaches to semi-synthetic minimal cells: a review. *Naturwissenschaften*, 93, 1–13.

Luisi, P. L., Allegretti, M., de Souza, T. P., Steiniger, F., Fahr, A., and Stano, P. (2010). Spontaneous protein overcrowding in liposomes: a new vista for the origin of cellular metabolism. *Chem. Biochem.*, 11, 1989–1992.

Luisi, P. L., Stano, P., and Chiarabelli, C., eds. (2014). Synthetic Biology. *Current Opinion in Chemical Biology*, 22, pp. v–vii (Editorial overview) and 1–162.

Luther, A., Brandsch, R., and von Kiedrowski, G. (1998). Surface promoted replication and exponential amplification. *Nature*, 396, 245–248.

Luzzati, V., Vargas, R., Mariani, P., Gulik, A., and Delacroix, H. (1993). Cubic phases of lipid-containing systems: elements of a theory and biological connotations. *J. Mol. Biol.*, 229, 540–551.

Ma, Q. G. and Remsen, E. F. (2002). Chemically induced supramolecular reorganization of triblock copolymer assemblies: Trapping of intermediate states via a shell-cross-linking methodology. *Proc. Natl. Acad. Sci. USA*, 99, 5058–5063.

Machy, P. and Leserman, L. (1987). *Liposomes in Cell Biology and Pharmacology*. London: John Libbey and Co., Ltd.

Madeira, V. M. C. (1977). *Biochim. Biophys. Acta*, 499, 202–211.

Maden, B. and Monro, R. E. (1968). Ribosome-catalyzed peptidyl transfer – effects of cations and pH value. *European Journal of Biochemistry*, 6, 309.

Mader, S. S. (1996). *Biology*, 5th edn. W. C. Brown Publisher.

Maestro, M. and Luisi, P. L. (1990). A simplified thermodynamic model for protein uptake by reverse micelles. In K. L. Mittal, ed., *Surfactants in Solution*, Vol. 9. Plenum.

Malyshev, D. A., Dhami, K., Quach, H. T., Lavergne, T., and Ordoukhanian, P. (2012). Efficient and sequence-independent replication of DNA containing a third base pair establishes a functional six-letter genetic alphabet. *Proc. Natl. Acad. Sci. USA*, 109 (30): 12005–12010.

Mandelbrot, B. (1982). *Fractal Geometry of Nature*. Fenn and Company.

Mandell, D. J. *et al.* (2015). Biocontainment of genetically modified organisms by synthetic protein design. *Nature*, 518, 55–60. http://dx.doi.org/10.1038/nature14121.

Mangiarotti, G. and Chiaberge, S. (1997). Reconstitution of functional eukaryotic ribosomes from Dictyostelium discoideum ribosomal proteins and RNA. *The Journal of Biological Chemistry*, 272, 19682–19687.

Mansy, S. S. and Szostak, J. W. (2009). Reconstructing the emergence of cellular life through the synthesis of model protocells. *Cold Spring Harbor Symp. Quant. Biol.*, 74, 47–54.

Mansy, S. S., Schrum, J. P., Krishnamurthy, M., Tobé, S., Treco, D. A., and Szostak, J. W. (2008). Template-directed synthesis of a genetic polymer in a model protocell. Nature. doi:10.1038/nature07018.

Margulis, L. (1993). *Symbiosis in Cell Evolution*. Freeman.

Margulis, L. and Sagan, D. (1995). *What is Life?* Weidenfeld and Nicholson.

Mariani, P., Luzzati, V., and Delacroix, H. (1988). Cubic phases of lipid-containing systems: structure analysis and biological implications. *J. Mol. Biol.*, 204, 165–189.

Marks-Tarlow, T., Robertson, R., and Combs, A. (2001). Varela and the Uroborus: the psychological significance of reentry. *Cybernetics Human Knowing*, 9, 31.

Marques, E. F., Regev, O., Khan, A., Miguel, M. D., and Lindman, B. (1998). Vesicle formation and general phase behavior in the catanionic mixture SDS-DDAB-water. The anionic-rich side. *J. Phys. Chem. B*, 102, 6746–6758.

Martinek, K. and Berezin, I. V. (1986). *Dokl. Akdam. Nauk. SSSR*, 289, 1271.

Martinek, K., Levashov, A. V., Pantin, V. I., and Berezin, I. V. (1978). Model of biological membranes or surface-layer (active center) of protein globules (enzymes) – reactivity of water solubilized by reversed micelles of aerosol OT in octane during neutral hydrolysis of picrylchloride. *Doklady Akademii Nauk SSSR*, 238, 626–629.

Martinek, K., Levashov, A. V., Klyachko, N. L., Pantin, V. I., and Berezin, I. V. (1981). The principles of enzyme stabilization. 6. Catalysis by water-soluble enzymes entrapped into reversed micelles of surfactants in organic solvents. *Biochem. Biophys. Acta*, 657, 277–295.

Martinek, K., Levashov, A. V., Klyachko, N., Khmelnttski, Yu. L., and Berezin, I. V. (1986). Micellar enzymology. *Eur. J. Biochem.*, 155, 453–468.

Martini, L. and Mansy, S. S. (2011). Cell-like systems with riboswitch controlled gene expression. *Chem. Commun.*, 47, 10734–10736.

Mascolo, R. (2011). *L'emergere della biologia della cognizione. La complessità della vita di Humberto Maturana Romecín*. Aracne Editrice, Rome.

Mason, S. F. and Tranter, G. E. (1983). The parity violating energy difference between enantiomeric molecules. *Chem. Phys. Lett.*, 94, 34.

Masters, J. R. (2002). HeLa cells 50 years on: the good, the bad and the ugly. *Nat. Rev. Cancer*, 2, 315–319.

Matsubayashi, H., Kuruma, Y., and Ueda, T. (December 2014). Cell-free synthesis of SecYEG translocon as the fundamental protein transport machinery. *Orig. Life Evol. Biosph.*, 44(4): 331–334.

Matsumura, S., Takahashi, T., Ueno, A., and Mihara, H. (2003). Complementary nucleo-base interaction enhances peptide–peptide recognition and self-replicating catalysis. *Chem. Eur. J.*, 9, 4829–4837.

Matthews, C. N. (1975). The origin of proteins, heteropolypeptides from hydrogen cyanide and water. *Origin of Life*, 6, 155–163.

Mattman, L. H. (1992). *Cell Wall Deficient Forms: Stealth Phatogens*. Boca Raton, FL: CRC Press.

Maturana, H. and Varela, F. (1980). *Autopoiesis and Cognition: The Realization of the Living*. Reidel.

Maturana, H. R. and Varela, F. J. (1998). *The Tree of Knowledge* (Based on revised edn. of 1992). Shambala. (First edn. 1984, *El árbol del conocimiento. Bases biológicas del entendimiento humano*. First English edn. 1987).

Maturana, H., Lettvin, J., McCulloch, W., and Pitts, W. (1960). Life and cognition. *Gen. Physiol.*, 43, 129–175.

Mavelli, F. (2004). *Theoretical investigations on autopoietic replication mechanisms*. ETH-Z Dissertation Nr. 15218, Zurich.

Mavelli, F. (2012). Stochastic simulations of minimal cells: the Ribocell model. *BMC bioinformatics*, 13(S 4): S 10. doi:10.1186/1471-2105-13-S4-S10.

Mavelli, F. and Luisi, P. L. (1996). Autopoietic self-reproducing vesicles: a simplified kinetic model. *J. Phys. Chem.*, 100, 16600–16607.

Mavelli, F., Altamura, E., Cassidei, L., and Stano, P., (2014). Recent theoretical approaches to minimal artificial cells. *Entropy*, 16, 2488–2511.

Maynard-Smith, J. and Szathmáry, E. (1995). *The Major Transitions in Evolution*. Oxford University Press.

(1999). *The Origins of Life*. Oxford University Press.

Mayr, E. (1974a). Teleological and teleonomic: a new analysis. *Boston Studies in the Philosophy of Science*, XIV, 91–117.

(1974b). *The multiple meanings of teleological*. In Ernst Mayr, *Toward a New Philosophy of Biology: Observations of an Evolutionist*. Cambridge, MA: Harvard University Press, 1988, pp. 44–45.

(1988). The limits of reductionism. *Nature*, 331, 475.

(1992). The idea of teleology. *Journal of the History of Ideas*, 53(1): 117–135.

McCollom, T. M., Ritter, G., and Simoneit, B. R. T. (1999). Lipid synthesis under hydrothermal conditions by Fischer-Tropsch-type reactions. *Orig. Life Evol. Biosph.*, 29, 153–166.

McLaughlin, B. P. (1992). The rise and fall of British emergentism. In A. Beckermann, H. Flohr and J. Kim, eds., *Emergence or Reduction: Essays on the Prospects of Nonreductive Materialism*, Library edn. de Gruyter, pp. 49–53.

Meier, C. A. (1992). *Wolfgang Pauli und C. G. Jung, Ein Briefwechsel*. Springer Verlag.

Menger, F. (1991). Groups of organic molecules that operate collectively. *Angew. Chem. Int. Ed. Engl.*, 30, 1086–1099.

Merleau-Ponty, M. (1967). *The Structure of Behaviour*. Beacon.

Michalodimitrakis, K. and Isalan, M. (2009). Engineering prokaryotic gene circuits. *FEMS Microbiol.*, 33(1): 27–37.

Mill, J. S. (1872). *System of Logic*, 8th edn. Longmans, Green, Reader and Dyer.

Miller, C., Cuendet, P., and Gratzel, M. (1991). Adsorbed omega-hydroxy thiol monolayers on gold electrodes – evidence for electron-tunneling to redox species in solution. *J. Phys. Chem.*, 95, 877–886.

Miller, D. M. and Gulbis, J. M. (2015). Engineering protocells: prospects for self-assembly and nanoscale production lines. *Life*, 5(2): 1019–1053.

Miller, M. B. and Basler, B. L. (2001). Quorum sensing in bacteria. *Ann. Rev. Microbiol.*, 55, 165–199.

Miller, S. L. (1953). Production of amino acids under possible primitive Earth conditions. *Science*, 117, 2351–2361.

  (1998). The endogenous synthesis of organic compounds. In A. Brack, ed., *The Molecular Origin of Life*. Cambridge University Press.

Miller, S. L. and Bada, J. (1988). Submarine hot springs and the origin of life. *Nature*, 334, 609–611.

  (1991). Extraterrestrial synthesis. *Nature*, 350, 388–389.

Miller, S. L. and Cleaves, H. J. (2007). Prebiotic chemistry on the primitive Earth. In I. Rigoutsos and G. Stephanopoulos, eds., *Systems Biology. Volume I, Genomics*. Oxford – New York: Oxford University Press.

Miller, S. L. and Lazcano, A. (1995). The origin and early evolution of life: prebiotic chemistry, the pre-RNA world, and time. *J. Mol. Evol.*, 41, 689–692.

Miller, S. L. and Parris, M. (1964). *Nature*, 204, 1248–1250.

Mingers, J. (1992). The problems of social autopoiesis. *Int. J. Gen. Syst.*, 21, 229–236.

  (1995). *Self-Producing Systems: Implications and Applications of Autopoiesis*. Plenum Press.

  (1997). A critical evaluation of Maturana's constructivist family therapy. *Syst. Practice*, 10(2): 137–151.

Miranda, M., Amicarelli, F., Poma, A., Ragnelli, A. M., and Arcadi, A. (1988). *Biochim. Biophys. Acta*, 966, 276–286.

Mojzsis, S. J., Harrison, T. M., and Pidgeon, R. T. (2001). Oxygen-isotope evidence from ancient zircons for liquid water at the Earth's surface 4,300 Myr ago. *Nature*, 409 (6817):178–181.

Monod, J. (1971). *Chance and Necessity*. A. A. Knopf.

Monro, R. and Marcker, K. A. (1967). Ribosome-catalysed reaction of puromycin with a formylmethionine-containing oligonucleotide. *Journal of Molecular Biology*, 25, 347–350.

Morgan, C. L. (1923). *Emergent Evolution*. William and Norgate.

Morigaki, K., Dallavalle, S., Walde, P., Colonna, S., and Luisi, P. L. (1997). Autopoietic self-reproduction of chiral fatty acid vesicles. *J. Am. Chem. Soc.*, 119, 292–301.

Morowitz, H. J. (1967). Biological self-replicating systems. *Prog. Theor. Biol.*, 1, 35–58.

  (1974). Manufacturing a living organism. *Hospital Practice*, 9, 210–215.

  (1992). *Beginnings of Cellular Life*. Yale University Press.

Morowitz, H. J., Deamer, D. W., and Smith, T. (1991). Biogenesis as an evolutionary process. *J. Mol. Evol.*, 33, 207–208.

Morowitz, H. J., Peterson, E., and Chang, S. (1995). The synthesis of glutamic acid in the absence of enzymes – implications for biogenesis. *Orig. Life Evol. Biosph.*, 25, 395–399.

Morowitz, H. J., Kostelnik, J. D., Yang, J., and Cody, G. D. (2000). The origin of intermediary metabolism. *Proc. Natl. Acad. Sci. USA*, 97, 7704–7709.

Mossa, G., Di Giulio, A., Dini, L., and Finazzi-Agrò, A. (1989). *Biochim. Biophys. Acta*, 986, 310–314.

Müller, D., Pitsch, S., Kittaka, A., Wagner, E., Wintner, C. E., and Eschenmoser, A. (1990). Chemie von α-Aminonitrilen. Aldomerisierung von Glycolaldehyd-phosphat zu racemischen Hexose-2,4,6-triphosphaten und (in Gegenwart von Formaldehyd) racemischen Pentose-2,4-diphosphaten: rac-Allose-2,4,6-triphosphat und rac-Ribose-2,4-diphosphat sind die Reaktionshauptprodukte. *Helvetica Chimica acta*, 73(5): 1410–1468.

Murtas, G., Kuruma, Y., Bianchini, P., Diaspro, A., and Luisi, P. L. (2007). Protein synthesis in liposomes with a minimal set of enzymes. *Biochem Biophys Res Comm.*, 363: 12–17.

Mushegian, A. (1999). The minimal genome concept. *Curr. Opin. Genetics Develop.*, 9, 709–714.

(2005). Protein content of minimal and ancestral ribosome. *RNA Society*, 11, 1400–1406.

Mushegian, A. and Koonin, E. V. (1996). A minimal gene set for cellular life derived by comparison of complete bacterial genomes. *Proc. Natl. Acad. Sci. USA*, 93, 10268–10273.

Nagel, E. (1961). *The Structure of Science*. Harcourt.

Nakajima, T., Yabushita, Y., and Tabushi, I. (1975). Amino acid synthesis through biogenic CO_2 fixation. *Nature*, 256, 60–61.

Nakashima, T., Toyota, H., Urabe, I., and Yomo, T. (2007). Effective selection system for experimental evolution of random polypeptides towards DNA-binding protein. *J. Bioscence and Bioengineering*, 103, 155–160.

Naoi, M., Naoi, M., Shimizu, T., Malviya, A. N., and Yagi, K. (1977). Permeability of amino acids into liposomes. *Biochim. Biophys. Acta*, 471, 305–310.

Nelson, K. E., Levy, M., and Miller, S. L. (2000). Peptide nucleic acids rather than RNA may have been the first genetic molecule. *Proc. Natl. Acad. Sci. USA*, 97(8): 3868–3871.

Neumann, J. von, and Burks, A., eds. (1966). *Theory of Self-Reproduction Automata*. University of Illinois Press.

Newport, J. (1987). Nuclear reconstitution in vitro: stages of assembly around protein–free DNA. *Cell*, 48, 205–217.

Nicolis, G. and Prigogine, I. (1977). *Self-Organization in Nonequilibrium Systems. From Dissipative Structures to Order Through Fluctuations*. New York: John Wiley & Sons.

Nierhaus, K. H. and Montejo, V. (1973). Protein involved in peptidyltransferase activity of Escherichia-coli ribosomes. *Proc. Natl. Acad. Sci. USA*, 70, 1931–1935.

Nierhaus, K. H. and Dohme, F. (1974). Total reconstitution of functionally active 50S ribosomal subunits from Escherichia coli. *Proc. Natl. Acad. Sci. USA*, 71, 4713–4717.

Nissen, P., Hansen, J., Ban, N., Moore, P. B., and Steitz, T. A. (2000). The structural basis of ribosome activity in peptide bond synthesis. *Science*, 289, 920–930.

Noble, D. (2006). *The Music of Life*. Oxford University Press.

Noireaux, V. and Libchaber, A. (2004). A vesicle bioreactor as a step toward an artificial cell assembly. *Proc. Natl. Acad. Sci. USA*, 101, 17669–17674.

Noireaux, V., Bar-Ziv, R., and Libchaber, A. (2003). Principles of cell-free genetic circuit assembly. *Proc. Natl. Acad. Sci. USA*, 100, 12672–12677.

Nomura, S. M., Yoshikawa, Y., Yoshikawa, K., *et al.* (2001). Towards proto-cells: "primitive" lipid vesicles encapsulating giant DNA and its histone complex. *Chem. Bio. Chem.*, 6, 457–459.

Nomura, S. M., Tsumoto, K., Yoshikawa, K., Ourisson, G., and Nakatani, Y. (2002). Towards proto-cells: "primitive" lipid vesicles encapsulating giant DNA and its histone complex. *Cell. Mol. Biol. Lett.*, 7, 245–246.

Nomura, S. M., Tsumoto, K., Hamada, T., *et al.* (2003). Gene expression within cell-sized lipid vesicles. *Chem. Bio. Chem.*, 4, 1172–1175.

Nooner, D. W., Gilbert, J. M., Gelpi, E., and Oró, J. (1976). Closed system Fischer-Tropsch synthesis over meteoritic iron, iron-ore and nickel-iron alloy. *Geochim. Cosmochim. Acta*, 40, 915–924.

Noyes, R. M. (1989). Some models of chemical oscillators. *J. Chem. Educ.*, 66, 190–191.

Nucara, L. (2014). *La Filosofia di Humberto Maturana*. Firenze: Casa Editrice Le Lettere.

Oberholzer, T. and Luisi, P. L. (2002). The use of lipsomes for constructing cell models. *J. Biol. Phys.*, 28, 733–744.

Oberholzer, T., Albrizio, M., and Luisi, P. L. (1995a). Polymerase chain reaction in liposomes. *Curr. Biol.*, 2, 677–682.

Oberholzer, T., Wick, R., Luisi, P. L., and Biebricher, C. K. (1995b). Enzymatic RNA replication in self-reproducing vesicles: an approach to a minimal cell. *Biochem. Biophys. Res. Commun.*, 207, 250–257.

Oberholzer, T., Nierhaus, K. H., and Luisi, P. L. (1999). Protein expression in liposomes. *Biochem. Biophys. Res. Commun.*, 261, 238–241.

O'Connor, T. (1994). Emergent properties. *Am. Phil. Q.*, 31, 91–104.

Okada, M. and Ohno, H. (1972). Assembly mechanism of tobacco mosaic virus particle from its ribonucleic acid and protein. *Molec. Gen. Genetics*, 114, 205–213.

Okasha, S. (2006). *Evolution and the Levels of Selection*. Oxford University Press.

Olsson, U. and Wennerstrom, H. (2002). On the ripening of vesicle dispersions. *J. Phys. Chem. B*, 106, 5135–5138.

O'Malley, M. A., Powell A., Davies J. F., and Calvert, J. (2007). *BioEssays*, 30, 57–65.

Oparin, A. I. (1924). *Proiskhozhdenie Zhisni*. Moskowski Rabocii.

    (1938). *Origin of Life*. McMillan.

    (1953). *The Origin of Life*. Dover Publications.

    (1957). *The Origin of Life on Earth*, 3rd edn. Academic Press.

    (1961). *Life: Its Nature, Origin and Development*. Oliver and Boyd.

Oppenheim, P. and Putnam, H. (1958). The unity of science as a working hypothesis. In H. Feigl, G. Maxwell and M. Scriven, eds., *Minnesota Studies in the Philosphy of Science*. University of Minnesota Press, pp. 3–36.

Orgel, L. E. (1968). Evolution of the genetic apparatus. *J. Mol. Biol.*, 38, 381–393.

    (1973). *The Origins of Life*. New York: John Wiley & Sons.

    (1994). The origin of life on the Earth. *Sci. Amer.*, 271(4): 53–61.

    (2000a). A simpler nucleic acid. *Science*, 290(5495): 1306–07.

    (2000b). Self-organizing biochemical cycles. *Proc. Natl. Acad. Sci. USA*, 97, 12503–12507.

    (2003). Some consequences of the RNA world hypothesis. *Orig. Life Evol. Biosph.*, 33, 211–218.

    (2004). Prebiotic chemistry and the origin of the RNA world. *Crit Rev Biochem Mol Biol.*, 39, 99–123.

Oró, J. (1960). Synthesis of adenine from ammonium cyanide. *Biochem. Bioph. Res. Commun.*, 2, 407–412.

    (1961). Amino acid synthesis from hydrogen cyanide under possible primitive Earth conditions. *Nature*, 190, 442–443.

    (1994). Early chemical stages in the origin of life. In S. Bengtson, ed., *Early Life on Earth: Nobel Symposium n. 84*. New York: Columbia University Press, pp. 48–59.

    (2002). Historical understanding of life's origin. In J. W. Schopf, ed., *Life's Origin, the Beginnings of Biological Evolution*. University of California Press, pp. 7–41.

Oró, J. and Kimball, A. P. (1961). Synthesis of purines under possible primitive Earth conditions. 1. Adenine from hydrogen cyanide. *Arch. Biochem. Biophys.*, 94, 221–227.

(1962). Synthesis of purines under possible primitive Earth conditions. 2. Purine intermediates from hydrogen cyanide. *Arch. Biochem. Biophys.*, 96, 293–313.

Ourisson, G. and Nakatani, Y. (1994). The terpenoid theory of the origin of cellular life: the evolution of terpenoids to cholesterol. *Chem. Biol.*, 1, 11–23.

(1999). Origin of cellular life: molecular foundations and new approaches. *Tetrahedron*, 55, 3183–3190.

Ousfouri, S., Stano, P., and Luisi, P. L. (2005). Condensed DNA in lipid microcompartments. *J. Phys. Chem. B.*, 109, 19929–19935.

Pääbo, S. (1993). Ancient DNA. *Scientific American*, 269(5): 60–66.

Palazzo, G. and Luisi, P. L. (1992). Solubilization of ribosomes in reverse micelles. *Biochem. Biophys. Res. Commun.*, 186, 1546–1552.

Paley, W. (1802; other sources report 1803). *Natural Theology, or Evidences of the Existence and Attributes of the Deity, Collected from the Appearances of Nature*, 12th edn (1986). Lincoln-Rembrandt Publishing.

Palyi, G., Zucchi, C., and Caglioti, L., eds. (2002). *Fundamentals of Life*. Elsevier.

Pantazatos, D. P. and McDonald R. C. (1999). Directly observed membrane fusion between oppositely charged phospholipid bilayers. *Membrane Biol.*, 170, 27–38.

Papahadjopoulos, D., Lopez, N., and Gabizon, A. (1989). Drug delivery by liposomes. In G. Lopez-Berenstein and I. J. Fidler, eds., *Liposomes in Therapy of Infectious Diseases and Cancer*. Alan Riss Inc., pp. 135–154.

Parens, E., Johnston, J., and Moses, J. (2008). Ethics. Do we need "synthetic bioethics"? *Science*, 321 (5895): 1449.

Park, J. H. and Lee, S. Y. (2008). Towards systems metabolic engineering of microorganisms for amino acid production. *Curr. Opin. Biotechnol.*, 19(5): 454–460.

Parker, E. T., Zhou, M., Burton, A. S., Glavin, D. P., Dworkin, J. P., *et al.* (2014). A plausible simultaneous synthesis of amino acids and simple peptides on the primordial Earth. *Proc. Natl. Acad. Sci. USA*, 108(12): 5526–4431.

Parsons, P. (1996). Dusting off panspermia. *Nature*, 383, 221–222.

Pascal, R., Boiteax, L., and Commeyras, A. (2005). From the prebiotic synthesis of amino acids towards a primitive translation apparatus. *Top Curr. Chem.*, 259–269.

Patel, B. H., Percivalle, C., Ritson, D. J., Duffy, C. D., and Sutherland, J. D. (2015). *Nature Chem.*, 7, 301–307.

Paul, N. and Joyce, G. F. (2002). A self-replicating ligase ribozyme. *Proc. Natl. Acad. Sci. USA*, 99, 12733–12740.

(2004). Minimal self-replicating systems. *Curr. Opin. Chem. Biol.*, 8, 634–639.

Paul, N., Springsteen, G., and Joyce, G. F. (2006). Conversion of a ribozyme to a deoxyribozyme through in vitro evolution. *Chem. & Biol.*, 13, 329–338.

Pereto, J., Lopez-Garcia, P., and Moreira, D. (2004). Ancestral lipid biosynthesis and early membrane evolution. *Trends Biochem. Sci.*, 29, 469–477.

Pfammatter, N., Guadalupe, A. A., and Luisi, P. L. (1989). Solubilization and activity of yeast cells in water-in-oil microemulsion. *Biochem. Biophys. Res. Commun.*, 161, 1244–1251.

Pfammatter, N., Hochköppler, A., and Luisi, P. L. (1992). Solubilization and growth of Candida pseudotropicalis in water-in-oil microemulsions. *Biotechnol. Bioeng.*, 40, 167–172.

Pfüller, U. (1986). *Mizellen, Vesikeln, Mikroemulsionen*. Springer Verlag.

Piaget, J. (1967). *Biologie et connaissance*. Gallimard.

Pietrini, A. V. and Luisi, P. L. (2002). Circular dichroic properties and average dimensions of DNA-containing reverse micellar aggregates. *Biochim. Biophys. Acta*, 1562, 57–62.

(2004). Cell-free protein synthesis through solubilisate exchange in water/oil emulsion compartments. *Chem. Bio. Chem*, 5, 1055–1062.

Pileni, M. P. (1981). Photoelectron transfer in reverse micelles – photo-reduction of cytochrome-c. *Chem. Phys. Lett.*, 81, 603–605.

Pizzarello, S. and Cronin, J. R. (2000). Non-racemic amino acids in the Murray and Murchison meteorites. *Geochim. Cosmochim. Acta*, 64, 329–338.

Pizzarello, S. and Weber, A. L. (2004). Prebiotic amino acids as asymmetric catalysts. *Science*, 303, 1151.

Plankensteiner, K., Righi, A., and Rode, B. M. (2002). Glycine and diglycine as possible catalytic factors in the prebiotic evolution of peptides. *Orig. Life Evol. Biosph.*, 32, 225–236.

Plasson, R., Biron, J. P., Cottet, H., Commeyras, A., and Taillades, J. (2002). Kinetic study of the polymerization of alpha-amino acid N-carboxyanhydrides in aqueous solution using capillary electrophoresis. *J. Chromatogr. A*, 952, 239–248.

Poerksen, B. (2004). *The Certainty of Uncertainty, Dialogues Introducing Constructivism*. Imprint Academic.

Pohorille, A. and Deamer, D. (2002). Artificial cells: prospects for biotechnology. *Trends Biotech.*, 20, 123–128.

Pojman, J. A., Craven, R., and Leard, D. C. (1994). Oscillations and chemical waves in the physical chemistry lab. *J. Chem. Educ.*, 71, 84–90.

Pollack, A. (2014). Scientists add letters to DNA's alphabet, raising hope and fear. *New York Times*, May 7.

Ponce de Leon, S. and Lazcano, A. (2003). Panspermia – true or false? *Lancet*, 362, 406–407.

Ponnamperuma, C. and Peterson, E. (1965). Peptide synthesis from aminoacids in aqueous solution. *Science*, 147, 1572.

Popa, R. (2004). *Between Necessity and Probability: Searching for the Definition and Origin of Life*. Springer Verlag.

Pope, M. T. and Muller, A. (1991). Polyoxometalate chemistry – an old field with new dimensions in several disciplines. *Angew. Chem. Int. Ed. Engl.*, 30, 34–48.

Portmann, M., Landau, E. M., and Luisi, P. L. (1991). Spectroscopic and rheological studies of enzymes in rigid lipidic matrices: the case of α-chymotrypsin in a lysolecithin/water cubic phase. *J. Phys. Chem.*, 95, 8437–8440.

Powner, M. W. and Sutherland, J. D. (2010). Phosphate-mediated interconversion of ribo- and arabino-configured prebiotic nucleotide intermediates. *Angew Chem. Int. Ed.*, 49, 4641–4643.

Powner, M. W., Gerland, B., and Sutherland, J. D. (2009). Synthesis of activated pyrimidine ribonucleotides in prebiotically plausible conditions. *Nature*, 459, 239–242.

Powner, M. W., Sutherland, J. D., and Szostak, J. W. (2010). Chemoselective multicomponent one-pot assembly of purine precursors in water. *J. Am. Chem. Soc.*, 24, 16677–16688.

Pozzi, G., Birault, V., and Werner, B. (1996). Single-chain polyprenyl phosphates form primitive membranes. *Angew. Chem. Int. Ed. Engl.*, 35, 177–179.

Prigogine, I. (1997). *The End of Certainty-Time, Chaos and the New Laws of Nature*. Free Press.

Prigogine, I. and Lefever, R. (1968). Symmetry breaking instabilities in dissipative systems. *J. Chem. Phys.*, 48, 1695–1700.

Prijambada, I.D., *et al.* (1996). Solubility of artificial proteins with random sequences. *Febs Letters*. 382, 21–25.

Primas, H. (1985). Can chemistry be reduced in physics? *Chem. Uns. Zeit*, 19, 160.
    (1993). In E. P. Fischer, ed., *Neue Horizonte 92/93: Ein Forum der Naturwissenschaften*. München: Piper.

Primas, H. (1998). Emergence in exact natural sciences. *Acta Politechnica Scand.*, 91, 86–87.

Pross, A. (2005). On the chemical nature and origin of teleonomy. *Origin of Life & Evol. Biosphere*, 35, 384–394.

Pryer, W. (1880). *Die Hypothesen über den Ursprung des Lebens*. Berlin.

Purrello, R. (2003). Lasting chiral memory. *Nature Mater.*, 2, 216–217.

Pyun, J., Zhou, X.-Z., Drockenmuller, E., and Hawker, C. J. (2003). *Mater. Chem.*, 13, 2653.

Qian, L. and Winfree, E. (2011). Scaling up digital circuit computation with DNA strand displacement cascades, *Science*, 332, 1196–1201. doi:10.1126/science.1200520.

Quack, M. (2002). *Angew. Chem.*, 41, 4618–4630.

Quack, M. and Stohner, J. (2003a). Combined multidimensional anharmonic and parity violating effects in CDBrClF. *J. Chem. Phys.*, 119, 11228–40.

(2003b). Molecular chirality and the fundamental symmetries of physics: influence of parity violation on rotovibrational frequencies and thermodynamic properties. *Chirality*, 15, 375–376.

Quack, M. and Stohner, J. (2014). The concept of law models in chemistry. *European review*, 22, S 50–86.

Raab, W. (1988). *Ärtzliche Kosmetologie*, 18, 213–224.

Radzicka, A., and Wolfenden, R. (1996). Rates of uncatalyzed peptide bond hydrolysis in neutral solution and the transition state affinities of proteases. *J. Am. Chem. Soc.*, 118, 6105–6109.

Rajamani, S., Vlassov, A., Benner, S., Coombs, A., Olasagasti, F., and Deamer, D. (2007). Lipid-assisted synthesis of RNA-like polymers from mononucleotides. Orig. Life Evol. Biosph. doi:10.1007/s11084-007-9113-2.

Ramundo-Orlando, A., Arcovito, C., Palombo, A., Serafino, A. L., and Mossa, G. (1993). *J. Liposome Res.*, 3, 717–724.

Ramundo-Orlando, A., Mattia, F., Palombo, A., and D'Inzeo, G. (2000). Effect of low frequency, low amplitude magnetic fields on the permeability of cationic liposomes entrapping carbonic anhydrase, Part II. *Bioelectromagnetics*, 21, 499–507.

Rao, M., Eichberg, J., and Oró, J. (1987). Synthesis of phosphatidylethanolamine under possible primitive earth conditions. *J. Mol. Evol.*, 25, 1–6.

Rasi, S., Mavelli, F., and Luisi, P. L. (2003). Cooperative micelle binding and matrix effect in oleate vesicle formation. *J. Phys. Chem. B*, 107, 14068–14076.

(2004). Matrix effect in oleat-micelles-vesicles transformations. *Orig. Life Evol. Bioph.*, 34, 215–24.

Rathman, J. F. (1996). Micellar catalysis. *Curr. Opin. Coll. Interf. Sci.*, 1, 514–518.

Rebek, J. (1994). A template for life. *Chem. Br.*, 30, 286–290.

Recordati, G. and Bellini, T. G. (2004). A definition of internal constancy and homeostasis in the context of non-equilibrium thermodynamics. *Experimental Physiology*, 89(1): 27–38.

Reichenbach, H. (1978). The aims and methods of physical knowledge. In M. Reichenbach and R. S. Cohen, eds., *Hans Reichenbach: Selected Writings 1909–53*. Translated by E. H. Schneewind. Reidel, pp. 81–225.

Reszka, R. (1998). Liposomes as drug carrier for diagnostics, cytostatics and genetic material. In J. E. Diederichs and R. H. Müller, eds., *Future Strategies for Drug Delivery with Particulate Systems*. Medpharm GmbH Scientific Publishers.

Ribo, J. M., Crusats, J., Sagues, F., Claret, J., and Rubires, R. (2001). Chiral sign induction during the formation of mesophases in stirred solutions. *Science*, 292, 2063–2066.

Riddle, D. S., Santiago, J. V., Bray-Hall, S. T., Doshi, N., Grantcharova, V. P., Yi, Q., and Baker, D. (1997). Functional rapidly folding proteins from simplified amino acid sequences. *Nature*, 4, 805–809.

Ringertz, N. R., Krondahl, U., and Coleman, J. R. (1978). Reconstitution of cells by fusion of cell fragments. *Experimental Cell Research*, 113, 233–246.

Rispens, T. and Engberts, J. B. F. N. (2001). Efficient catalysis of a Diels-Alder reaction by metallo-vesicles in aqueous solution. *Org. Lett.*, 3, 941–943.

Riste, T. and Sherrington, D., eds. (1996). *Physics of Biomaterials: Fluctuations, Selfassembly and Evolution* (Nato Science Series, Series E, Applied Sciences). Kluwer.

Rizzotti, M., ed. (1996). *Defining Life*. University of Padua.

Robertson, M. P. and Joyce, G. F. (2012). The origins of the RNA world. *Cold Spring Harb Perspect Biol.*, 4, 5. doi:10.1101/cshperspect.a003608.

Robertson, R. N. (1983). *The Lively Membrane*. Cambridge University Press.

Rode, B. M., Son, H. L., and Suwannachot, Y. (1999). The combination of salt induced peptide formation reaction and clay catalysis: a way to higher peptides under primitive earth conditions. *Orig. Life. Evol. Biosph.*, 29, 273–286.

Rodrigo, G., Landrain, T. E., and Jaramillo, A. (2012). De novo automated design of small RNA circuits for engineering synthetic riboregulation in living cells. *Proc. Natl. Acad. Sci. USA.*, 109, 15271–15276. doi:10.1073/pnas.1203831109.

Rodrigo, G., Landrain, T. E., Majer, E., Daròs, J.-A., and Jaramillo, A. (2013). Full design automation of multi-state RNA devices to program gene expression using energy-based optimization. *PLoS Comp. Biol.*, 9, e1003172. doi:10.1371/journal.pcbi.1003172.

Rogerson, M. L., Robinson, B. H., Bucak, S., and Walde, P. (2006). Kinetic studies of the interaction of fatty acis with phosphatidylcholine vesicles (Liposomes). *Colloid and Surfaces B: Biointerfaces*, 48, 24–34.

Rohl, C. A., Strauss, C. E., Misura, K. M., and Baker, D. (2004). Protein structure prediction using Rosetta. *Methods Enzymol.*, 383, 66–93.

Rojas, N. R. L., *et al.* (1997). De novo heme proteins from designed combinatorial libraries. *Protein Science*, 6, 2512–2524.

Rolle, F. (1863). *Ch. Darwin's Lehre von der Entstehung der Arten, in ihrer Anwendung auf die Schöpfunggeschichte*. J. C. Hermann.

Roseman, A., Lentz, B. R., Sears, B., Gibbes, D., and Thompson, T. E. (1978). Properties of sonicated vesicles of three synthetic phospholipids. *Chem. Phys. Lipids*, 21, 205–222.

Rotello, V., Hong, J. I., and Rebek, J. (1991). Sigmoidal growth in a self-replicating system. *J. Am. Chem. Soc.*, 113, 9422–9423.

Runion, G. E. (1990). *The Golden Section*. Dale Seymour Publications.

Rushdi, A. I. and Simoneit, B. R. (2001). Lipid formation by aqueous Fischer-Tropf-type synthesis over a temperature range 100 to 400 °C. *Orig. Life Evol. Biosph.*, 31, 103–118.

Sackmann, E. (1978). Dynamic molecular-organization in vesicles and membranes. *Ber. Bunsen-Gesell. Phys. Chem.*, 82, 891–909.

Sada, E., Katoh, S., Terashima, M., and Tsukiyama, K.-I. (1988). Entrapment of an ion-dependent enzyme into reverse-phase evaporation vesicles. *Biotechnol. Bioeng.*, 32, 826–830.

Sada, E., Katoh, S., Terashima, M., Shiraga, H., and Miura, Y. (1990). Stability and reaction characteristics of reverse-phase evaporation vesicles (revs) as enzyme containers. *Biotechnol. Bioeng.*, 36, 665–671.

Saetia, S., Liedl, K. R., Eder, A. H., and Rode, B. M. (1993). Evaporation cycle experiments: a simulation of salt-induced peptide synthesis under possible prebiotic conditions. *Orig. Life Evol. Biosph.*, 3, 167–176.

Sagan, C. (1985). *Cosmos*. Ballantine Publishing.

(1994). The search for extraterrestrial life. *Sci. Amer.*, 271(4): 71–77.

Saito, H., Yamada, A., Ohmori, R., Kato, Y., Yamanaka, T., *et al.* (2007). Towards constructing synthetic cells: RNA/RNP evolution and cell-free translational systems in giant liposomes. *Micro-NanoMechatronics and Human Science*, 2007. MHS '07 (International Symposium on), 286–291.

Samoylov, A. M., Samoylova, T. I., Pustovyy, O. M., *et al.* (2005). Novel metal cluster isolated from blood are lethal to cancer cells. *Cells Tissues Organs*, 179 (3): 115–124.

Sanchez, R. A., Ferris, J. P., and Orgel, L. E. (1966). Conditions for purine synthesis: did prebiotic synthesis occur at low temperature? *Science*, 153, 72–73.

(1968). Studies in prebiotic synthesis. IV, The conversion of 4-aminoimidazole-5-carbonitrile derivatives to purines. *J. Mol. Biol.*, 38, 121–28.

Scartazzini, R. and Luisi, P. L. (1988). Organogels from lecithins. *J. Phys. Chem.*, 92, 829–833.

Schaerer, A. A. (2002). Conceptual conditions for conceiving life – a solution for grasping its principle, not mere appearances. In G. Palyi, C. Zucchi, and L. Caglioti, eds., *Fundamentals of Life*. Elsevier, pp. 589–624.

Schafmeister, C. E., *et al.* (1997). A designed four-helix bundle protein with native like structure. *Nature*, 4, 1039–1046.

Schmidli, P. K., Schurtenberger, P., and Luisi, P. L. (1991). Liposome-mediated enzymatic synthesis of phosphatidylcholine as an approach to self-replicating liposomes. *J. Am. Chem. Soc.*, 113, 8127–8130.

Schopf J. W. and C. Klein, eds. (1992). In *The Proterozoic Atmosphere*. Cambridge University Press.

Schopf, J. W. (1993). Microfossils of the early archean apex chert: new evidence of the antiquity of life. *Science*, 260, 640–646.

(1998). Chemical evolution and the origin of life. In A. Brack, ed., *The Molecular Origin of Life*. Cambridge University Press.

(2002). *Life's Origin*. University of California Press.

Schröder, J. (1998). Emergence: non-deducibility or downward causation? *Phil. Q.*, 48, 434–452.

Schulze, H. and Nierhaus, K. H. (1982). Minimal set of ribosomal components for reconstitution of the peptidyltransferase activity. *Embo Journal*, 1, 609–613.

Schurtenberger, P., Scartazzini, R., Magid, L. J., Leser, M. E., and Luisi, P. L. (1990). Structural and dynamic properties of polymer-like reverse micelles. *J. Phys. Chem.*, 94, 3695–3701.

Schurtenberger, P., Magid, L. J., King, S. M., and Lindner, P. (1991). Cylindrical structure and flexibility of polymerlike lecithin reverse micelles. *J. Phys. Chem.*, 95, 4173–4176.

Schuster, P. and Swetina, J. (1988). Stationary mutant distributions and evolutionary optimization. *Bull. Math. Biol.*, 50, 636–660.

Schwabe, C. (2001). *The Genomic Potential Hypothesis, a Chemist's View on the Origin and Evolution of Life*. Landes Bioscience.

Schwabe, C. and Warr, G. W. (1984). A polyphyletic view of evolution. The genetic potential hypothesis. *Persp Biol. Med.*, 27, 465–485.

Seddon, J. M. (1990). Structure of the inverted hexagonal (HII) phase, and non-lamellar phase transitions of lipids. *Biochim. Biophys. Acta*, 1031, 1–69.

Seddon, J. M., Hogan, J. J., Warrender, N. A., and Pebay-Peyroula, E. (1990). *Prog. Coll. Polym. Sci.*, 81, 189–197.

Sela, M., White, F. H. Jr., and Anfinsen, C. B. (1957). Reductive cleavage of disulfide bridges in ribonuclease. *Science*, 125, 691–692.

Selsis, F. (2000). *Modèle d'évolution physico-chimique des atmosphères de planètes telluriques. Application à l'atmosphère primitive terrestre et aux planètes extrasolaires*. Ph.D. thesis, Université Bordeaux 1 (France).

Severin, K., Lee, D. H., Kennan, A. J., and Ghadiri, M. R. (1997). A synthetic peptide ligase. *Nature*, 16(389): 706–709.

Shapiro, R. (1984). The improbability of prebiotic nucleic acid synthesis. *Orig. Life*, 14(1–4): 565–570.

(1986). *Origins: a Skeptic's Guide to the Creation of Life on Earth*. Summit Books.

(1988). Prebiotic ribose synthesis: a critical analysis. *Orig. Life Evol. Biosph.*, 18, 71–85.

(1995). The prebiotic role of adenine: a critical analysis. *Orig. Life Evol. Biosph.*, 25, 83–98.

(2000). A replicator was not involved in the origin of life. *IUBMB Life*, 49, 173–176.

Shen, C., Lazcano, A., and Oró, J. (1990a). The enhancement activities of histidyl-histidine in some prebiotic reactions. *J. Mol. Evol.*, 31, 445–452.

Shen, C., Mills, T., and Oró, J. (1990b). Prebiotic synthesis of histidyl-histidine. *J. Mol. Evol.*, 31, 175–179.

Shen, C., Yang, L., Miller, S. L., and Oró, J. (1990c). Prebiotic synthesis of histidine. *J. Mol. Evol.*, 31, 167–174.

Sheng, J., Li, L., Engelhart, A.E., Gan, J., Wang, J., and Szostak, J.W. (2014). Structural insights into the effects of 2'- 5' linkages on the RNA duplex. *Proc. Natl. Acad. Sci. USA*, 111(8): 3050–3055. doi:10.1073/pnas.1317799111.

Shermer, M. (2003). Is the universe fine-tuned for life? *Sci. Amer.*, Jan. 23.

Shimizu, Y., Inoue, A., Tomari, Y., Suzuki, T., Yokogawa, T., Nishikawa, K., and Ueda, T. (2001). Cell-Free Translation Reconstituted with Purified Components. *Nat. Biotechnol.*, 19, 751–755.

Shimizu, Y., Kanamori, T., and Ueda, T. (2005). Protein Synthesis by Pure Translation. *Systems Methods*, 36, 299–304.

Shimoyama, A. and Ogasawara, R. (2002). Dipeptides and diketopiperazines in the Yamato-791198 and Murchison carbonaceous chondrites. *Orig. Life Evol. Biosph.*, 32(2): 165–179.

Shimkets, L. J. (1998). Structure and sizes of genomes of the archaea and bacteria. In F. J. De Bruijn, J. R. Lupskin, and G. M. Weinstock, eds., *Bacterial Genomes: Physical Structure and Analysis*. Kluwer, pp. 5–11.

Shiner, E. K., Rumbaugh, K. P., and Williams, S. C. (2005). Interkingdom signaling: deciphering the language of acyl homoserine lactones. *FEMS Microbiol. Rev.*, 29, 935–947.

Shohda, K. and Sugawara, T. (2006). DNA polymerization on the inner surface of a giant liposome for synthesizing an artificial cell model. *Soft Matter*, 2, 402–408.

Sievers, D. and von Kiedrowski, G. (1994). Self-replication of complementary nucleotide-based oligomers. *Nature*, 369, 221–224.

Sievers, D., Achilles, T., Burmeister, J., *et al.* (1994). Molecular replication – from minimal to complex systems. In G. Fleischacker, S. Colonna and P. L. Luisi, eds., *Self-Production of Supramolecular Structures*. Kluwer Publishers.

Silin, V. I., Wieder, H., Woodward, J. T., *et al.* (2002). The role of surface free energy on the formation of hybrid bilayer membranes. *J. Am. Chem. Soc.*, 124, 14676–14683.

Silverman, J. A., Balakrishnan, R., and Harbury, P. B. (2001). Reverse engineering the $(\beta/\alpha)_8$ barrel fold. *Proc. Natl. Acad. Sci. USA*, 98, 3092–3097.

Simpson, G. G. (1973). Added comments on "The non-prevalence of humanoids." In C. Sagan, ed., *Communication with Extraterrestrial Intelligence*. MIT Press, pp. 362–364.

Smith, H. O., Hutchison, C. A. III, Pfannkoch, C., and Venter, J. C. (2003). Generating a synthetic genome by whole genome assembly: phiX174 bacteriophage from synthetic oligonucleotides. *Proc. Natl. Acad. Sci. USA*, 100, 15440–15445.

Smith, R. S. and Iglewski, B. H. (2003). P. aeruginosa quorum sensing systems and virulence. *Curr. Opin. Microbiol.*, 6, 56–60.

Soai, K., ed. (2008). *Amplification of Chirality* (Topics in Current Chemistry, vol. 284). Springer.

Soga, H., Fuji, S., Yomo, T., Kato, Y., Watanabe, H., and Matsuura, T. (2014). In vitro membrane protein synthesis inside cell-sized vesicles reveals the dependence of membrane protein integration on vesicle volume. *ACS Synth. Biol.*, 3(6): 372–379.

Solomon, B. and Miller, I. R. (1976). Interaction of glucose oxidase with phospholipid vesicles. *Biochim. Biophys. Acta*, 455, 332–342.

Sommer, A. P. and Wickramasinghe, N. C. (2005). Functions and possible provenance of primordial proteins-part II: Microorganism aggregation in clouds triggered by climate change. *J. Proteome Res.*, 4, 180–184.

Spang, A., Saw, J. H., Jørgensen, S. L., Zaremba-Niedzwiedzka, K., Martijn, J., *et al.* (2015). Complex archaea that bridge the gap between prokaryotes and eukaryotes. *Nature*, 521(7551): 173–179. doi: 10.1038/nature14447.

Sperry, R. W. (1945). *Journal of Neurophysiology*, 8, 15.

(1986). Discussions: macro- versus microdeterminism. *Phil. Sci.*, 53, 265–270.

Spirin, A. (1986). *Ribosome Structure and Protein Synthesis*. Benjamin Cummings Publishing.

Sprinzak, D. and Elowitz, M. B. (2005). Reconstruction of genetic circuits. *Nature*, 438 (7067): 443–448.

Stano, P. and Luisi, P. L. (2007). Basic questions about the origins of life: proceedings of the Erice International School of Complexity. *Origins of Life and Evolution of Biospheres*, 37, 303–307.

Stano, P., Bufali, S., Pisano, C., *et al.* (2004). Novel campotothecin analogue (Gimatecan)-containing liposomes prepared by the ethanol injection method. *J. Lipos. Res.*, 14, 87–109.

Stano, P., Wehrli, E., and Luisi, P.L. (2006). Insights on the oleate vesicles self-reproduction. *J. Physics Condensed Matter*, 18 S2231–S2238.

Stano, P., Carrara, P., Kuruma, Y., de Souza, T. P., and Luisi, P. L. (2011). Compartmentalized reactions as a case of soft-matter biotechnology: synthesis of proteins and nucleic acids inside lipid vesicles. *J. Mater. Chem.*, 21, 18887–18902.

Stano, P., Rampioni, G., Damiano, L., D'Angelo, F., Carrara, P., Leoni, L., and Luisi, P. L. (2014). Towards the engineering of chemical communication between semi-synthetic and natural cells. In S. Cagnoni, M. Mirolli and M. Villani, eds., *Evolution, Complexity and Artificial Life*. Dordrecht: Springer, pp. 91–104.

Stetter, K. O. (1998). Hyperthermophiles and their possible role as ancestors of modern life. In A. Brack, ed., *The Molecular Origin of Life*. Cambridge University Press.

Stocks, P. G. and Schwartz, A. W. (1982). Basic nitrogen-heterocyclic compounds in the Murchison meteorite. *Geochim. Cosmochim. Acta*, 46, 309–315.

Strogatz, S. H. (1994). *Non Linear Dynamics and Chaos, With Applications to Physics, Biology, Chemistry, and Engineering*. Perseus Book Group.

Strogatz, S. (2001). Exploring complex networks. *Nature*, 410, 268–276.

Stryer, L. (1975). *Biochemistry*. Freeman and Co.

Summers, D. P and Chang, S. (1993). Prebiotic ammonia from iron(II) reduction of nitrite on the early earth. *Nature*, 365, 630–633.

Summers, D. P and Lerner, N. R. (1998). Ammonia from iron(II) reduction of nitrite and the Strecker synthesis: do iron(II) and cyanide interfere with each other? *Orig. Life Evol. Biosphere*, 28, 1–11.

Sunami, T., Sato, K., Matsuura, T., Tsukada, K., Urabe, I., and Yomo, T. (2006). Femtoliter compartment in liposomes for in vitro selection of proteins. *Analytical Biochemistry*, 357, 128–136.

Sunami, T., Caschera, F., Morita, Y., Toyota, T., Nishimura, K., Matsuura, T., Suzuki, H., Hanczyc, M. M., and Yomo, T. (2010). Detection of association and fusion of giant vesicles using a fluorescence-activated cell sorter. *Langmuir*, 26, 15098–15103.

Susskind, L. (2005). *The Cosmic Landscape: String Theory and the Illusion of the Intelligent Design*. Little Brown.

Sutherland, J. D. (2007). Looking beyond the RNA structural neighborhood for potentially primordial genetic system. *Angewandte Chemie Int. Ed.*, 46, 2354–2356.

Sutherland, J. D., Anastasi, C., Buchet, F. F., Crower, M. A., Parkes, A. L., Powner, M. W., and Smith, J. M. (2007). RNA: prebiotic product, or biotic invention. *Chemistry & Biodiversity*, 4(4): 721–739.

Swairjo, M. A., Seaton, B. A., and Roberts, M. F. (1994). Biochem. *Biophys. Acta*, 1191, 354–361.

Szathmáry, E. (2002). Units of evolution and units of life. In G. Palyi, L. Zucchi and L. Caglioti, eds., *Fundamentals of Life*. Elsevier SAS, pp. 181–195.

Szostak, J. W., Bartel, D. P., and Luisi, P. L. (2001). Synthesizing life. *Nature*, 409, 387–390.

Taillades, J., Cottet, H., Garrel, L., *et al.* (1999). *N*-Carbamoyl amino acid solid–gas nitrosation by NO/NOx: a new route to oligopeptides via α-amino acid *N*-carboxyanhydride. Prebiotic implications. *J. Mol. Evol.*, 48, 638–645.

Takahashi, Y. and Mihara, H. (2004). Construction of a chemically and conformationally self-replicating system of amyloid-like fibrils. *Bioorg. Med. Chem.*, 12, 693–699.

Takakura, K., Toyota, T., and Sugawara, T. (2003). A novel system of self-reproducing giant vesicles. *J. Am. Chem. Soc.*, 125, 8134–8140. [See also: Takakura, K., Toyota, T., Yamada, K., *et al.* (2002). Morphological change of giant vesicles triggered by dehydrocondensation reaction. *ChemLett.*, 31, 404–405.]

Takakura, K., Yamamoto, T., Kurihara, K., Toyota, T., Ohnuma, K., and Sugawara, T. (2014). Spontaneous transformation from micelles to vesicles associated with sequential conversions of comprising amphiphiles within assemblies. *Chem. Commun. (Camb)*., 50(17): 2190–2192. doi:10.1039/c3cc47786j.

Tanford, C. (1978). The hydrophobic effect and the organization of living matter. *Science*, 200, 1012–1018.

Tegmark, M. (2003). Parallel universes. *Scientific American*, May, 41–51.

Teramoto, N., Imanishi, Y., and Yoshihiro, I. (2000). In vitro selection of a ligase ribozyme carrying alkylamino groups in the side chains. *Bioconjugate Chem.*, 11, 744–748.

Thomas, C. F. and Luisi, P. L. (2004). Novel properties of DDAB: matrix effect and interaction with oleate. *J. Phys. Chem. B*, 108, 11285–11290.

(2005). RNA selectively interacts with vesicles depending on their size. *J. Phys. Chem. B.*, 109, 14544–14550.

Thompson, E. (2007). *Mind in Life*. The Belknap Press of the Harvard University Press.

(2014). *Waking, Dreaming, Being: Information and Consciousness in Neuroscience*. Columbia University Press.

Thompson, E. and Varela, F. J. (2001). Radical embodiment: neural dynamics and consciousness. *Trends Cog. Sci.*, 5, 418–425.

Torre, P., Keating, C. D., and Mansy, S. S. (2014). Multiphase water-in-oil emulsion droplets for cell-free transcription-translation. *Langmuir*, 30, 5695–5699.

Traub, P. and Nomura, M. (1968). Structure and function of E. coli ribosomes, V. reconstitution of functionally active 30S ribosomal particles from RNA and proteins. *Proc. Nat. Acad. Sci. USA.*, 59, 737–741.

Tsumoto, K., Nomura, S. M., Nakatani, Y., and Yoshikawa, K. (2002). Giant liposome as a biochemical reactor: transcription of DNA and transportation by laser tweezers. *Langmuir*, 17, 7225–7228.

Turing, A. (1952). The chemical basis of morphogenesis. *Phil. Trans. Royal. Soc. London B*, 237, 37.

Ulbricht, W. and Hoffmann, H. (1993). Physikalische Chemie der Tenside. In K. Kosswig, and H. Stache, ed., *Die Tenside*. Carl Hanser Verlag, pp. 1–114.

Ulman, A. (1996). Formation and structure of self-assembled monolayers. *Chem. Rev.*, 96, 1533–54.

Uster, P. S. and Deamer, D. W. (1981). Fusion competence of phosphatidylserine-containing liposomes quantitatively measured by a fluorescence resonance energy transfer assay. *Arch. Biochem. Biophys.*, 209(2): 385–395.

Uwin, P. J. R., Webb, R. I., and Taylor, A. P. (1998) Novel nano-organisms from Australian sandstones. *Am. Mineralogist*, 83, 1541–1550.

Valenzuela, C. Y. (2002). Does biotic life exist? In G. Palyi, C. Zucchi, and L. Cagiliati, eds., *Fundamentals of Life*. Elsevier, pp. 331–334.

Vancanneyt, M., Schut, F., Snauwaert, C., Goris, J., Swings, J., and Gottschal, J. C. (2001). *Sphingopyxis alaskensis* sp. Nov, a dominant bacterium from a marine oligotrophic environment. *Int. J. Syst. Evol. Microbiol.*, 51, 73–79.

Van der Gulik, P., Massar, S., Gilis, D., Buhrman, H., and Rooman, M. (2009). The first peptides: the evolutionary transition between prebiotic amino acids and early proteins. *Journal of Theoretical Biology*, 261(4): 531–553.

Varela, F. J. (1979). *Principles of Biological Autonomy*. North Holland/Elsevier. Translated in French, *see infra* (Varela, 1989b).

   (1989a). Reflections on the circulation of concepts between a biology of cognition and systemic family therapy. *Family Process*, 28, 15–24.

   (1989b). *Autonomie et Connaissance*. Seuil.

   (1999). *Ethical Know-How: Action, Wisdom, and Cognition*. Stanford University Press.

   (2000). *El Fenómeno de la Vita*. Dolmen Ensayo.

Varela, F. J., Maturana, H. R., and Uribe, R. B. (1974). Autopoiesis: the organization of living system, its characterization and a model. *Biosystems*, 5, 187–196.

Varela, F. J., Thompson, E., and Rosch, E. (1991). *The Embodied Mind: Cognitive Science and Human Experience*. Cambridge, MA: MIT Press.

Veomett, G., Prescott, D. M., Shay, J., and Porter, K. R. (1974). Reconstruction of mammalian cells from nuclear and cytoplasmic components separated by treatment with cytochalasin B. *Proc. Nat. Acad. Sci. USA.*, 71, 1999–2002.

Vilanova, C. and Porcar, M. (2014). Table 1: Finalist projects in the iGEM competition, 2006–2013. *Nature Biotechnology*, 32, 420–424. doi:10.1038/nbt.2899.

Villarreal, L. P. (2009). The source of self: genetic parasites and the origin of adaptive immunity. *Ann. N. Y. Acad. Sci.*, 1178, 194–232.

   (2011). Viral ancestors of antiviral systems. *Viruses*, 3(10): 1933–1958.

Villarreal, L. P., and Witzany, G. (2010). Viruses are essential agents within the roots and stem of the tree of life. *J. Theor. Biol.*, 262(4): 698–710.

Wächtershäuser, G. (1988). Before enzymes and templates: theory of surface metabolism. *Microbiol. Rev.*, 52, 452–484.

   (1990a). Evolution of the first metabolic cycles. *Proc. Natl. Acad. Sci. USA*, 87, 200–204.

   (1990b). The case for the chemoautotrophic origin of life in the iron–sulfur world. *Origin Life Evol. Biosph.*, 20, 173–176.

   (1992). Groundworks for an evolutionary biochemistry: the iron–sulfur world. *Prog. Biophys. Mol. Biol.*, 58, 85–201.

   (1997). The origin of life and its methodological challenge. *J. Theor. Biol.*, 187, 483–494.

   (2000). Life as we don't know it. *Science*, 289, 1307–1308.

Waks, M. (1986). Proteins and peptides in water-restricted environments. *Proteins*, 1, 4–15.

Waks, Z. and Silver, P. A. (2009). Engineering a synthetic dual-organism system for hydrogen production. *Appl. Environ. Microbiol.*, 75(7): 1867–1875.

Walde, P. (2000). Enzymatic reactions in giant vesicles. In P. L. Luisi and P. Walde, eds., *Giant Vesicles, Perspectives in Supramolecular Chemistry*. John Wiley & Sons, pp. 297–311.

Walde, P. and Ishikawa, S. (2001). Enzymes inside lipid vesicles: preparation, reactivity and applications. *Biomol. Eng.*, 18, 143–177.

Walde, P. and Mazzetta, B. (1998). Bilayer permeability-based substrate selectivity of an enzyme in liposomes. *Biotechnol. Bioeng.*, 57, 216–219.

Walde, P., Goto, A., Monnard, P.-A., Wessicken, M., and Luisi, P. L. (1994a). Oparin's reactions revisited: enzymatic synthesis of poly(adenylic acid) in micelles and self-reproducing vesicles. *J. Am. Chem. Soc.*, 116(17): 7541–7547.

Walde, P., Wick, R., Fresta, M., Mangone, A., and Luisi, P. L. (1994b). Autopoietic self-reproduction of fatty acid vesicles. *J. Am. Chem. Soc.*, 116(26): 11649–11654.

Walde, P., Cosentino, K., Hengel, H., and Stano, P. (2010). Giant vesicles: preparations and applications. *ChemBioChem*, 11, 848–865.

Walter, K. U., Vamvaca, K., and Hilvert, D. (2005). An active enzyme constructed from a 9-amino acid alphabet. *The Journal of Biological Chemistry*, 280, 37742–37746.

Wang, J. and Wang, W. (1999). A computational approach to simplifying the protein folding alphabet. *Nature Structural Biology*, 6, 1033–1038.

Weber, A. (2002). The "surplus of meaning." Biosemiotic aspects in Francisco J. Varela's philosophy of cognition. *Cybernetics Human Knowing*, 9, 11–29.

Weber, W., Lienhart, C., El-Baba, M. D., and Fussenegger, M. (2009). A biotin-triggered genetic switch in mammalian cells and mice. *Metabolic Engineering*, 11(2): 117–124.

Wei, Y. and Hecht, M. H. (2004). Enzyme-like proteins from an unselected library of designed amino acid sequences. *Protein Engineering, Design & Selection*, 17, 67–75.

Wei, Y., Liu, T., Sazinskiy, S. L., Moffet, D. A., Pelczer, I., and Hecht, M. H. (2003). Stably folded de novo proteins from a designed combinatorial library. *Protein Science*, 12, 92–102.

Wenneström, H. and Lindmann, B. (1979). *Phys. Rev.*, 52, 1–86.

Westhof, E. and Hardy, N., eds. (2004). *Folding and Self-Assembly of Biological Macromolecules*. World Scientific Publishing Company.

Whitesides, G. M. and Boncheva, M. (2002). Beyond molecules: self-assembly of mesoscopic and macroscopic components. *Proc. Natl. Acad. Sci. USA*, 99, 4769–4774.

Whitesides, G. M. and Grzybowski, B. (2002). Self-assembly at all scales, *Science*, 295, 2418–2421.

Whitesides, G. M., Mathias, J. P., and Seto, C. T. (1991). Molecular self-assembly and nanochemistry – a chemical strategy for the synthesis of nanostructures. *Science*, 254, 1312–1319.

Whitfield, J. (2006). *In the Beat of a Heart: Life, Energy, and the Unity of Nature*. National Academies Press.

Wick, R., Walde, P., and Luisi, P. L. (1995). Autocatalytic self-reproduction of giant vesicles. *J. Am. Chem. Soc.*, 117, 1435–1436.

Wick, R., Angelova, M., Walde, P., and Luisi, P. L. (1996). Microinjection into giant vesicles and light microscopy investigations of enzyme mediated vesicle transformations. *Chemistry and Biology*, 3, 105–111.

Wieczorek, R., Dorr, M., Chotera, A., Luisi P. L., and Monnard, P.-A. (2013). Formation of RNA phosphodiester bond by histidine containing dipeptides. *ChemBioChem*, 14(2): 217–223.

Williams, T. A., Foster, P. G., Cox, C. J., and Embley, T. M. (2013). An archaeal origin of eukaryotes supports only two primary domains of life. *Nature*, 504, 231–236.

Willimann, H. and Luisi, P. L. (1991). Lecithin organogels as matrix for the transdermal transport of drugs. *Biochem. Biophys. Res. Commun.*, 177, 897–900.

Wilschut, J., Duzgunes, N., Fraley, R., and Papahadjopoulos, D. (1980). Studies on the mechanism of membrane-fusion – kinetics of calcium-ion induced fusion of phosphatidylserine vesicles followed by a new assay for mixing of aqueous vesicle contents. *Biochemistry*, 19, 6011–6021.

Wilson, T. L. (2001). The search for extraterrestrial intelligence. *Nature*, 409, 1110–1114.

Wimsatt, W. C. (1972). Complexity and organization. In K. F. Schaffner and R. S. Cohen, eds., *Boston Studies in the Philosophy of Science, Proceedings of the Philosphy of Science Association*. Reidel, pp. 67–86.

(1976a). Reductionism, levels of organization, and the mind-body problem. In G. Globus, G. Maxwell, and I. Savodinik, eds., *Consciousness and the Brain*. Plenum Press, pp. 205–266.

(1976b). Reductive explanation, a functional account. In C. A. Hooker, G. Pearse, A. C. Michealos, and J. W. van Evra, eds., *Proceedings of the Meetings of the Philosophy of Science Association 1974*. Reidel, pp. 671–710.

Winfree, A. T. (1984). The prehistory of the Belousov-Zhabotinsky oscillator. *J. Chem. Educ.*, 61, 661–663.

Woese, C. (1967). *The genetic code*. New York: Harper & Row.

Woese, C. R. (1979). A proposal concerning the origin of life on the planet Earth. *J. Mol. Evol.*, 13, 95–101.

Wolynes, P. G. (1997). As simple as can be? *Nature Structural Biology*, 11, 871–874.

Wong, J. T. (1975). A co-evolution theory of the genetic code. *Proc. Natl. Acad. Sci. USA*, 72, 1909–1912.

Wong, J. T. and Xue, H. (2002). Self-perfecting evolution of heteropolymer building blocks. In *Fundamentals of Life, Editions scientifiques et medicales Elsevier* SAS. Paris.

Wood, W. B. (1973). Genetic control of bacteriophage T4 morphogenesis. In F. J. Ruddle, ed., *Genetic Mechanisms of Development*. Academic Press, pp. 29–46.

Woodle, M. C. and Lasic, D. D. (1992). *Biochim. Biophys. Acta*, 1113, 171–199.

Yao, S., Ghosh, I., Zutshi, R., and Chmielewski, J. (1997). A pH-modulated self-replicating peptide. *J. Am. Chem. Soc.*, 119, 10559–10560.

Yao, S., Ghosh, I., and Chmielewski, J. (1998). Selective amplification by auto- and cross-catalysis in a replicating peptide system. *Nature*, 396, 447–450.

Yaroslavov, A. A., Udalyk, O. Y., Kabanov, V. A., and Menger, F. M. (1997). Manipulation of electric charge on vesicles by means of ionic surfactants: effects of charge on vesicle mobility, integrity, and lipid dynamics. *Chem. Eur. J.*, 3, 690–695.

Yarus, M. (2011). Getting past the RNA world: the initial Darwinian ancestor. *Cold Spring Harb Perspect Biol.*, 3(4). doi:10.1101/cshperspect.a003590.

Yonath, A. (2010). Polar bears, antibiotics, and the evolving ribosome (Nobel Lecture). *Angew Chem Int Ed Engl*, 49, 4341–4354.

(2012). Ribosomes: Ribozymes that Survived Evolution Pressures but Is Paralyzed by Tiny Antibiotics. In M. A. Carrondo and P. Spadon, eds., *NATO Science for Peace and Security Series A: Chemistry and Biology, Macromolecular Crystallography*, pp. 195–208.

Yoshimoto, M., Walde, P., Umakoshi, H., and Kuboi, R. (1999). Conformationally changed cytochrome c-mediated fusion of enzyme- and substrate-containing liposomes. *Biotechnol. Prog.*, 15, 689–696.

Yu, W., Sato, K., Wakabayashi, M., *et al.* (2001). Synthesis of functional protein in liposome. *J. Biosc. Bioeng.*, 92, 590–593.

Yuen, G. U. and Knenvolden, K. A. (1973). Monocarboxylic acids in Murray and Murchison carbonaceous meteorites. *Nature*, 246, 301–302.

Yuen, G. U., Lawless J. G., and Edelson, E. H. (1981). Quantification of monocarboxylic acids from a spark discharge synthesis. *J. Mol. Evol.*, 17, 43–47.

Zaher, H. S. and Unrau, P. J. (2007). Selection of an improved RNA polymerase ribozyme with superior extension and fidelity. *RNA*, 13, 1017–1026.

Zamarev, K. I., Romannikov, V. N., Salganik, R. I., Wlassoff, W. A., and Khramtsov, V. V. (1997). Modelling of the prebiotic synthesis of oligopeptides: silicate catalysts help to overcome the critical stage. *Orig. Life Evol. Biosph.*, 27, 325–337.

Zampieri, G. G., Jäckle, H., and Luisi, P. L. (1986). Determination of the structural parameters of reverse micelles after uptake of proteins. *J. Phys. Chem.*, 90, 1849.

Zeleny, M. (1977). Self-organization of living systems formal model of autopoiesis. *Int. J. Gen. Syst.*, 4, 13–28.

Zelinski, W. S. and Orgel, L. E. (1987). Autocatalytic synthesis of a tetranucleotide analogue. *Nature*, 327, 346–347.

Zeng, F. W. and Zimmermann, S. C. (1997). Dendrimers in supramolecular chemistry: from molecular recognition to self-assembly. *Chem. Rev.*, 97, 1681–1712.

Zepik, H. H., Bloechliger, E., and Luisi, P. L. (2001). A chemical model of homeostasis. *Angew. Chem. Int. Ed. Engl.*, 40, 199–202.

Zhang, B. and Cech, T. R. (1998). Peptidyl-transferase ribozymes: trans reactions, structural characterization and ribosomal RNA-like features. *Chem. Biol.*, 5, 539–553.

Zhao, M. and Bada, J. L. (1989). Extraterrestrial amino acids in cretaceous/tertiary boundary sediments at Stevns Klint, Denmark. *Nature*, 339, 463–465.

Zhao, Y., Liu, X., Wu, M., Tao, W., and Zhai, Z. (2000). In vitro nuclear reconstitution could be induced in a plant cell-free system. *FEBS Letters*, 480 (2–3): 208–212.

Zhu, J., Zhang, L., and Reszka, R. (1996). Liposome-mediated delivery of genes and oligonucleotides for the treatment of brain tumors. In G. Gregoriadis and B. McCormack, eds., *Targeting of Drugs: Strategies for Oligonucleide and Gene Delivery in Therapy*. Plenum Press, pp. 169–187.

Zhu, J., Zhang, L., Hanisch, U. K., Felgner, P. L., and Reszka, R. (1996). In vivo gene therapy of experimental brain tumors by continuous administration of DNA-liposome complexes. *Gene Therapy*, 3, 472–476.

Ziegler, M., Davis, A. V., Johnson, D. W., and Raymond, K. N. (2003). Supramolecular chirality: a reporter of structural memory. *Angew. Chem. Int. Ed. Engl.*, 42, 665–668.

Names index

Subject index

abiogenesis, 3, 414
acetic acid, 38
actin, 192, 197, 203–204
activation energy, 63, 191, 201,
  271, 284
active surface, 199
active transport, 172
adenine, 44, 47, 316, 327, 329
adenosine, 251
adenosine diphosphate (ADP), 320–321, 324,
  327, 387
adenosine triphosphate (ATP), 91–92, 126, 329, 337,
  382, 389
aggregate, 5, 57, 83, 84, 145, 149, 193–196, 199,
  202, 205, 217–218, 232, 266–268, 270, 277, 279,
  282, 285–287, 289, 290, 296, 300–301, 310,
  315–316, 320, 327, 377, 387
amide-bond condensation, 252
aminoacyl-phosphate, 91
ammonia, 3, 5, 33, 38–41, 53
ammonia salts, 3
amphiphilic, 42–43, 84, 94, 109, 110, 193–194, 196,
  265–266, 279, 283, 292, 316, 369
anthropic principle (AP), 24–26, 28
aqueous micelles, 146, 256, 267, 270–271, 273–274,
  289, 292
Archaea, 18, 205, 376, 411
artificial chemical life, 21, 61
artificial life, 224, 248, 261, 336–337, 400,
  408, 417
aspartate transcarbamoylase, 197
astrobiology (or bio-astronomy, or exobiology), 31,
  36–38, 293, 400, 406
attractors, 87, 238–239, 390
autopoiesis, i, xiii, xiv, 6, 60, 81, 95, 119, 120,
  122–129, 132–139, 144–148, 150–151, 153–174,
  176, 178, 180–181, 183–184, 186–187, 246, 256,
  322, 400–401
autopoietic system, 125–129, 134, 148, 151, 154,
  159–167, 172–173, 181, 187
auto-regulation, 158
autotrophic, 89, 410
average micellar size, 273
axoneme, 204, 205

Bacteria, 205, 376, 411
basic autopoiesis, 151
benzene, 231, 236
big bang, 25, 243
big-bang theory, 10
bilayer, 57, 84, 111, 148, 153–154, 194–196, 199,
  258, 260, 267, 269, 283, 286–287, 289, 297, 312,
  315–316, 323, 325, 380, 392–395
black smokes, 38
blood, 158, 260
breaking of symmetry, 202, 407
British emergentism, 229, 236
Buddhism, 26, 131–132, 142–143, 165, 179

Cambrian, 12
Cambrian explosion, 12, 16
carbon dioxide, 39, 50, 88–89, 98, 231, 257, 341
carbonic anhydride, 53
carbonyl compounds, 53
catalytic peptides, 75, 100, 414, 418
cell reproduction, 250, 261
chemical autopoiesis, 145, 147, 155, 173, 188, 255
Chemical Autopoiesis approach, 136, 138
chemical information, 65, 99, 248
chemoton, 46, 153–155
chloroplasts, 18, 206, 212, 376
clay, 64, 88–89, 100, 102
coacervates, 5, 83
co-evolution, 47, 75, 91, 169, 240, 255
cognition, i, xiii, 6, 119, 124, 126, 130–131,
  134–137, 139–142, 149, 152, 154–155, 157–178,
  180–181, 183, 186–187, 246, 401, 441
compartmentalistic approach, 72, 82–83, 86, 88, 95,
  114, 395, 402–403
compartmentalistic hypothesis, 68, 95
compartmentalized metabolism, 85
compartmentation, 46–47, 84, 194, 196, 232, 256,
  265, 274–275, 279, 282, 286, 320
compartments, 12, 42, 47, 53, 67, 71, 78, 80, 82–84,
  87, 89, 94, 260, 265, 269–270, 274, 279, 280, 289,
  290, 292, 294–295, 324, 373, 375, 376–377, 379,
  380, 385, 387–388, 392–394, 411, 414, 418
complementary liposomes, 316
compositional information, 93, 94

461

Printed in the United States
By Bookmasters